QD
561
D6813
1972

Dorfner, Konrad.

Ion exchangers; properties and applications. Edited by Andrée Fé Coers. [3d ed. Ann Arbor, Mich.] Ann Arbor Science [1972]

317 p. illus. 24 cm.

Translation of Ionenaustauscher.
Bibliography: p. 281-309.

1. Ion exchange. I. Title.

QD561.D6813 1972 541'.3723 77-141232
ISBN 0-250-97501-7 MARC

Library of Congress 72

Ion Exchangers
Properties and Applications

Ion Exchangers

Properties and Applications

by Konrad Dorfner, Ph. D.

Edited By Andrée Fé Coers

Second Printing 1973

P.O. Box 1425, Ann Arbor, Michigan 48106

Library of Congress Catalog Card No. 77–141232
SBN 250 97501 7

Printed in the United States of America

Translated from the original German by
STS Incorporated,
Ann Arbor, Michigan.

Preface to the First Edition

Ion exchangers have been in wide use in the laboratory and production plant, in research and technology for quite some time. The first available exchange materials had served for the study of a number of fundamental exchange phenomena when this new direction of analysis became generally known following the discovery of Adams and Holmes of the production of ion exchangers based on synthetic resins.

On the other hand, the diverse applications of these materials are so confusing that the need for a survey adapted to practical application by students as well as by the technician became evident. The author therefore attempted to summarize all questions concerning ion exchange in the present book and thus to add an independent general introduction to his monograph on ion exchange chromatography published by the Akademie Verlag of Berlin.

The author is hopeful that all those who must work with ion exchangers will be saved long searches and valuable time. In view of the large number of existing original studies, this survey cannot claim completeness; but it is hoped that all fields of application have been indicated. The author takes this opportunity to express his appreciation to all colleagues with an interest in this field for their communications and data which may be useful in a further expansion of the study.

Ludwigshafen, 1962. KONRAD DORFNER

Preface to the Second Edition

The astonishingly kind reception given to the present monograph on ion exchange by those interested in the field soon showed that a new edition would be appropriate. It is herewith offered in a revised and expanded form.

In this revision it was possible to consider numerous suggestions from readers. Without a change in the entire structure, the individual chapters were expanded with consideration made of some new developments. Many new literature references could be added to offer the reader a still simpler access to original sources.

A new chapter on the theory of ion exchange has been added, so that this aspect has also become more nearly complete. This was done with the purpose of presenting a general theoretical framework without giving details, since only in this manner is it possible to offer a condensed version of the theory.

The author is sincerely indebted to many colleagues for their notices and suggestions which have assisted him in the preparation of the revised edition.

Special acknowledgment is expressed to Prof. Dr. G. Dickel of Munich for making partly unpublished data available on the theory of ion exchange and for his critical review of this chapter, as well as to Prof. Dr. K. Fischbeck of Heidelberg for numerous discussions and suggestions. Moreover, appreciation is due to a number of firms for placing documenting data at the author's disposal.

Ludwigshafen, 1964. KONRAD DORFNER

Preface to the Third Edition

The continued development of ion exchangers and their applications since the publication of the second edition of this monograph and persisting interest in a survey of this sector have indicated that a revision is appropriate. It is offered herewith as an enlarged third edition.

To give the state of the art and technology of ion exchangers their just due, it was necessary to reclassify the material. After a longer introduction, the various types of ion exchangers as well as their properties and test methods are described more fully. The application of ion exchangers in industry has received particularly detailed discussion to cover new developments. Thus, a mathematical treatment and the special techniques of the fixed-bed process, the most recent development of continuous processes, water treatment with ion exchangers, and the use of ion exchangers for waste water treatment, metal refining, sugar production, and other industrial applications have been described as fully as possible within the available space. The remaining chapters have been reviewed, improved and enlarged on the basis of recent findings.

It is not easy to give an appropriate survey of such a broad field as ion exchange. The author frequently had to depend on the willingness of his colleagues to offer assistance. His gratitude for advice and information as well as for data and contributions is expressed particularly to Prof. Dr. K. Fischbeck, Heidelberg, Prof. Dr. G. Dickel, Munich, Prof. Dr. E. Blasius, Saarbrücken, and R. Brunner, Engineer, Frankfurt, in addition to a number of colleagues and friends who are dealing with this field. Sincere appreciation is also expressed to the publishers for their rapid publication and consideration of the author's delayed wishes.

Mannheim, 1971. KONRAD DORFNER

Contents

Chapter 1

Introduction

1.1 History

In 1850, Thompson [584] and Way [621] reported on the observation that cultivated soil can exchange various ions, such as ammonium ions, for calcium or magnesium ions. Spence, a pharmacist from York, working under contract to Thompson, had prepared a bed of sandy clay treated with ammonium sulfate in a glass column and allowed water to flow through it. Instead of ammonium sulfate, he obtained gypsum in the eluate. This phenomenon, which was first called base exchange, was interpreted as a chemical process by Henneberg and Stohmann [266] in 1858, when they succeeded in confirming the reversibility and equivalence of such ion exchange processes. Finally, in 1870, Lemberg [386, 387] demonstrated that a number of natural minerals, particularly the zeolites, are capable of ion exchange when he succeeded in converting lucite into analcime. These and similar observations remained the subject of basic research for about 50 years until Gans [206, 207] in 1905 synthesized inorganic ion exchangers, among which sodium perutite found practical application in water treatment.

One of the most important events in the history of ion exchangers is represented by the discovery of Adams and Holmes [3, 4, 85, 86] (1935) that synthetic resins have ion-exchanging properties. Recognizing the importance of the discoveries of Adams and Holmes, IG-Farbenindustrie AG purchased the patents on the subject and since 1936, has continued development work in this field with the aim of a systematic production of ion exchange resins with the desired properties (Griessbach [235, 236]).

The ion exchangers which were first obtained by polycondensation came to be replaced increasingly by polymerization products after 1945, when d'Alelio [599] succeeded in incorporating sulfonic acid groups into a cross-linked polystyrene resin. Further development dealt with improvements of ion exchange resins and the production of special resins with specific ion exchange properties. Skogseid [556] developed a potassium-specific exchanger and Cassidy [115] prepared exchangers with a reducing and oxidative action as described earlier by Griessbach, Lauth and Meier [213]. The stormy development of ion exchanger research after 1945 is evident from the graph of Figure 1, in which the number of scientific papers is plotted as a function of the year of publication [373]. The method of ion exchange thus became an important modern tool which was accepted in the laboratory and in industry [142].

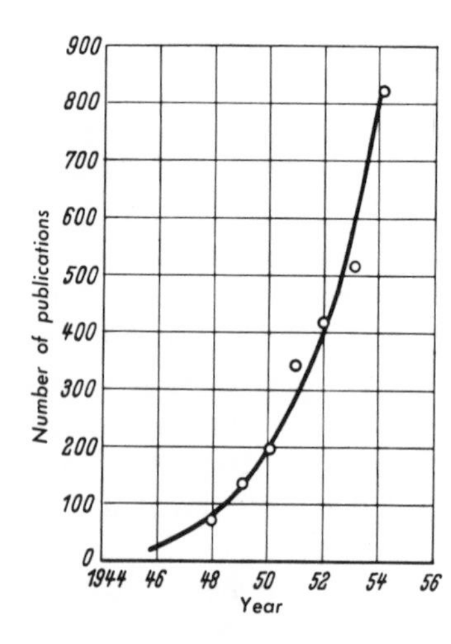

Figure 1 Number of publications on ion exchange as a function of time [373].

The advances of the most important types of ion exchangers, *i.e.*, synthetic resin exchangers, over the last 30 years of research and development can no longer be expressed in simple numbers of publications, since the fields in which ion exchange is employed have become extremely diverse [581]. The future prospects for ion exchangers indicate a scope exceeding all original expectations [365].

1.2 Principles

Ion exchangers are solid and suitably insolubilized high molecular weight polyelectrolytes which can exchange their mobile ions for ions of equal charge from the surrounding medium. The resulting ion exchange is reversible and stoichiometric with the displacement of one ionic species by another on the exchanger. Viewed in a different light, ion exchangers can be considered high molecular weight acids or bases with a high molecular weight cation, which can exchange their hydrogen or hydroxyl ions for equally charged ions and thus are converted into high molecular weight salts. If such a solid acid is neutralized with a base into the salt, however, the cations bound to the polyelectrolytes can again be displaced by other cations. The resulting process is known as cation exchange and the polyelectrolyte is the cation exchanger. In the second case, a solid base is obtained which is capable of OH^- exchange and can be neutralized with an acid, and the anion which was bound first can again be displaced by another anion—a process then known as anion exchange. The polyelectrolyte taking place in this process is called the anion exchanger.

The most widely used modern ion exchangers are organic materials based on synthetic resins. The fundamental processes can be described most easily with these products and on the other hand, all—even general—phenomena of ion exchange can be more easily understood on the basis of this principle. Figure 2 shows some types of ion exchangers.

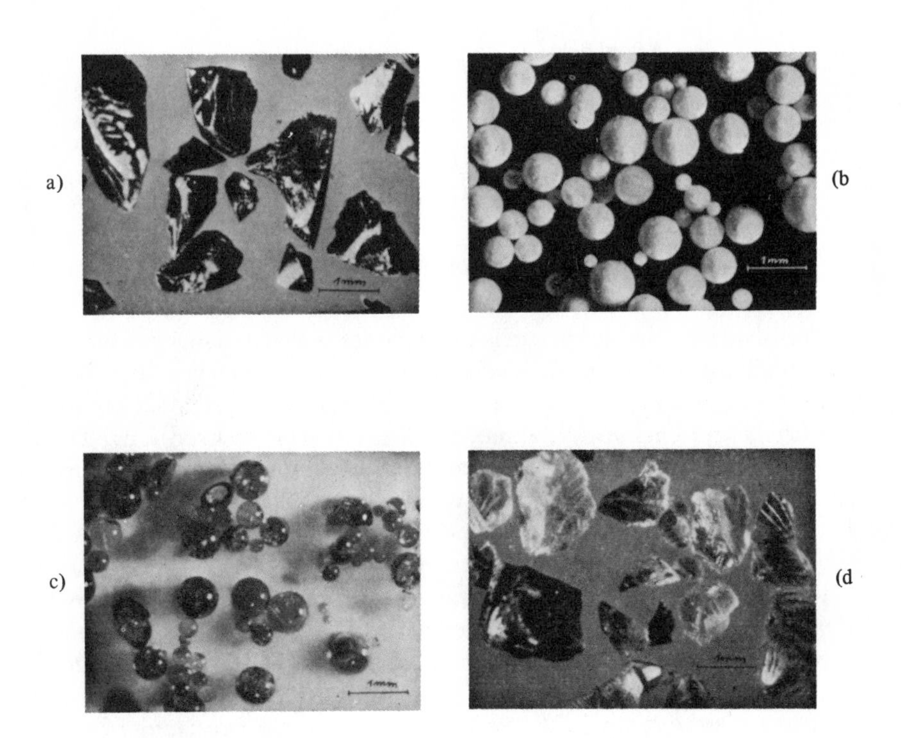

Figure 2 Some common types of ion exchange: (a) weak acid cation exchanger in granulate form; (b) weak base anion exchanger in spherical form; (c) strong acid cation exchanger in spherical form; (d) strong base anion exchanger in granulate form.

The macromolecule of the ion exchanger in the most general case represents a three-dimensional network with a large number of attached ionizable groups. Such a network, a strong acid cation exchanger containing sulfonic acid groups as active sites, is shown in Figure 3.

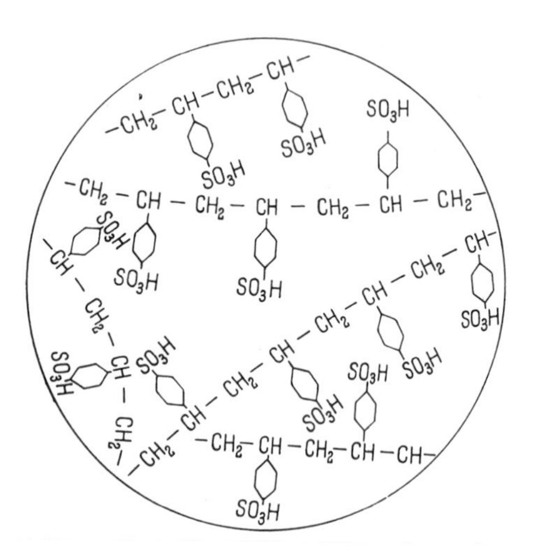

Figure 3 Structure of a highly acid polystyrene–sulfonic acid resin (not cross-linked).

The nomenclature of ion exchange includes some fundamental terms. The high molecular weight skeleton composed of different building blocks is known as the matrix to which the so-called functional groups are firmly and chemically bound; these are known as fixed ions. The exchangeable ions in a heteropolar bond are called the counterions. In contrast, co-ions are all those ionic species which can be present in an exchanger but have the same charge as the fixed ions, *i.e.*, the functional group without its counterion.

Polyfunctional ion exchangers are those with different types of functional groups, while monofunctional exchangers contain only one type of functional group. The channels which enclose the ion exchanger skeleton are usually called pores.

If we give the symbol CE^{-n} to a cation exchanger consisting of any matrix carrying a functional group with any number n of negative charges and the symbol C_1^+ to a monovalent cation, ion exchange can generally be described as follows:

$$CE^{-n} \cdot nC_1^+ + nC_2^+ + nX^- \rightarrow CE^{-n} \cdot nC_2^+ + nC_1^+ + nX^-$$

and for hydrogen ion exchange, which already represents a special case, this simplifies to:

$$CE^{-n} \cdot nH^+ + nC^+ + nX^- \rightarrow CE^{-n} \cdot nC^+ + nH^+ + nX^-$$

becoming a process analogous to neutralization when the cation intended for the exchange is used in the hydroxide form, *i.e.*, as a base:

$$CE^{-n} \cdot nH^+ + nC^+ + nOH^- \rightarrow CE^{-n} \cdot nC^+ + nH_2O.$$

Using the symbol AE^{+m} for an anion exchanger in the same manner, we obtain the general formulation for anion exchange:

$$AE^{+m} \cdot mC_1^- + mC_2^- + mX^+ \rightarrow AE^{+m} \cdot mC_2^- + mC_1^- + mX^+$$

for the hydroxyl exchange which also represents a special case:

$$AE^{+m} \cdot mOH^- + mC^- + mX^+ \rightarrow AE^{+m} \cdot mC^- + mOH^- + mX^+$$

and again, for the process analogous to neutralization:

$$AE^{+m} \cdot mOH^- + mC^- + mH^+ \rightarrow AE^{+m} \cdot mC^- + mH_2O.$$

As we will see later, these fundamental formulations can be applied to all ion exchange processes and can be useful as a generalizing scheme if certain processes are to be interpreted as ion exchange events.

Ion exchange processes require a mediating agent, generally water, in which the exchanging ions are dissolved. The ion exchanger present in water contains water in its pores, so that when the exchange process is initiated, the solute ions penetrate the pores at the same time, exchange places, so-to-speak, with the existing counterions, and—since electron neutrality must be maintained—join the co-ions which have penetrated previously. From the physicochemical standpoint, therefore, diffusion is an important process during ion exchange.

If the system produced in this manner is left to itself, an equlibrium forms as in other chemical reactions and for the general case of an ion exchanger IE with counterions C_1 of valence x and a counterion C_2 of valence z, this equilibrium can be written as follows:

$$zIE \cdot C_1 + xC_2^{z+} \rightleftarrows zIE \cdot C_2 + xC_1^{z+}.$$

By formal application of the law of mass action, the following is obtained for the equilibrium state:

$$K = \frac{a_{rC_2}^z \cdot a_{C_1}^x}{a_{rC_1}^z \cdot a_{C_2}^x}$$

where a_{C_1} and a_{C_2} are the chemical activities of particles C_1 and C_2 in the solution, a_{rC_1} and a_{rC_2} are those in the resin, and K is the thermodynamic equilibrium constant. The activities are obtained from the concentration $[C_i]$ and the activity coefficients f_i. If this is solved, we obtain:

$$K = \frac{f^z_{rC_2} \cdot f^x_{C_1}}{f^z_{rC_1} \cdot f^x_{C_2}} \cdot \frac{[C^z_{r_1}] \cdot [C^x_1]}{[C^z_{r_1}] \cdot [C^x_2]}$$

By combining K with the activity coefficients, we obtain an empirical equilibrium constant:

$$K_{emp.} \equiv \frac{[C^z_{r_2}] \cdot [C^x_1]}{[C^z_{r_1}] \cdot [C^x_2]} = K \frac{f^z_{rC_1} \cdot f^x_{C_2}}{f^z_{rC_2} \cdot f^x_{C_1}}$$

containing values which are analytically measurable.

However, this empirical equilibrium constant depends on the properties of the ion exchanger and the nature and concentration of the electrolytes, so that finally, different empirical equilibrium constants are obtained for different ions and the end result has been a new way of consideration which has led to the concept of selectivity.

Ion exchange has frequently been compared to adsorption, but in view of the aforementioned stoichiometry this is incorrect, even though ion exchangers may and do have adsorbent properties and even though materials known as adsorbents, such as aluminum oxide or silica gel, may act as ion exchangers. It is not always simple to establish the boundaries in this case.

1.3 Procedures

Three techniques are used for the practical application of ion exchange: (1) batch operation, (2) column processes, and (3) continuous processes.

1.3.1 *Batch Operation*

Batch operation is the simplest ion exchange process. The ion exchanger is contacted with the electrolyte solution in any desired vessel until an exchange equilibrium has been established between the counterions of the exchanger and the ions of equal charge of the electrolyte:

$$IE \cdot C_1 + C_2X \rightleftarrows IE \cdot C_2 + C_1X.$$

The degree to which this process takes place depends on the equilibrium constant of the ion exchange system. After equilibrium has been attained, the ion exchanger is filtered. If additional ions are to be exchanged from the electrolyte solution by the exchanger, fresh ion exchanger must be added and filtration must be performed again after adjustment of the equilibrium. This process therefore can be compared to discontinuous extraction and adsorption techniques.

The applicability of a batch technique is limited to exchange processes in which the equilibrium has been shifted strongly to the right of the above equation. This is always true when a cation exchanger in the H-form is contacted with a metal hydroxide solution or an anion exchanger in the OH-form with an acid solution. In both reactions, ion exchange is followed by the formation of water.

1.3.2 *Column Processes*

The column process is the most important and most frequently used laboratory technique. The ion exchanger is packed in a glass column and all necessary operations are carried out in this bed. Fundamentally, two techniques can be distinguished here, *i.e.*, the descending and the ascending flow process. In the first case, the liquid moves down and in the second, up through the exchanger bed.

If an exchanger column contains the ion exchanger with counterion C_1, we are dealing with the equipment shown schematically in Part I of Figure 4. The counterion C_1 of an exchanger is to be exchanged for

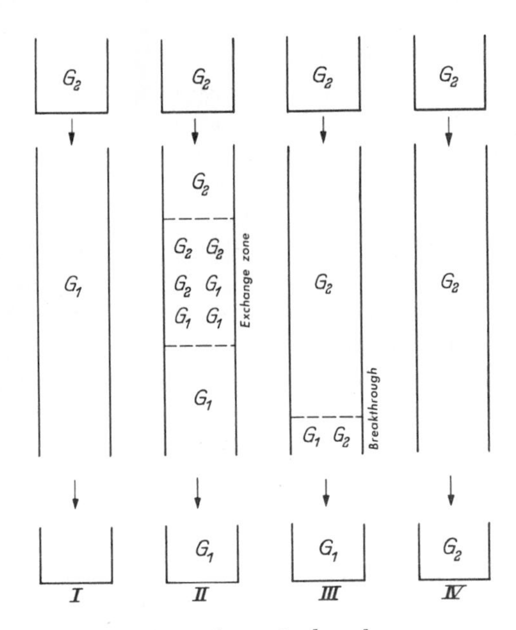

Figure 4 Ion exchange in the column process.

the counterion C_2 of a solution in an overhead reservoir. As soon as solution with C_2 has entered the exchanger bed (Figure 4, II), C_2 ions are exchanged by the exchanger. After a short time, the exchanger in the upper section of the column is completely loaded with C_2 as counterions. Additional C_2 ions flow unhindered through this part of the bed and reach the exchange zone farther down, where the C_1 counterions exchange sites quantitatively with the C_2 ions. The liberated C_1 ions are eluted at the lower end of the column in a stoichiometric ratio. If this process is continued, the exchange zone in the column continues to migrate until it reaches the lower end and the overall process has come to the point where C_1 and C_2 ions are simultaneously eluted from the column. Breakthrough takes place, at which the concentration of C_2 ions in the flow begins to increase prominently until it finally has reached the same concentration as in the solution which was initially charged on the column (Figure 4, III). If the C_2 ions are charged once more (Figure 4, IV), no further ion exchange can take place, since the entire exchanger already has the C_2 form; the C_2 ions flow through the column without hindrance.

During flow through the column, the ions which are to be exchanged continuously contact fresh ion exchanger, so that the equilibrium is increasingly shifted in the desired direction. Compared to a batch technique, ion exchange here becomes a complete and simple process.

Ion exchange columns. The ion exchange literature describes numerous columns either because practical considerations led to new designs or certain techniques made these necessary. In principle, the ratio of diameter to height in laboratory columns should amount to 1:10–1:20. In any case, they must be designed such that the liquids can flow through them easily.

A simple ion exchange column can be homemade with materials which are usually available in any laboratory. As shown by Figure 5, it consists of a simple glass tube provided with a bored stopper on both ends. The lower stopper is equipped with a glass tube attached to the tip of a capillary as a dropping attachment via a flexible tube connection. The pinch cock permits a regulation of the dropping rate. The upper end of the column is provided with an ordinary dropping funnel as a reservoir. Cotton balls inserted on both ends of the ion exchanger bed (cellulose, synthetic fiber, quartz, or glass fiber balls are used) prevent plugging of the discharge tube by ion exchange particles on one hand and turbulence of the ion exchanger bed on the other. It is also possible to have a glass blower produce a column with ground joints (Figure 6) on which reservoirs of different size can be attached. An overflow tube can be connected to the flexible tube provided under

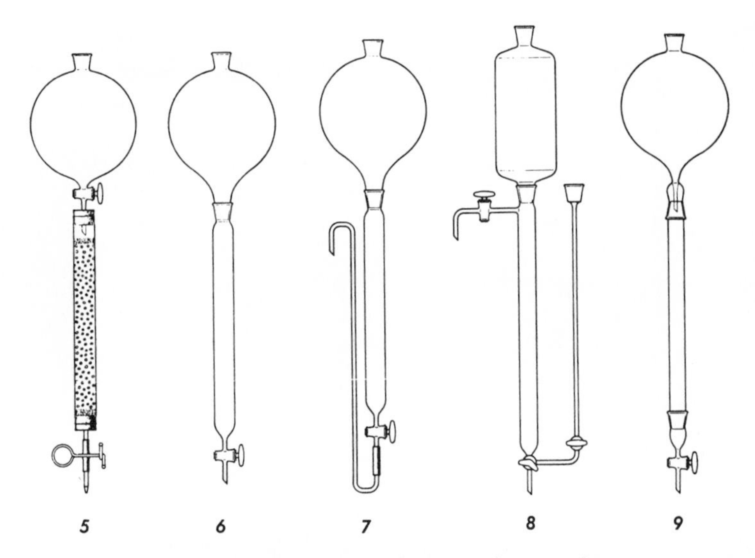

Figure 5 Simple homemade ion exchange column.
Figure 6 Ion exchange column with ground joint.
Figure 7 Ion exchange column with overflow.
Figure 8 Countercurrent ion exchange column.
Figure 9 Wickbold inversion column

the discharge cock (Figure 7), thus preventing the drainage of the column. Several advantages are offered by a so-called countercurrent column (Figure 8) which can be used with an ascending flow technique as well as with the simple descending flow technique. This is obtained by a suitable adjustment of the three-way stopcock. The liquid flows through the column from the right attachment and is discharged through the left stopcock. The effect obtained is also known as reverse flow in adsorption and exchange processes (counterflow). The materials which have been exchanged near the top of the column do not need to cover the distance through the entire column bed and can thus be eluted more easily. The same end is achieved very simply by the Wickbold inversion column (Figure 9). A column version which is used very frequently in laboratories is the burette type. As shown by Figure 10, it consists of a simple burette with the necessary connection at the top for loading of solutions. For ion exchange processes at certain high or low temperatures, the column and reservoir are equipped with a temperature-regulating jacket (Figure 11).

In series studies with ion exchangers, a column system as shown in Figure 12 has proved useful. Five columns of the type shown in Figure 12a are interconnected into one system by means of a distribu-

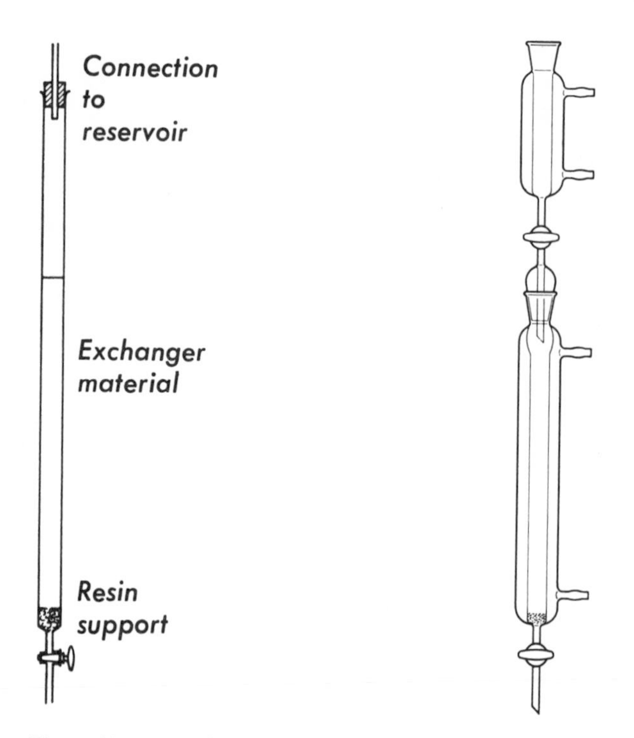

Figure 10 Ion exchange column of the burette type.

Figure 11 Heatable Column.

ting system. The columns can be separately equipped with reservoirs and after the preceding common treatment can continue to operate individually (Figure 12c; for the sake of simplicity the ancillary equipment has been omitted).

Larger columns to test an ion exchange process can in principle be constructed in the same manner as described using glass as the ma-

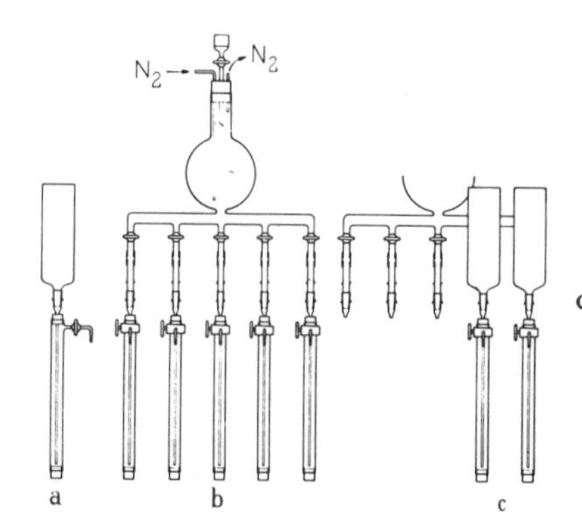

Figure 12 Schematic diagram of the column system.

terial. Dimensions of 50-mm diameter and 800-mm length are recommended. Acid-washed sand or quartzite is frequently used as the resin support. Experiments in columns of such dimensions usually already furnish results which can serve as the basis for the construction of production units.

Charging the ion exchangers into a column can be easily done with some practice. The exchanger, present in any form, is first treated with distilled water in a beaker. The resulting swelling must be carried out under all circumstances to prevent rupture of the column or too close a packing due to swelling in the column. Two hours are usually sufficient for swelling. The exchanger is subsequently poured into the column (Figure 13) with care that rapid charging leads to a uniform

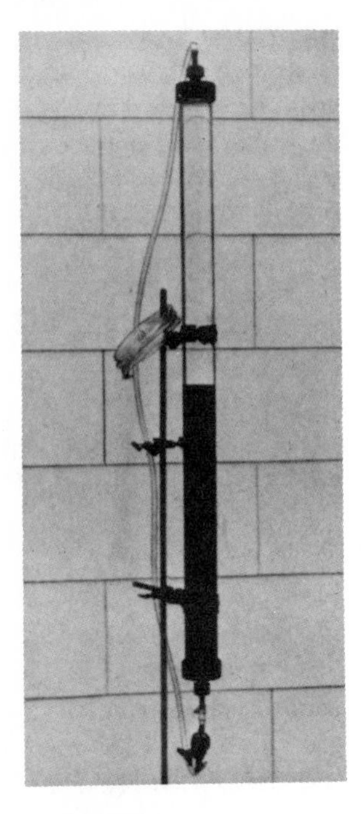

Figure 13 Larger column for preliminary industrial tests.

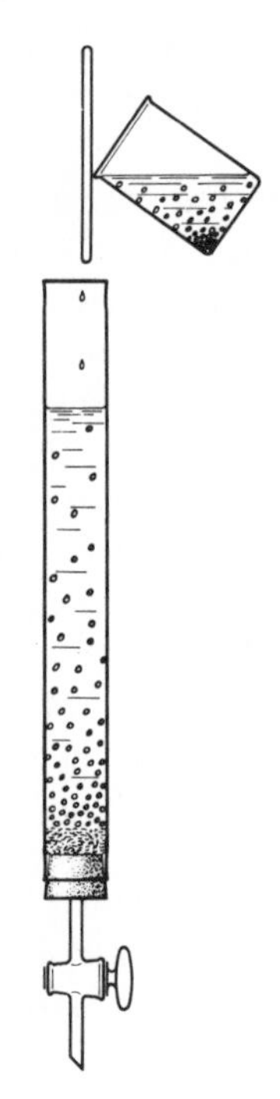

Figure 13a Charging of an ion exchange column.

packing of the different sized exchanger particles and that the exchanger is always covered with water to prevent the inclusion of air bubbles. Excess water is continuously suctioned from the column. If too much water has been removed accidentally from the column and air has entered between the particles, it usually suffices to add water and to swirl the ion exchanger bed by tipping the column. Finally, a cotton ball is inserted on the top rim of the ion exchanger bed and the column is washed a few times with distilled water.

The particle processes of ion exchange which take place in an exchange column during one cycle are:

1. Ion exchange
2. Washing of the exchanger bed.
3. Regeneration or elution.

Ion exchange proper may take place in the most diverse forms. It depends on the problem involved; several examples concerning this subject will be given later in applications in laboratory and industry. Washing of the exchanger bed is necessary between individual runs to remove excess reagent solution remaining in the bed. During regeneration, the exchanger is transformed into its original form; if the ion which has been exchanged in the first run is to be recovered, it is eluted from the exchanger with a suitable liquid and collected in the eluate.

Three parameters serve to describe the dynamic and chemical processes taking place in exchange columns, *i.e.*, flow rate, pressure drop and breakthrough capacity.

Although the quantity of liquid flowing through a column can be expressed in drops per second and used again as a comparison criterion for one and the same column, a precise indication of the flow rate is given in $ml \cdot cm^{-2} \cdot min^{-1}$. If the flow rate is expressed only in $ml \cdot min^{-1}$, the diameter of the column must also be indicated. The linear flow rates in $cm \cdot min^{-1}$ or $cm \cdot sec^{-1}$ are used less frequently.

The resistance produced by friction leads to a pressure drop as a liquid flows through an ion exchange column, so that the flow rate and thus the volume flowing through per unit of time are limited. The pressure drop depends on the particle size of the ion exchangers, as demonstrated in Figure 14 with the example of the cation exchanger Dowex 50X8 in the particle sizes 20–50 mesh and 50–100 mesh. The pressure drop is a function of the apparent density, particle size, and shape. This is equally applicable to cation exchangers and anion exchangers.

As indicated briefly by the explanations of Figure 4, the exchange of one species of counterion for another in an ion exchange column finally reaches a point where breakthrough occurs. This breakthrough process of new counter ions which are no longer exchanged can also be represented by a diagram. Breakthrough curves in this diagram generally permit a good description of the processes in ion exchange columns since a breakthrough capacity characteristic for a column

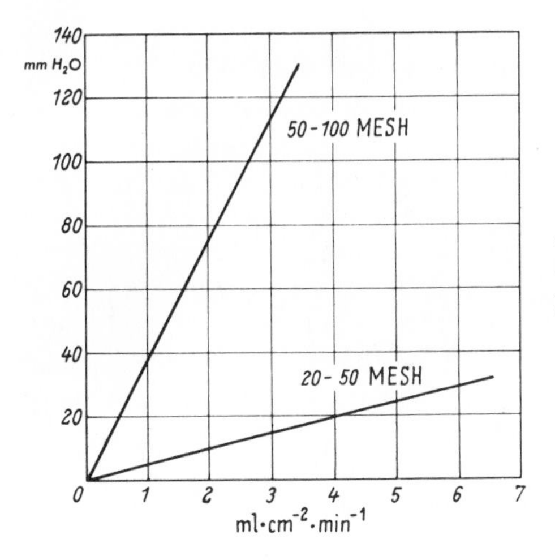

Figure 14 Pressure loss and flow rate of the same Dowex 50X8 ion exchanger of different particle sizes.

under given conditions can be assigned to these curves. The concentration, column length or time or the equivalent fraction of the latter, the column volume, or the volume of solution flowing through the column at the same flow rate can be used as the coordinates of breakthrough curves. As an example, Figure 15 shows a breakthrough diagram which is in frequent use: the dependence of the concentration of ions retained in the loaded solution, *i.e.*, ions which were not exchanged as a function of the volume eluted from the column.

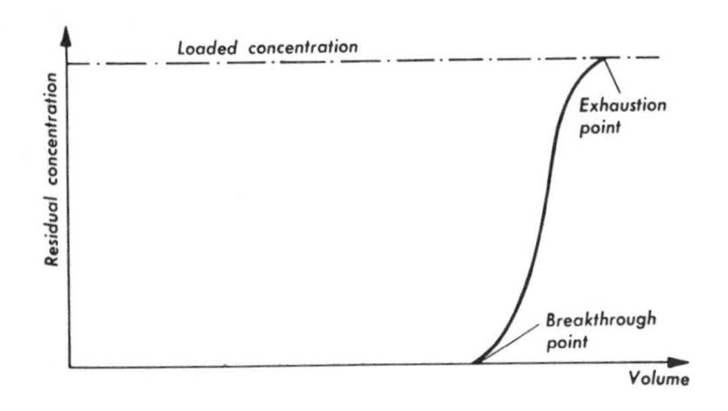

Figure 15 Breakthrough curve.

1.3.3 *Continuous Processes*

In the continuous ion exchange processes which are essentially reserved for industrial applications, the exchanger and liquid usually move in countercurrent columns. It is a characteristic and disadvantage of column processes that a large part of the upper ion exchange bed remains in the column without utilization during the run, while the lower part performs the ion exchange. The logical theoretical consequence of this fact is that the exhausted part of an ion exchange column should be continuously removed and regeneration performed immediately, and this already establishes the characteristic feature of the continuous or fluid-bed process.

The technical difficulties which needed to be overcome to realize such processes were great [388]. A number of systems are described, which appeared to be usable and which had the characteristic that ion exchange proper and regeneration or elution were carried out simultaneously in different parts of the equipment. However, this appeared to hold little promise for a long time, since the equipment was too costly and the ion exchange particles too unstable. When it becomes possible to overcome the problems of process engineering, continuous ion exchange processes will find broad fields of application and will be the future of ion exchange technology. This is already indicated by three well-advanced techniques.

Chapter 2

Ion Exchanger Types

Ion exchange has become a general operating technique today. As indicated by the brief historical review in Chapter 1, ion exchange was investigated first with inorganic materials; these even found industrial use but were almost completely displaced by synthetic ion exchange resins. The demand for ion exchange materials that would satisfy the most diverse requirements led to the development of new products which must be considered as separate ion exchanger types according to their special properties.

Among the available types, synthetic ion exchange resins have gained the greatest significance. In addition, the inorganic ion exchangers should be discussed, since these have again aroused great interest because of their mechanical, thermal and chemical properties and because, in view of their mineralogical parallels, they offer information about numerous processes in soil science. The cellulose ion exchangers represent a separate type with their own characteristics related most closely with another group, *i.e.*, the dextran ion exchangers. Both types have been rapidly accepted in ion exchange chromatography where they have become indispensable for analytical as well as preparative work. In addition, ion exchangers based on carbon have been produced and, together with a number of other materials with ion-exchanging properties, have been tested for their applicability in ion exchange.

In the case of ion exchange membranes, interest was aroused not so much by questions of their structure as by their physicochemical mechanism of action and its utilizations; this will continue only until sufficiently sophisticated materials for general industrial applications have become available. Liquid ion exchangers as another type probably are important primarily because of the technological possibilities offered by them.

Finally we should note the redox exchangers, which have become known as electron exchangers, redoxites, or redox ion exchangers. In analogy with the reversible exchange of ions in ion exchangers, a reversible redox reaction can be carried out with these substances. Redox exchangers are frequently discussed in monographs dealing with ion exchange. However, since their analogy with ion exchangers has only a formal context while ion exchange in principle is unrelated to a redox reaction, we shall dispense with a description of redox exchangers and refer the reader to the specialized literature [116, 161, 405].

2.1 Synthetic Resin Ion Exchangers

For a description and an understanding of synthetic ion exchange resins and their properties, three factors are decisive with regard to their production: The raw material which was used for the construction of the skeleton or the matrix, bridging agents for cross-linking and insolubilization, and the type and number of so-called functional groups. Synthetic ion exchange resins chemically are insoluble polyelectrolytes with a limited moisture regain capacity. On the basis of their function, we speak of cation exchangers, anion exchangers, and amphoteric ion exchanger types.

Polymerization and polycondensation in principle can be used as the synthesis routes to form the matrix. At the present time, polymer resins have become more important than polycondensation resins. They have a higher chemical and thermal stability.

The most important starting material, which is used nearly exclusively in exchangers of the strongly acid and strongly basic types produced on a large scale, is styrene. The styrene is cross-linked with itself and with divinylbenzene into a polymeric network:

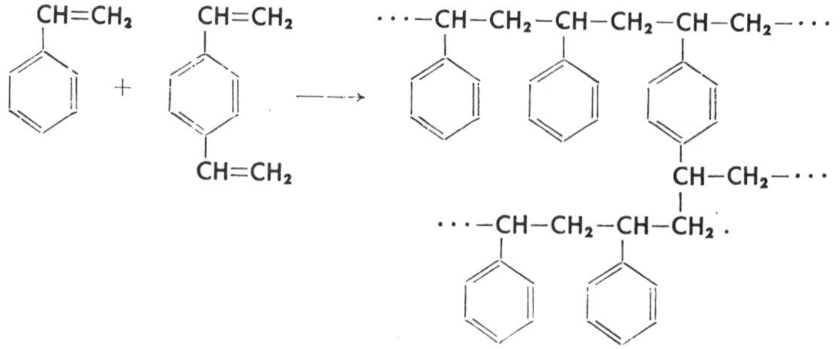

$$CH{=}CH_2\text{-}C_6H_5 + CH_2{=}CH\text{-}C_6H_4\text{-}CH{=}CH_2 \longrightarrow \cdots\text{-}CH\text{-}CH_2\text{-}CH\text{-}CH_2\text{-}CH\text{-}CH_2\text{-}\cdots$$

On the other hand, matrices for primarily weakly acid cation exchangers are formed, also with divinylbenzene, from acrylic acid or methacrylic acid:

CH_3–$C(=CH_2)$–$COOH$ + $CH=CH_2$ / $CH=CH_2$ (divinylbenzene) ⟶ ···–$C(CH_3)(COOH)$–CH_2–CH–CH_2–··· / ···–CH–CH_2–···

In polycondensation resins, the matrix is usually constructed of phenol and formaldehyde:

OH + HCHO ⟶ OH, OH, OH, CH_2, CH_2, CH_2, CH_2

where the properties of the exchanger depend on the nature of the phenols, the quantity of starting materials used, and their side groups [613].

As far as the structure of styrene–divinylbenzene copolymers is concerned, the reaction leading to their formation produces matrices which are not uniform in degree of cross-linking. Such ion exchange particles accordingly must be considered heterogeneous substances in which regions with longer network chains formed at the beginning of polymerization are found together with relatively closely cross-linked domains [81, 167]. From the standpoint of high polymer synthesis, it is probably impossible for practical purposes to obtain a uniform bridge distribution through the structure of a cross-linked copolymer. Naturally, the determination of significant thermodynamic values becomes difficult as a result and this is always evident in all theoretical considerations. The homogeneity of ion exchange structures depends on the purity, nature and properties of starting materials used for their production as well as on optimum conditions of polymer synthesis. In spite of the irregular structure of the matrix, however, it is possible to produce ion exchangers with a uniform distribution of ionic groups [506].

Cross-linking naturally depends first of all on the quantity of divinylbenzene used as the cross-linking agent in the production. Commercial ion exchangers of the gel type contain between 2 and 12% divinylbenzene. The nature and degree of cross-linking have a decisive influence on the properties of ion exchange particles; additional details on this subject are described in Section 2.1.1.3.

Regardless of the influence of the matrix on properties of an ion exchanger, the decisive factor is represented by the functional group. Up until now, the following have been incorporated in cation exchangers:

$$-SO_3^-, -COO^-, -O^-, PO_3^{2-}, -PO_3H^-, -AsO_3^{2-}, -SeO_3^-$$

and in anion exchangers:

$$-\overset{+}{N}H_3, =\overset{+}{N}H_2, \equiv\overset{+}{N}, -\overset{+}{N}R_3,$$

$$-C_6H_4-\overset{+}{N}(CH_3)_3, \equiv S^-, \equiv P^-, -C_5H_4\overset{+}{N}-OH$$

The functional groups confer the property of ion exchange to the matrix. Depedning on the acidity or basicity of the functional groups, we distinguish between strong and weak acid and between strong and weak basic ion exchangers. It has become accepted practice to distinguish between two types of strong exchangers: Type I contains a trimethylamine group and Type II a dimethyl-β-hydroxyethylamine group.

$$-\overset{+}{N}(CH_3)_3 \qquad -\overset{+}{N}(C_2H_4OH)(CH_3)_2$$

Type I Type II

Type I is more strongly basic than Type II but more difficult to regenerate. Type II has a higher thermal stability but is more sensitive to oxidants.

The conditions set for a matrix, especially insolubility, are satisfied by a number of polymers. However, many of these polymers do not contain chemical groups with an ion exchange capacity. Although such groups can be introduced into most polymers, the end products still cannot be used in ion exchange, since their water solubility increases together with the ionic and thus the hydrophilic groups. Consequently, only the cited special polymers and copolymers are suited for the synthesis of ion exchangers.

In addition to cation and anion exchangers, amphoteric exchangers exist containing both acid and base groups. For example, condensation products of amines and phenols contain both very weakly acid phenolic OH-groups and basic amino groups. Bipolar exchange resins have also been produced by the introduction of acid and basic groups into the same skeleton of a styrene–divinylbenzene copolymer [431]. Amphoteric ion exchangers of the aminophosphonic acid and amino-carboxylic acid type have also been investigated [408–410].

Cation exchanger also can contain two different functional groups with the same charge, such as sulfonic acid and carboxyl groups, and they are then known as polyfunctional. For example, copolymers of acrylic acid or methacrylic acid with divinylbenzene have been sulfonated [601, 602] and mixed phosphoric acid–sulfonic acid resins have been produced [603].

Strong acid cation exchangers of the sulfonic acid type. The strong acid cation exchangers with functional groups consisting of sulfonic acid have attained the greatest importance among ion exchangers produced from a matrix of styrene and divinylbenzene as the cross-linking agent since they can be used industrially for water softening. They are produced by the sulfonation of suspension copolymer beads with sulfuric acid, sulfur trioxide, fuming sulfuric acid or chlorosulfonic acid. The SO_3-groups, which are the functional groups and produce the cation exchange function, probably are primarily in paraposition and a double sulfonation is probably impossible for stearic reasons. The ion exchange capacity of 5 meq/g which is usually obtained confirms that only one aromatic ring on the average has been sulfonated.

The production of a strong acid cation exchanger of the sulfonic acid type involves a large number of problems of detail which have only been solved with time. It is important to obtain undamaged crack-free beads. From the chemical standpoint, the sulfonation reaction leads to exchangers which contain hydrogen ions as the counterions. By treatment with sodium hydroxide solution, the exchangers are then converted into the Na^+-form and further utilized as such. Complete conversion into the Na-form is important because the hydrogen ions remaining in an exchanger may lead to equipment corrosion—for example, with their use in water softening.

Weak acid cation exchangers of the carboxyl type. In the most frequently used forms, the functional group represented by carboxyl in weak acid cation exchangers consists of one of the copolymer components, so that another production step becomes unnecessary. With dissociation constants of between 10^{-5} and 10^{-7}, they are weak acids which can be very effectively converted from the salt into the acid form by means of strong acids. Because of their high selectivity for Ca^{2+} and Mg^{2+} ions, a regeneration with sodium chloride is not very effective and practically not feasible. The carboxyl cation exchangers are primarily suited for the removal of cations from basic solutions and for the cleavage of weakly alkaline salts of polyvalent cations. However, structural modifications can also increase the acidity of carboxyl cation exchangers to such a degree that a cleavage of sodium and potassium salts is possible [118, 119].

Phosphorus- and arsenic-containing cation exchangers. Although these types of cation exchangers are not of very great significance for applications, they are of interest as medium-strong acid exchangers because of their selectivity and high capacities. This applies to arsenic-containing types, particularly with regard to their high affinity to uranium cations.

Weak basic anion exchangers of the amine type. This group of synthetic ion exchange resins includes materials with functional groups of $-NH_2$, $=NH$, and $\equiv N$, both individually and mixed. Polycondensate and polymers serve for their production. In styrene–divinylbenzene polymers, a particular role is played by the chloromethylation reaction and subsequent amination which leads to the weak basic anion exchanger. Because of their low basicity only anions of strong acids, such as HCl or H_2SO_4, are exchanged sufficiently; the anions of weaker acids, such as SiO_3^{2-} or HCO_3^-, are not extensively exchanged. For the same reason, however, these anion exchangers can also be converted into the OH-form by weak bases, such as Na_2CO_3 or NH_4OH. The customary commercial Cl-form is easily hydrolyzed. The weak basic functional groups cannot exchange neutral salt anions.

Strong basic quaternary ammonium anion exchangers. These anion exchangers are obtained by a relatively simple method from the chloromethylation products of styrene–divinylbenzene copolymers by their conversion with tertiary amines. The exchangers obtained are extremely stable and have a high exchange capacity. As strong basic products they are also capable of exchanging silicate and carbonate. They are easily converted from the chloride form into the OH-form by treatment with NaOH. However, with Na_2CO_3 this regeneration is difficult, and with NH_4OH it is almost impossible. By conversion with trimethylamine one obtains the strong basic anion exchanger known as Type I, which is the most common in general. For reasons of regeneration feasibility, a Type II was finally developed with dimethylamino ethanol which proved to be less stable, however, and did not become as important as expected.

Anion exchangers of the pyridine type. Polymers with pyridine as the group with exchange activity should be weak basic anion exchangers. Little has become known about their production and application. It is expected that products of this type will have a good chemical, thermal, and radiation resistance as well as good kinetic characteristics [463].

Depending on the counterion attached to the functional group to maintain electron neutrality, the individual types can be handled in different forms. In the case of cation exchangers, we then speak of an H-form or Na-form and in anion exchangers, of an OH-form or a Cl-form, etc. Either the ion exchangers can be used in their delivered ionic form or they need to be converted first into another form. The composition of the solution to be treated and the selectivity of an exchanger for certain ions generally determine the choice of an exchanger and its particular form.

The reconversion of an ion exchanger into its working form is known as regeneration. Different regenerants are used, depending on the exchanger—whether it is a weak or strong acid or a weak or strong basic type. The regenerants must be easily managed and inexpensive. Table 1 shows a survey of the most common regenerants for synthetic ion exchange resins from the standpoint that a given ion exchange type is to be converted into a given ionic form. The regenerant must be used with an approximate concentration of 1 N in a quantity greater than would correspond to a stoichiometric conversion; this is also apparent in Table 1.

This survey offers general working information, mainly intended for laboratories, but it must be kept in mind that different regenerants in

Table 1
Regenerating agents for the various resin ion exchanger types for conversion into a desired ionic form

Ion exchanger type	*Desired ionic form*	*Regenerating agent (approx. 1 N solution)*	*Required amount*	
			Regenerating agent, meq	*Ion exchanger, meq*
Strong acid cation exchanger	H^+-	HCl	3 –5	1
	H^+-	H_2SO_4	3 –5	1
	Na^+-	NaCl	3 –5	1
Weak acid cation exchanger	H^+-	HCl	1.5–2	1
	H^+-	H_2SO_4	1.5–2	1
	Na^+-	NaOH	1.5–2	1
Stong base anion exchanger, type I	OH^--	NaOH	4 –5	1
	Cl^--	NaCl	4 –5	1
	Cl^--	HCl	4 –5	1
	SO_4^{--}-	Na_2SO_4	4 –5	1
	SO_4^{--}-	H_2SO_4	4 –5	1
Strong base anion exchanger, type II	OH^--	NaOH	3 –4	1
Weak base anion exchanger	free base	NaOH	1.5–2	1
	free base	NH_4OH	1.5–2	1
	free base	Na_2CO_3	1.5–2	1
	Cl^-	HCl	1.5–2	1
	SO_4^{--}-	H_2SO_4	1.5–2	1

different concentrations may lead to better results with certain types from various manufacturers or for certain purposes. For example, we may note that a 6% nitric acid solution should be used for the removal of heavy metals from a strong acid cation exchanger based on polystyrene with 8% divinylbenzene. Special caution in the choice of a regenerant is indicated when the ions to be exchanged may form precipitates with the regenerant on the resin. In such a case, it is most suitable to use a 1–2% regenerant solution first and to convert the resin intermittently into the Na^{+}-form by treatment with a 4–10% NaCl solution. This leads to the gradual removal of the ion, which gives rise to precipitates until the regeneration can be completed as usual without involving a risk for the resin.

Special developments of synthetic ion exchange resins require separate consideration since they differ from customary resin types either because of their special matrix or their particular functional groups.

Macroreticular ion exchange resins. Macroreticular ion exchangers are types in which a solvent was used during their production from the monomers, so that a macroporous matrix structure is formed in the course of polymerization [366, 374, 541]. These macroporous structures with large internal surfaces can be sulfonated very easily and completely and are much more resistant to osmotic shock. Furthermore, they are extremely uniform in external shape and, in contrast to the gel types, they are opaque. Pore sizes of several thousand angstrom diameter and surfaces of up to 100 m^2/g and more can be obtained. To prevent collapse of the structures, a larger portion of cross-linking agent needs to be used. However, in connection with the large internal porosity, this leads to a number of advantages which become evident in a smaller swelling difference in polar and nonpolar solvents, a smaller loss of volume during drying, and a higher oxidation resistance. Because of their higher porosity, larger molecules can also penetrate the interior. The economy of macroreticular ion exchangers is somewhat limited by their lower capacities and higher regeneration costs. However, their suitability for catalytic purposes is unique. Macroreticular ion exchangers are fully developed products which are in frequent use because of their advantages [268].

Isopor ion exchange resins. The isopor ion exchangers are a group in which cross-linking and pore structure are modified to obtain a large-pore structure during the chloromethylation with chloromethyl–methylether and aluminum chloride in the preformed styrene–divinylbenzene beads [355]. In practical application, they show little sensitivity to organic fouling. In some cases, they are particularly suited as anion exchangers for the removal of silicate. Probably the maximum structural homogeneity attainable up until now has been achieved in isopor ion exchangers, since highly cross-linked regions are relatively rare and cross-linking is more or less homogeneous [356].

Specific and chelate ion exchangers. Specific ion exchangers are those types in which functional groups were introduced which have the properties of a specific reagent. The specificity is based on the chemical structure of the ion exchanger itself and may not be confused with the phenomenon of selectivity. Because of its specificity, an ion exchanger can sorb one ionic species to the exclusion of others under a broad range of conditions. Their mechanism of action will be explained by an example:

Skogseid [556] subjected polystyrene to nitrogenation, reduction, conversion of the formed polyaminostyrene with picrylic chloride, and renewed nitrogenation, and obtained an exchanger containing the following building blocks (I):

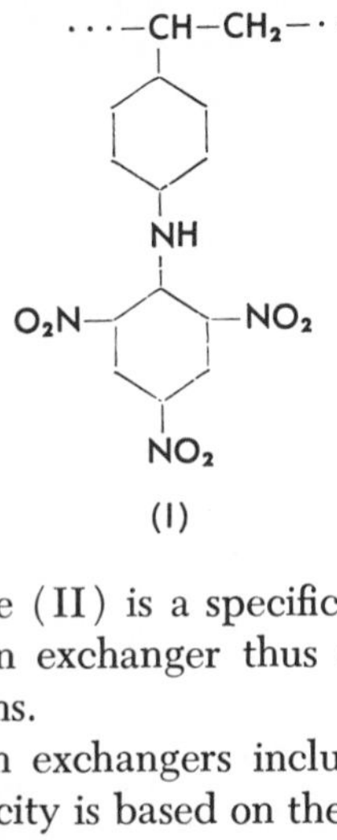

$$\cdots-CH-CH_2-\cdots \quad NO_2$$
$$O_2N- \quad -NO_2$$
$$NH \quad NH$$
$$O_2N- \quad -NO_2 \quad O_2N- \quad -NO_2$$
$$NO_2 \quad NO_2$$

(I) 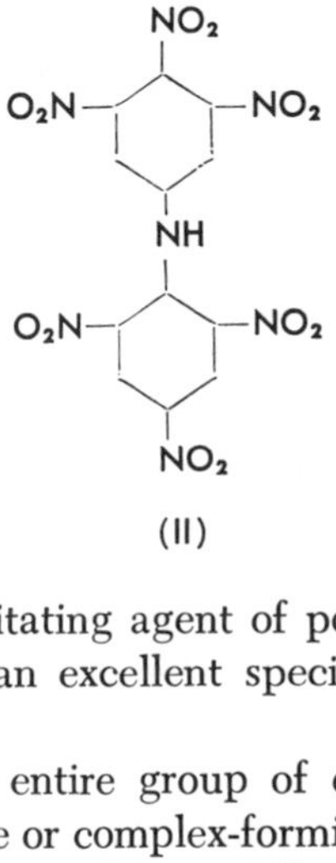 (II)

Dipicrylamine (II) is a specific precipitating agent of potassium, and the above ion exchanger thus shows an excellent specific uptake of potassium ions.

Specific ion exchangers include the entire group of chelate resins whose specificity is based on the chelate or complex-forming functional group. The best known commercial type is the iminodiacetate ion exchanger (Dowex A-1), in which the iminodiacetate groups are directly attached to the styrene matrix:

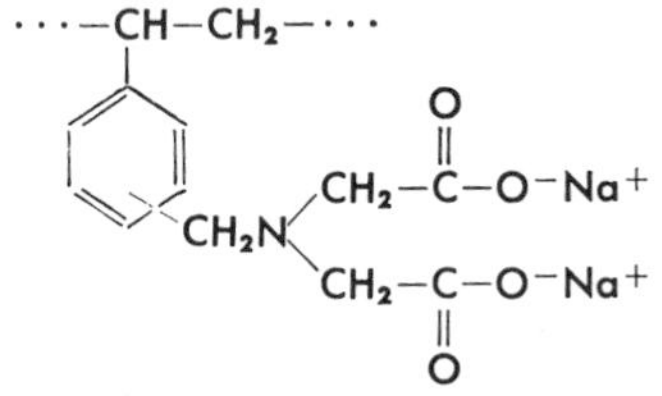

$$\cdots-CH-CH_2-\cdots$$
$$CH_2N\begin{cases} CH_2-\overset{O}{\overset{\|}{C}}-O^-Na^+ \\ CH_2-\underset{O}{\underset{\|}{C}}-O^-Na^+ \end{cases}$$

and which can fix polyvalent ions with a high affinity by the formation of heterocyclic metal chelate complexes:

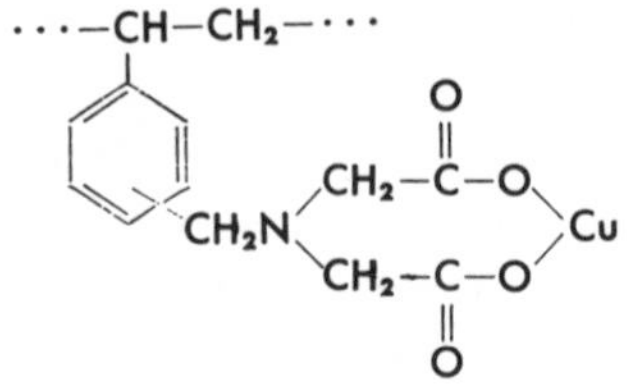

$$\cdots-CH-CH_2-\cdots \quad (C_6H_4)-CH_2N(CH_2-C(=O)-O)_2Cu$$

A series of specific and chelate-forming ion exchangers has been produced, but their main application probably will be in the field of analysis [63, 267, 453].

Snake-cage ion exchangers. The so-called snake-cage ion exchangers were developed as a separate group of ion exchange materials to permit the removal of salts from a single polyelectrolyte and regenerate the exhausted resin with water. In these snake-cage resins, a polymeric linear "snake" is formed by the polymerization of suitable monomers with a given charge in such a way that it will be located in a polymeric cross-linked "cage" of opposite charge. Since the distance between charged groups determines their effect, some efforts were needed to obtain a controlled structure of the effective interstices between negative and positive functional groups. This was achieved by the polymerization of acrylic or methacrylic acid into different quaternary ammonium anion exchange resins and led to the desired structures [256, 606]:

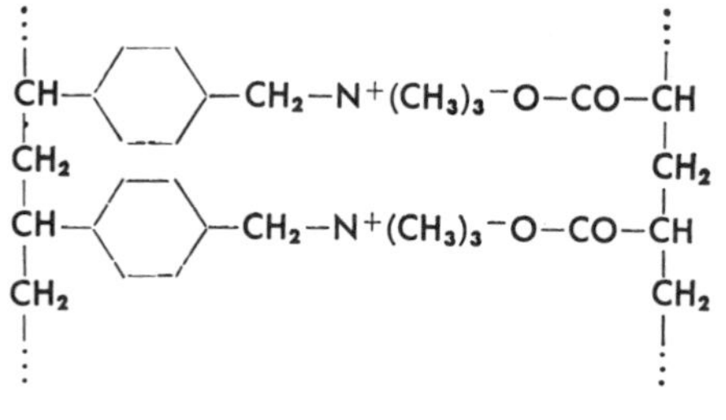

$$\begin{array}{l} \vdots \\ CH-(C_6H_4)-CH_2-N^+(CH_3)_3\,^-O-CO-CH \\ CH_2 \qquad\qquad\qquad\qquad\qquad\qquad\qquad\qquad CH_2 \\ CH-(C_6H_4)-CH_2-N^+(CH_3)_3\,^-O-CO-CH \\ CH_2 \qquad\qquad\qquad\qquad\qquad\qquad\qquad\qquad CH_2 \\ \vdots \end{array}$$

Such a snake-cage ion exchanger can exchange salts from aqueous solutions of organic compounds, such as glycerol, by simple sorption and can be regenerated subsequently by washing with water [639]. Since the snake-cage ion exchangers can also have a selective action, they also permit the separation of electrolytes, such as NaCl and NaOH. The ion retardation process is carried out with snake-cage ion exchangers.

Among the synthetic ion exchange resins described above, those of the gel type still have the greatest importance since this group continues to be used most frequently in industry. The production of an ion exchanger requires some specialized knowledge, so that it will hardly be worthwhile to attempt homemade preparation. Ion exchangers also assume a sort of intermediate position between that of a chemical product which is manufactured and further processed, and equipment serving for the manufacture of a certain product. They are highly sophisticated chemical materials but are used only as processing aids and thus are only the transfer agents of a chemical process without being completely changed or turned to another purpose. Because of the possibility for a reversible conversion into different forms, they have diverse applications and should have an infinite life. Naturally, this is made impossible by the fact that like any material they are subject to chemical and mechanical wear and thus have only a limited life.

The ion exchangers available today are commercial products from several manufacturers. They have registered tradenames, such as Allasion, Amberlite, Chempro, De-Acidite, Diaion, Dowex, Duolite, Imac, Ionac, Kastel, Lewatit, Liquonex, Mykion, Permutit, Wofatit, Zeo-Karb, Zerolit, and others. In Russian and other Eastern bloc countries, ion exchangers are usually referred to as cationites or anionites. Since it has frequently become customary to designate individual types only by letters or numbers, such as SBS, 10-P, etc., this somewhat confusing dual nomenclature must be retained [462, 585]. The Appendix lists a selection of important commercial ion exchangers frequently cited in the literature. The number of available types is so large that it has become almost confusing. Data on their characteristics have been derived from sales literature of the manufacturers—the best source of information concerning the applicability of an available synthetic ion exchange resin.

These designations give the specification of an ion exchanger by covering the necessary data resulting from its properties. The acidity and basicity follow from the functional group with its exchange activity and thus are immediately understandable. The production process determines the external form of an ion exchanger. While polymerization resins are generally delivered in the form of beads, polycondensation resins are available as milled granulates. The color of individual types ranges from white through yellow, brown, and dark brown to black. Important additional characteristics are their total exchange capacity, moisture-holding capacity, particle size distribution, physical structure, and stability as well as data on elutable components. In practice, a knowledge is needed of additional data which depend on the intended application and frequently need to be determined specifically for special uses.

In conclusion, it may be said that the criteria which have led to the present significance of synthetic ion exchange resins are their great mechanical strength and chemical resistance, high exchange rates, high capacity and the possibilities of varying their properties as a result of their synthetic nature. Their only disadvantage is their limited range of operating temperatures, although this range is also sufficient for most purposes.

2.1.1 *Properties of Synthetic Ion Exchanger Resins*

The properties of ion exchangers—and the discussion—might be simplified to capacity, equilibrium properties, and kinetic behavior since these criteria mainly determine the ion exchange process; however, they are not sufficient by themselves to characterize synthetic products. It is precisely the possibility of predetermining the properties of exchanger materials by their synthesis which has been the decisive factor in giving synthetic ion exchange resins their superior position compared to natural ion exchange materials. These synthetic possibilities in themselves, however, have the purpose of manufacturing products which always satisfy the objective, are of high quality, and, equally important, can be used economically.

The objective of an ion exchanger is satisfied by a sufficiently high capacity and rapid adjustment of equilibrium. The selectivity which appears at this stage immediately demonstrates, however, that a number of other factors which influence the properties of an ion exchanger would perhaps remain neglected at first. The quality to be demanded of an ion exchanger includes its physical and chemical resistance and the characteristic to release none of its own components to the surrounding media; this, in turn, can be given a positive rating only if oxidative, hydrolytic, thermal, or radioactive influences neither produce nor accelerate such a release. Synthetic ion exchange resins offer sufficient economic advantages when they can be manufactured inexpensively and have a long life.

The purpose of describing the properties, in addition to completeness, is to offer practical instructions for the measurement of criteria required in practice. It can be easily seen that the equipment needed for this experimental work is simple and inexpensive—a characteristic which applies generally to all ion exchange equipment.

As yet no criterion exists for the kinetic behavior of ion exchange resins; however, in the future an inclusion of this property in the specifications will be indispensible.

2.1.1.1 Moisture content and density

Commercial ion exchange materials have a certain moisture content in the form of bound water resulting from the hygroscopic properties of the resin. Beyond this, ion exchangers can take up free water or surface water which can be removed as an unbound amount of water by centrifuging and has no influence on the exchanger properties.

The quantity of bound water depends on the nature, amount and form of the functional groups, which have a positive influence on moisture uptake, and on the density of network formed by cross-linking between matrix molecules, which permits a moisture uptake only until the osmotic forces are compensated. The resulting moisture content is expressed in per cent of moisture per weight of wet resin, in per cent of moisture per weight of dry resin, or in weight or mole number of water per equivalent of exchange capacity. With a higher degree of cross-linking (measured by the divinylbenzene content), less water is bound by the exchanger, as shown by Figure 16. The uptake of moisture from the atmosphere with different relative humidities depends on the ionic form of the exchanger, as demonstrated by Figure 17.

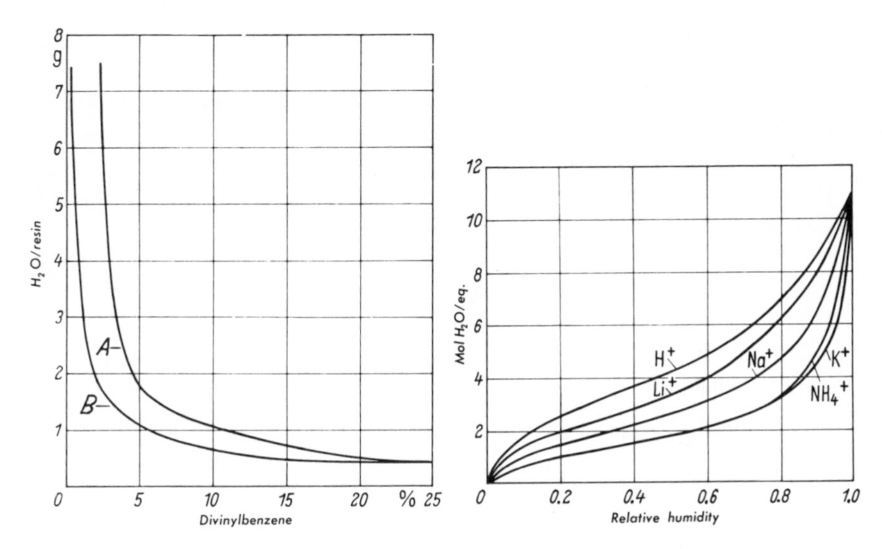

Figure 16 Moisture regain of cation exchangers of different degrees of cross-linking. A = strong acid sulfonic acid exchanger; B = weak acid carboxyl exchanger.

Figure 17 Moisture regain of a strong acid cation exchanger as a function of the relative humidity and ionic form.

A number of methods can be used to determine the bound water in ion exchange resins. Oven drying or azeotropic distillation are techniques which are generally used for high polymers. Their disadvantages can be circumvented by titration with Karl Fischer reagent in which the bound water can be determined rapidly and satisfactorily in methanol or pyridine [478]. However, all of these techniques fail when the residual water content is less than 100 ppm. Since a knowledge of such low moisture contents is important for ion exchange in nonaqueous solvents, a method was developed with the use of tritium-labeled water as the indicator. This method has a limit of detection of 10^{-10} mg H_2O [66]. A comparison with the Karl Fischer method showed that titration is of greater advantage for more than 0.5 wt.% of water.

The density of dry water-free ion exchange resins is about 1.2 for anion exchangers and about 1.4 g/ml for cation exchangers. A moisture uptake produces a characteristic density change in the various types, so that it amounts to about 1.3 in the strong acid polystyrene cation exchangers and to about 1.1 g/ml in strong basic anion exchangers. For practical purposes, the apparent or bulk density is decisive, amounting to 0.6–0.8 kg/1 as a rule.

2.1.1.2 Particle size

Ion exchange beads and granulates. Ion exchange beads or granulates generally are marketed in particle sizes of between 0.04 and 1 mm. Particle sizes of more than 1 mm tend to fragment during production. The listed measure of particle size is the diameter in mm or, as in the Anglo-Saxon literature, according to standard screen sizes in "mesh" values. The American standard screen size (US mesh) can be converted into millimeters by the following rule of thumb:

$$\frac{16}{\text{mesh}} = \text{particle diameter in mm}$$

Table 2 offers a comparison of these two units of measurement.

For British standard screens (BSS mesh) the following conversion formula applies:

$$\frac{12.2\text{–}15.5}{\text{BSS mesh}} = \text{particle diameter in mm}$$

Table 3 compares BSS mesh values and particle diameters.

The ion exchange particles have a different volume in the dry and the wet states and therefore also have different particle sizes. This difference is due to the moisture-holding capacity of the exchange resins and depends on the nature of the functional group and the degree of cross-linking. If the functional group is known, swelling factors of a certain exchanger can be stated as a function of the percentage content of cross-linking agent. A microscopic technique has been developed to investigate the particle volume [183]; it offers information on the volume in the swollen state but as a function of the loading state.

To determine the particle size of a dry exchanger, a screen analysis with the use of a standard set of screens is used in the simplest case. If customary screen sets are used, no other discussion is necessary.

Table 2
Particle size in US mesh and mm

US mesh	*mm diameter*
16– 20	1.2 –0.85
20– 50	0.85–0.29
50–100	0.29–0.15
100–200	0.15–0.08
200–400	0.08–0.04

A different method, which is more suitable for a comparison of particle sizes, makes use of the effective particle size and the similarity coefficient. The effective size is a screen size which passes 10% of the total quantity while 90% is retained. The similarity coefficient is obtained as the ratio of the mesh size in mm of the screen which passes 60% and the mesh size of the screen which passes 10%. This value offers information on the sharpness of the particle size distribution, for with a smaller similarity coefficient, the distribution is narrower.

In many cases, however, the particle size of the wet resin must be considered, since wet resins are used in most ion exchange applications. Wet screening is carried out so that about 50 ml of an exchanger, after preswelling in distilled water for 2 hours, is loaded on the largest mesh screen on which it is to be classified. All particles of smaller particle size are now flushed through the screen with a flow of distilled water. This separates the upper limit of a certain particle size range. The fraction which has passed through the screen is loaded on the next

Table 3
Particle size in BSS mesh and mm

BSS Mesh	*Particle diameter in mm*	
10	1.500	
20	0.735	
30	0.485	
40	0.360	
50	0.286	
60	0.235	Linearly
70	0.200	decreasing
80	0.173	factor
90	0.155	from
100	0.136	15.0 to
120	0.125	12.2
140	0.093	
160	0.079	
180	0.069	
200	0.061	

screen, which represents the lower limit of the desired particle size range, and is again treated as before. The particles retained on this screen are flushed off with water; it is advisable to make use of a brush to remove particles adhering to the screen. If this procedure is used with several screens, the result of the screen analysis expresses in per cent the amount of wet exchanger retained by each screen.

The use of exchange resins in ion exchange chromatography has demonstrated that the particle size of the exchanger material and its uniformity are the most important conditions to be satisfied for the attainment of a sharp separation. Hamilton [253] has described a method for the preparation of uniform particle size fractions which is

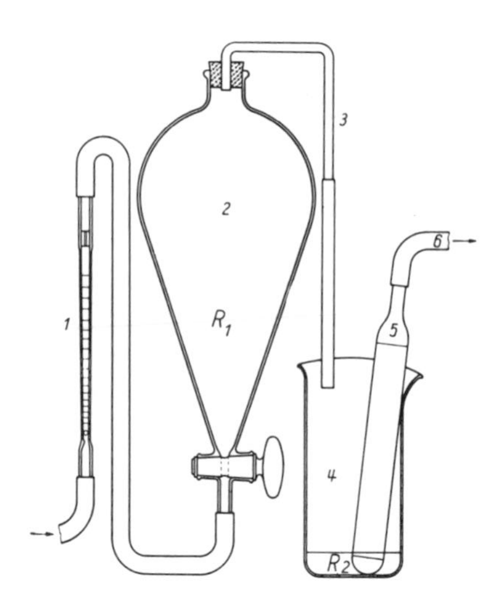

Figure 18 Apparatus of Hamilton [253] for the recovery of uniform particle size fractions. 1 = flow meter; 2 = separatory funnel; 3 = overflow; 4 = beaker; 5 = ball filter; 6 = pump connection; R_1 = charged exchanger mixture; R_2 = separated exchanger fraction.

carried out in the apparatus shown in Figure 18. The ion exchanger is charged into the separatory funnel and distilled water is passed through in an ascending flow. Since the rate of settling in water is proportional to the effective particle size, all fines are removed first, while the desired particle sizes and the oversize fractions remain in the separatory funnel. After changing the collecting beaker, the flow rate is increased and the desired particle size is flushed out. With careful manipulation, it is possible to obtain fractions in which the sizes of 60–80% of the particles do not deviate by more than $\pm 3\ \mu$ from the average. About 40 liters of water is needed for the separation of narrow cuts from a 2-liter separatory funnel.

A modification of this method was developed by Vassiliou and Kunin [614]. Two separatory funnels of different size are connected in series, and with the use of a circulating pump and a flowmeter a more rapid as well as simultaneous separation into different size fractions is obtained.

The required particle size depends on the operation to be performed and, if necessary, needs to be determined to optimize the ion exchange process. In the laboratory, a particle size of 0.3–0.5 mm is correct in most cases. This is the lowest limit for industrial installations. In investigations of the distribution coefficients of metal cations as a basis for chromatographic separations, particle sizes of 100–200 mesh are usually employed. Smaller particles are mechanically more stable than coarse fractions, a fact that needs to be kept in mind if the exchanger bed is moved or if it is subject to large volume changes. In any case, it is of advantage to maintain a uniform granulometry. For example, if ion exchange resins are to be milled for analytical purposes, a moisture content of about 40% is best. After swelling, the particles are milled in a meter mill at 13,000–15,000 rpm, then are screened or fractionated by the method of Hamilton.

The particle size may have a purely hydraulic or kinetic influence on the ion exchange process. In the column process, the dependence of the flow rate on particle size is most apparent. Since the frictional resistance is higher with smaller particle sizes, the flow rate also decreases with decreasing particle size. To prevent a complete holdup of the liquid to be filtered in the column, an overpressure must be applied on it by a suitable device. In an operation with a descending liquid front, a volume loss which is greater with small than with large particle sizes occurs at the head of the columns. Inversely, when working with an ascending liquid front, a volume expansion of the exchanger bed which is also greater with small than with large particle sizes can be observed. The influence of particle size on the exchange rate can be seen in Figure 19. The exchange rate will increase as the

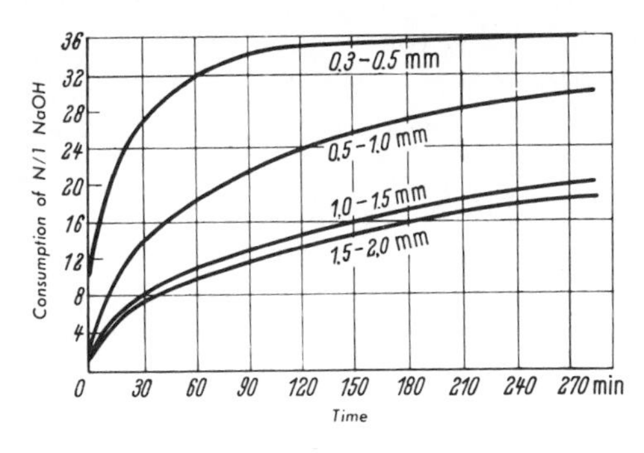

Figure 19 Influence of particle size on the exchange rate [245]). Particle sizes given in mm.

particle size decreases. The diffusion path lengths of the exchanging ions to and from the active sites are shorter with smaller particle sizes, so that exchange will be more rapid.

Ion exchange powders. Milling of ion exchange beads or granulates leads to an ion exchange powder with a considerably larger surface, an aspect which makes it a special type of ion exchanger. Powdered ion exchangers are finely divided resins of about 30 μ with an increased reaction rate and high adsorption power; they can be formed into compressed cakes or pellets. Powdered ion exchangers have consequently found diverse applications in industrial processes and as supporting materials [416]. Powdered ion exchangers cannot be regenerated and therefore can be used only once, but as a rule this is of subordinate importance in their intended fields of application.

Ultrafine ion exchange resins. So-called ultrafine ion exchange resins with a particle size of 0.5–1.5 μ are also available in the form of microspheres or agglomerates of microspheres. Because of their larger surface, the exchange rate is higher due to the greater accessibility of ionic sites. Their application in columns is limited. However, these particle sizes seem to have their own field of application in the production of ion exchange papers and for incorporation in plastics, films, coatings, and fibers [535].

2.1.1.3 Cross-linking

The degree of spatial interlinkage of ion exchangers is determined by the production process. Commercial synthetic ion exchange resins of the gel type usually contain 2–12% divinylbenzene as a cross-linking agent. Since cross-linking has a controlling influence on several properties of an exchanger, this step offers an opportunity in synthesis to produce special types of exchangers for special purposes.

Cross-linking influences not only the solubility but the mechanical stability, exchange capacity, water uptake and swelling behavior, volume changes in different forms of loading, selectivity, and chemical as well as oxidation resistance of ion exchangers. Exchangers with a low degree of cross-linking are soft and mechanically unstable (in the swollen state), while highly cross-linked products are hard and brittle with an increased sensitivity to osmotic influences. The exchange capacity of a dry exchanger increases with a smaller degree of cross-linking and vice versa, while the exchange capacity per volume decreases with decreasing cross-linking because of more extensive swelling. With a higher degree of cross-linking, the moisture content and swelling behavior of exchangers decrease. The volume change of the loading form which changes during regeneration is greater with a lower than with a high degree of cross-linking. The selectivity increases with higher degrees of cross-linking. It is also of interest that a higher degree of cross-linking improves chemical and oxidation resistance. In particular, however, one should draw attention to the exchange rate which decreases with a higher divinylbenzene content, as demonstrated by Figure 20 for a strong acid cation exchanger based on polystyrene.

In the literature, the degree of cross-linking usually is characterized by the value "X% cross-linked." This expresses in a percentage the amount of cross-linking agent which was used to produce the ion exchanger. In the Dowex types, the degree of cross-linking can be recognized from the type identification. For example, Dowex 50X4 means that this exchanger consists of 96% styrene and other monovinyl monomers, and 4% divinylbenzene as cross-linking agent.

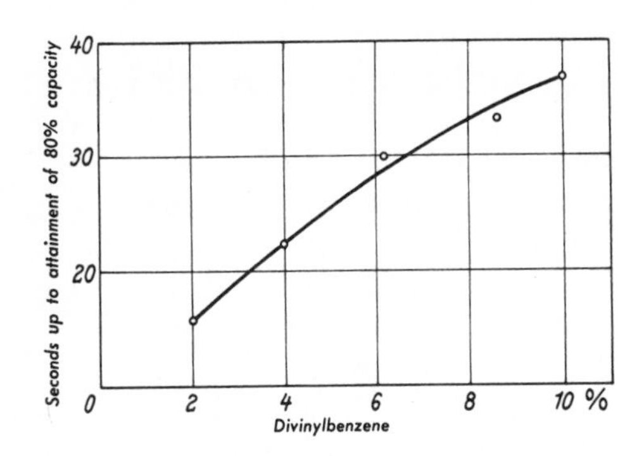

Figure 20 Exchange rate as a function of cross-linking.

However, in an evaluation of cross-linking in cation and anion exchangers, it must be kept in mind that the cross-linking data do not have the same meaning for both types of ion exchangers. The degree of cross-linking of a polystyrene–sulfonic acid exchanger (cation exchanger) is primarily determined by the concentration of bridge formerly in the monomer. Although additional cross-linkages can theoretically form during sulfonation by the formation of sulfur bridges, this effect is of less practical significance (although it cannot be overlooked) from the standpoint of production of highly uniform products [75]. In the case of anion exchangers, however, additional cross-linkages can be introduced by chloromethylation—which is the usual technique for the production of these exchangers—with the formation of methylene bridges between the benzene rings of parallel hydrocarbon chains. This new formation of cross-linkages takes place approximately as follows, according to Friedel-Crafts:

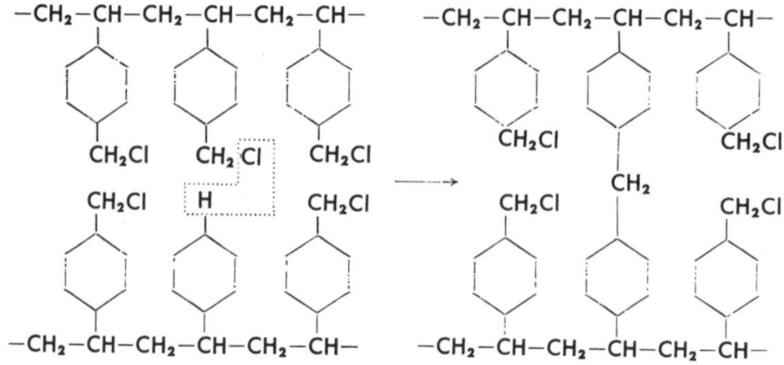

$-CH_2-CH-CH_2-CH-CH_2-CH-$ → $-CH_2-CH-CH_2-CH-CH_2-CH-$

CH_2Cl CH_2Cl CH_2Cl → CH_2Cl CH_2 CH_2Cl

CH_2Cl H CH_2Cl → CH_2Cl CH_2Cl

$-CH_2-CH-CH_2-CH-CH_2-CH-$ → $-CH_2-CH-CH_2-CH-CH_2-CH-$

This additional cross-linking can represent a considerable part of the total cross-linking in the exchanger and cannot be considered solely as a supplementary contribution [18]. Consequently, giving the degree of cross-linking in per cent of cross-linking agent used in anion exchangers is not of real significance even though this is found in the literature.

Thus, by modifying the degree of cross-linking, *i.e.*, by a change in the quantity of bridge forming agent with respect to monomer utilized, it is possible to obtain a range from linear structures through those with a low (0.5–1% divinylbenzene) and intermediate (4–10% divinylbenzene) degree of cross-linking to a close packed network through which the larger ions can no longer pass. This results from the decreasing mesh width of a spatial network with an increasing degree of cross-linking. The "pores" formed in the network of the exchanger thus have different sizes; in a normally cross-linked resin, these are assumed to be 10–20 Å. The property of ion exchangers—having a separation effect on ions as a result of their pore size—is the basis for the use of ion exchangers as ionic sieves [67].

If the unknown degree of cross-linking of a complete cation exchanger is to be determined, the method of Boyd and Soldano may be used. It has been found that an empirical relation exists between the equivalent volume of the dry (V_t) and the equivalent volume of the swollen (V_q) ion exchange resin:

$$\frac{V_q}{V_t} - 1 = \frac{k}{x}$$

In the above, x is the degree of cross-linking and k is a constant for a given salt form. The following k-values can be used for styrene–sulfonic acid exchangers:

Cation	H^+	Li^+	NH_4^+	K^+	$(C_2H_5)_4N^+$
k	10.7	11.8	9.1	9.1	5.2

from which x is immediately obtained in per cent divinylbenzene provided that the cross-linking falls into a range of 1–25%. The extent to which this or another relation applies to other cation exchanger types and anion exchangers has not yet been sufficiently investigated.

Attempts have repeatedly been made to determine cross-linking with physicochemical methods of analysis. In the infrared spectrum, absorption band differences were found at 830 cm^{-1} and 797 cm^{-1} compared to pure m-DVB and p-DVB[191], indicating the degree of cross-linking with these isomers. Corresponding measurements should make it possible to determine most cross-linking.

In addition to this relationship between cross-linking and swelling and the above-indicated effects on other properties of ion exchange resins, the neutral salt adsorption is also influenced and will be described more fully here.

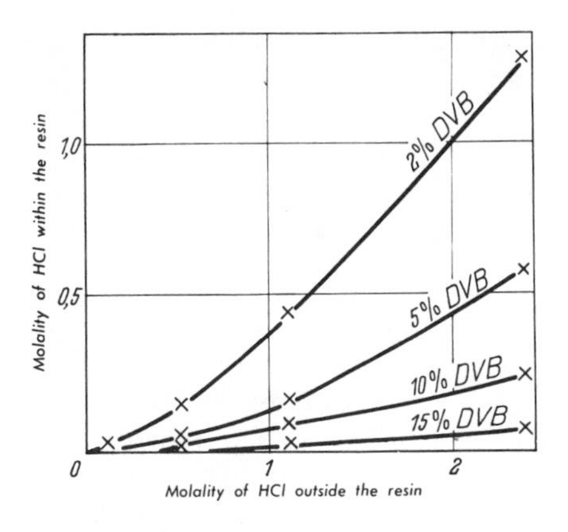

Figure 21 Neutral salt adsorption of HCl on a cation exchanger of different degrees of cross-linking [468].

Neutral salt adsorption is the phenomenon of adsorptive binding of neutral salts by ion exchangers, so that a cation exchanger in the presence of a high concentration of the external electrolyte can take up additional anions which entrain the corresponding cations. This is shown in Figure 21 for the adsorptive fixation of hydrochloric acid on a polystyrene–sulfonic acid resin as a function of the degree of cross-linking.

2.1.1.4 Porosity

As noted above, the porosity of ion exchange resins is related to the degree of cross-linking and the network formed as a result. The size of these capillary channels depends on the degree of cross-linking but is not uniform in one and the same ion exchange particle.

Detailed studies of the "porosity" of ion exchangers have shown that some caution is indicated in the use of this term. As noted earlier, the network of exchanger resins results in voids which have a greater of smaller moisture regain capacity depending on the degree of cross-linking. In contrast, it has been possible to produce ion exchangers which represent true phase-dispersed systems. The two types already differ in appearance. While the former are clear and transparent, the latter are opaque, and when combined with water show a cloudy opalescence. But this results in problems of nomenclature. Since no uniform terminology has become accepted for the distinction, and since the custom of speaking of the "porosity" of the exchanger network with reference to the mesh width is widespread, the former case shall be identified as apparent porosity and the latter as the real porosity whenever necessary. The really porous ion exchangers show relatively little swelling with a higher degree of cross-linking; but because they are permeated by fine pores, they have a relatively high capacity. The pore sizes may be measured on the basis of experiences in the determination of pore radii in swollen gels, among other things. Since exchange resins that have been swollen in aqueous solution exhibit diffusion coefficients of radioactively labeled counterions below 0°C with a more than exponential decrease with the reciprocal value of the absolute temperature, the theory of mixtures permits us to use these results to determine the fraction of pores in which the swelling agent has solidified [527].

The porosity also influences other ion exchanger properties, mainly the capacity and selectivity. The capacity would be much lower than it really is if the exchange resins had no pores and if only the functional groups at the surface were active for ion exchange. The high ion exchange capacities are obtained only because active functional groups located in the interior of the pores also contribute to the capacity. A limiting factor here is the ratio of pore size to the size of ions or molecules which are to be exchanged, since the resulting sieve effect leads to a selection of counterions on the basis of size. Consequently, larger organic ions are found to have an effective capacity which frequently is far below that obtained with small ions or calculated theoretically.

Since the "porosity" has such far-reaching effects on the properties and thus on the application of ion exchangers, it is understandable that discussions concerning its nature and degree in individual types of ion exchangers have not ceased. To obtain a better understanding of the pore character, Mikes [430] observed the formation of the disperse system from the start of polymerization. In syntheses leading to true porous products, the structure is built up in individual steps in which a molecular arrangement forms first, followed by intermolecular rearrangement and, finally, regrouping with the formation of visible pores. Kunin [369] again pointed out the difficulties in nomenclature and the confusions which have been repeatedly mentioned in the present discussion. The distinction between "macroreticular" and "microreticular" for a description of the pore character therefore appears very useful. "Macroreticular" pores or true porosity refers to structures in which the pores are larger than the atomic distances and are not a part of the gel structure. Their size and appearance depend little on the surrounding conditions. The expression "microreticular" refers to the apparent porosity of atomic dimensions which depends on the swelling behavior of the gel and, with regard to its size, on the surrounding conditions.

Measurement of pore sizes must be carried out to characterize pore structure. In addition to techniques for the determination of the total pore volume from the true density determined with a helium densitometer and the apparent density measured with a modified mercury porosimeter, a study of pore sizes requires a distinction between the specific surface O, pore volume V and pore radius r [260]. On the basis of model theory, the following relation exists between these parameters:

$$r = 2.7 \frac{V}{O}$$

The surface is determined by the Brunauer-Emmett-Teller method (BET method) or by the emanation method of Hahn. The pore volumes are measured by the saturation values of carbon tetrachloride adsorption [52] or by a titration method [443], either of which leads to values which differ by less than 10%. The pore radii can then be calculated from the measured surfaces and pore volumes on the basis of the above equation. With a method of van Bemmelen, Bachmann, and Maier, the pore radii and their distribution can also be determined directly. In addition to other techniques for the investigation of porosity [403], Kun and Kunin [367] developed a modified mercury method in which the exchangers can be investigated in the hydrated state. Additional information of the pore structure is offered by electron microscopy [467], which furnishes results in good agreement with the indirect methods and gives data on the appearance, position and character of the pores.

2.1.1.5 Swelling

The volume of an ion exchanger depends on several factors: (1) the surrounding medium (air, water, organic solvents), (2) the nature of the resin skeleton (type of matrix and cross-linking), (3) the charge density (nature and concentration of ionic groups), and (4) the type of counterions.

The volume change which takes place during transition from one medium into another and which is influenced by the other factors is known as swelling. This swelling is produced by osmotic pressure in the interior of the ion exchangers against the external, more dilute solution, so that the internal ion concentration decreases by a swelling which represents a solvent uptake. A distinction must be made between absolute swelling, in terms of the dried exchanger, and the difference in swelling volumes (breathing difference), *i.e.*, the volume change of a swollen exchanger under different loads.

Absolute swelling takes place when an air-dried resin becomes wet; this must be kept in mind when an ion exchanger sample is prepared. This volume adaptation, which is also known as preswelling, can be measured in a graduated cylinder; however, in laboratory practice it plays only a subordinate role since ion exchange resins are slurried into the columns only in the preswollen state. In industrial filter beds, preswelling must be performed carefully, possibly with the use of steam at elevated temperature, since undesirable malfunctions may occur otherwise as a result of fragmentation of the ion exchanger beads or too close packing.

During absolute swelling a certain quantity of water is taken up by the exchanger. Pepper, Reichenberg, and Hale [468] have described a centrifuge method for the determination of this quantity, the apparatus for which is shown in Figure 22. About 1 g air-dried resin is loaded into a glass filter attachment with a glass frit, and the assembly is immersed in distilled water for 1 hour at 25°C. The glass filter attachment is then inserted into a 15-ml centrifuge tube, closed with a rubber cap, and centrifuged for 30 minutes at 2,000 rpm. The filter attachment is weighed, immersed once more in water, and again centrifuged. It is then dried overnight at 110°C and brought to constant weight in vacuum at 1 mm Hg over phosphorus pentoxide at 110°C. Subsequently, or better yet before, the empty glass filter attachment is dried at 110°C and weighed, and after centrifuging, it is weighed with water as a blank. The dry weight and wet weight of the sample can thus be easily calculated from these data, so that the quantity of water taken up by absolute swelling can be determined [459].

In this case, swelling depends on the quantity of regained water. This is influenced first of all by the matrix structure, *i.e.*, the degree of cross-linking. While uncross-linked polystyrene sulfonic acid undergoes unlimited swelling in water, the moisture regain capacity of the exchanger decreases with increasing cross-linking. If the water contains an electrolyte, swelling furthermore depends on the electrolyte concentration. When this concentration increases, the moisture uptake will decrease, since the osmotic pressure difference between the external and internal solution is then smaller.

The swelling processes become more difficult to interpret when one changes from water to other solvents. Water with its strong dipole character interacts with the highly hydrophilic functional groups. As the dipole character of a solvent becomes weaker, swelling of the exchanger will be less pronounced since its electrostatic solvation tendency decreases. The ion exchange skeleton of hydrocarbon chains, on the other hand, has a tendency to take up less polar solvents, so that weak acid cation exchangers in the H-form, for example, swell more in alcohol than in water. The weakly ionic character of the functional groups and the resulting possibility of electrostatic solvation become subordinate to the affinity of alcohol and carboxyl groups [289].

The breathing difference is important in working with ion exchange columns. Its dependence on the species of counterion with a certain degree of cross-linking of the particular ion exchanger can be easily determined with the apparatus described by Blasius (Figure 23) [64]. For this purpose, a cation exchanger—for example, Dowex 50X1—can be used. A quantity of 4 g is charged into the 200-ml reservoir of the apparatus. The ground joints are then connected, and the ion exchanger is fed on the glass frit with opened stopcocks and is transferred into the desired state of loading. After complete loading, the

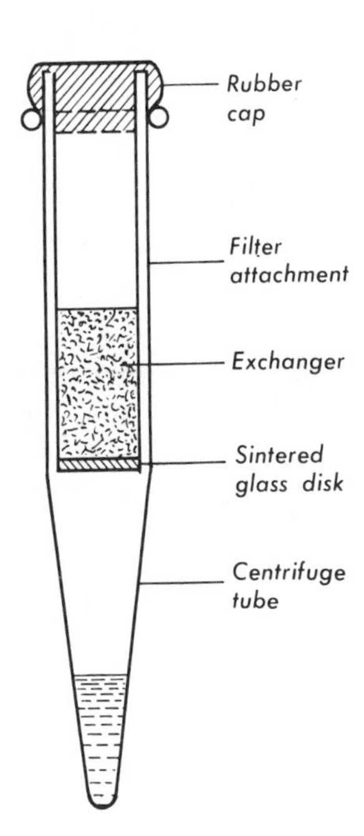

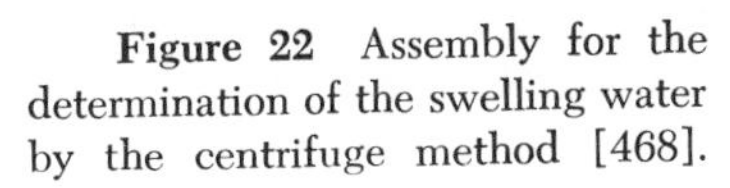
Figure 22 Assembly for the determination of the swelling water by the centrifuge method [468].

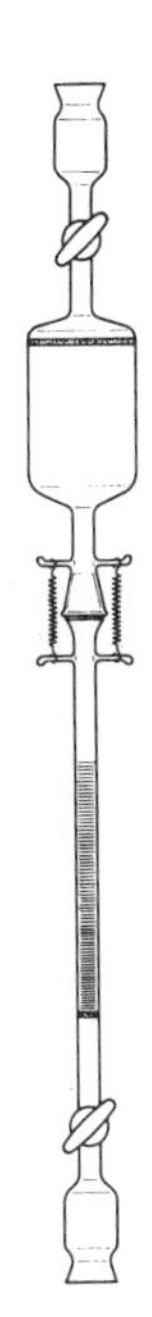

Figure 23 Apparatus for the determination of the breathing difference according to Blasius.

assembly is inverted and the resin is flushed into the graduated tube with distilled water; the tube is also closed by a fused-in sintered glass disk. As soon as the resin volume has become constant with further washing, it is read off from the scale. For example, if K^+ has been used as the counterion in a first experiment and Fe^{3+} in a second, a volume decrease will be observed. This phenomenon follows the general rule that the moisture holding capacity of this exchanger with a low degree of cross-linking decreases extensively when loaded with mono-, bi-, and trivalent ions. In a more highly cross-linked exchanger, the breathing difference in various loading states is smaller.

The conditions of the relation of swelling to the counterions is not always as simple as described in the above demonstration experiments. Although the rule has been accepted that the swollen volume of an exchanger is smaller the higher the valence of the counterion, the quality of equivalent ions also plays a role, since equivalent ions may differ in size and solvation power [224]. If the counterions exhibit a higher degree of dissociation, swelling will be more pronounced, while ion pair production and ion association will lead to a decrease of swelling. The weak acid cation exchangers have a smaller volume in the H-form than in the Na-form and the volume of weak basic anion exchangers is larger in the Cl-form than in the OH-form.

2.1.1.6 Capacity

The capacity is the most important property of an ion exchanger, since it permits a quantitative determination of how many counterions can be taken up by an exchanger.

In the course of development of ion exchangers, various definitions and units of measure were formulated for the capacity. It is important, first of all, to distinguish between total capacity and effective capacity. Total capacity is obtained from the total quantity of counterions capable of exchange. Effective capacity is that which can be fully utilized in an exchange column under the selected conditions. Depending on whether the total capacity—which alone is of interest in laboratory experiments—refers to the weight or the volume of dry or swollen resin, we obtain a capacity by weight or by volume. Since the weight and volume of an exchanger can differ greatly in the dry and the swollen state, as is readily evident from the discussions on cross-linking and swelling, very different values are obtained for the capacity as a function of the parameter to which its determined value is referred. It thus becomes necessary to indicate the specific units and conditions when capacity values are cited.

For laboratory and research purposes it has become increasingly popular to state the capacity in milliequivalents per gram (meq/g) or in milliequivalents per 100 g (meq/100 g). This is understood to refer to cation exchangers in the H-form and to anion exchangers in the Cl-form. In the case of production problems, the effective capacity is indicated in gram CaO per liter of exchanger (g CaO/l) or in kilogram CaO per cubic meter (kg CaO/m^3). This again demonstrates the greater tendency to use weight capacity in one case and volume capacity in the other.

To determine the weight capacity in meq/g exchanger, it is probably most suitable to use the method described by Fisher and Kunin [180]. This procedure is particularly appropriate for routine tests, since it is relatively rapid and furnishes reliable values while other methods, which undoubtedly furnish more precise values for the total capacity, may require a period of one week to six months for a determination.

The assembly shown in Figure 24, which was derived from the original literature, serves as the equipment; it consists of simple com-

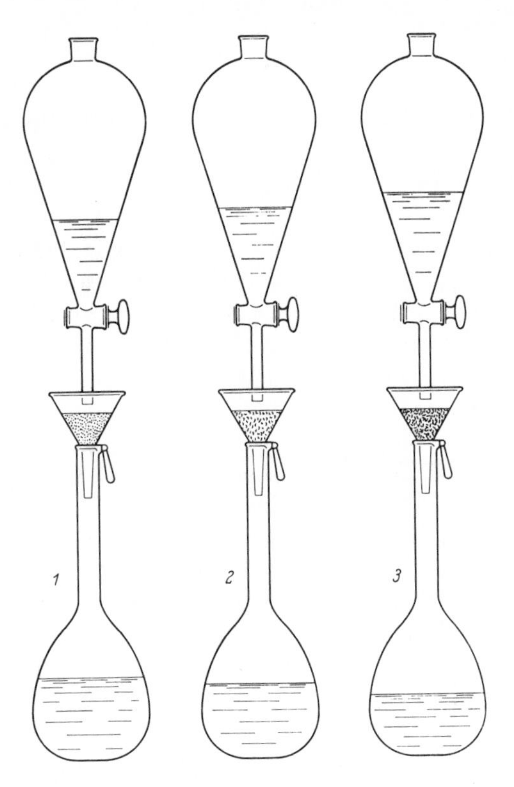

Figure 24 Equipment for the capacity determination according to Fisher and Kunin [180].

ponents available in every laboratory. The practical procedure consists of the steps which follow.

Determination of the capacity of a cation exchanger. The cation exchanger (5 g) is transformed into the H-form by slow treatment with about 1 liter N HNO_3 in the funnel of Figure 24. Subsequently, it is washed to neutrality with distilled water, suction-filtered, and dried in air. Of this quantity, 1.000 ± 0.005 g is weighed into a weighing bottle and is allowed to stand overnight in a dry 250-ml Erlenmeyer flask with 200 ml 0.1 N NaOH in 5% sodium chloride solution. An exchanger sample (1 g) of the same material is separately weighed into a weighing bottle, dried at 110°C overnight, and weighed again to determine the percentage of dry substance. Of the supernatant solution in the Erlenmeyer flask, 50 ml is back-titrated with 0.1 N H_2SO_4 against phenolphthalein. The capacity is then calculated by the formula:

$$\text{Capacity (meq/g)} = \frac{(200 \cdot \text{normality}_{\text{NaOH}}) - 4(\text{ml}_{\text{acid}} \cdot \text{normality}_{\text{acid}})}{\text{sample weight} \cdot \dfrac{\%\ \text{solids}}{100}}$$

and represents the total weight capacity of the exchanger in the dry H-form.

Nitric acid serves for exchanger transformation into the H-form, since hydrochloric acid and sulfuric acid may lead to precipitation in a reaction with the heavy metal cations present in the exchanger. The resin must be completely in the H-form before weighing of the sample, since differences in equivalent weights of different ions would lead to errors. The adjusted sodium hydroxide solution is treated with 5% sodium chloride to obtain a complete exchange equilibrium by the excess of Na-ions. A reproducibility of ± 1% can consequently be obtained.

Determination of the capacity of an anion exchanger. Air-dried anion exchanger (10 g) is transformed into the Cl-form by slow treatment with 1,000 ml N HCl in the funnel of Figure 24. The chloride is subsequently washed with alcohol until the filtrate is neutral to Methyl Orange, and is air-dried. Of this sample, 5.000 ± 0.005 g is weighed into a fresh funnel and the chlorine ion is eluted with 1,000 ml 4% sodium sulfate solution; the eluate is collected in a 1-liter graduated cylinder. Using aliquots of 100 ml of the throughput, the chloride is titrated with 0.1 N silver nitrate solution and potassium chromate as indicator. Separately, 1 g of the same exchanger sample is weighed into a weighing bottle, dried overnight at 110°C, and weighed back to determine the percentage of dry solids. The capacity is then calculated by the formula:

$$\text{Capacity (meq/g)} = \frac{\text{ml}_{\text{AgNO}_3} \cdot \text{normality}_{\text{AgNO}_3} \cdot 10}{\text{sample weight} \cdot \dfrac{\%\ \text{solids}}{100}}$$

and represents the total weight capacity of the exchanger in the dry Cl-form.

The exchanger is converted into the Cl-form to prevent errors due to the different equivalent weights of different anion forms and to prevent the possibility of a decomposition of the free base form during drying. The exchanger is washed with alcohol instead of water to avoid a possible hydrolysis of the salt form of weakly basic ion exchangers.

The capacities which are obtained by this method are not in complete agreement with a so-called analytical capacity. Analytical capacity refers to that calculated by a sulfur determination for cation exchangers of the sulfonic acid type and by a nitrogen determination for anion exchangers. The deviations can be explained by the fact that active sites are no longer accessible to ion exchange due to pore plugging or constriction. In sulfonic acid exchangers, moreover, a part of the sulfur may be bound in the sulfone form. That steric hindrance generally influences the measured capacity is evident from the fact that highly cross-linked exchangers exhibit different capacities when loaded with ions of different sizes.

For industrial purposes the effective volume capacity is most frequently determined, and this is appropriately done in a semi-industrial or industrial filtering test. It is performed with calcium chloride solution, the capacity in g CaO/l being calculated directly from the quantity of calcium determined in the form of oxalate.

The capacity depends on various external conditions, among which the influence of the pH is most important. Since the ionic groups represent strong or weak acids and strong or weak bases, they are naturally pH-dependent. Thus, carboxyl groups or phenolic OH-groups remain partially or completely undissociated at low pH values, so that the exchange capacity of exchangers with such groups with respect to metal cations from acid and neutral solutions is lower than from alkaline solutions. Anion exchangers with primary, secondary, and tertiary amino groups exchange acid anions from alkaline and neutral solutions to a lesser degree than from acid solutions. Ion exchangers with strong acid or strong basic groups (in the latter case, quaternary ammonium groups), however, are capacity-independent over a broad pH range. Accordingly, they are also active over a wide pH range and can be used under acid, neutral, and alkaline conditions. Table 4 shows a summary of the effective pH ranges of various types of ion exchangers.

Besides the exchange of ions, an additional ion adsorption may take place in ion exchangers, because the exchanger material adsorbs electrolytes beyond the capacity of the fixed ions (functional groups). Although the adsorption capacity in most cases is substantially smaller

Table 4
Effective pH range of different types of ion exchangers

Type of ion exchanger	*Approximate pH range*
Strong acid cation exchanger	4-14
Weak acid cation exchanger	6-14
Weak base anion exchanger	0- 7
Strong base anion exchanger	1-12

than the exchange capacity, it would, strictly speaking, have to be determined and expressed separately from the total capacity. It is admittedly difficult to distinguish between the exchange and the adsorption capacity.

In connection with the capacity and its dependence on the pH, it is important to note a method of pH titration of ion exchangers which permits a precise characterization of exchangers on the basis of their active groups. This experimental method, which was introduced by Griessbach [236], is based on the fact that cation exchangers in the H-form and anion exchangers in the OH-form can be titrated with alkalies and acids like ordinary acids and bases. The neutralization process is observed by measurement of the pH.

To perform the titration, a known quantity of 5–10 meq of the cation or anion exchanger is first transformed into the H- or OH-form with hydrochloric acid or sodium hydroxide, respectively, then washed with distilled water, and finally treated with N/10 NaCl solution in a beaker. Small portions of 2 N sodium hydroxide or 2 N hydrochloric acid are added with vigorous stirring, and the pH adjustment is observed on the glass electrode. For an evaluation of the test results, the measured pH-values are plotted as a function of the consumed quantity of alkali or acid.

Figure 25 shows the titration curves of some cation exchangers according to Griessbach. The capacity of the exchanger can be read from these curves at the intersection of the curve with the line for pH 7. Moreover, these curves can be used to evaluate the exchangers for their acid strength, operating interval, and buffering power.

2.1.1.7 Selectivity

The selectivity of ion exchangers is defined as the property of certain ion exchangers to exhibit a preferential activity for different ions. In

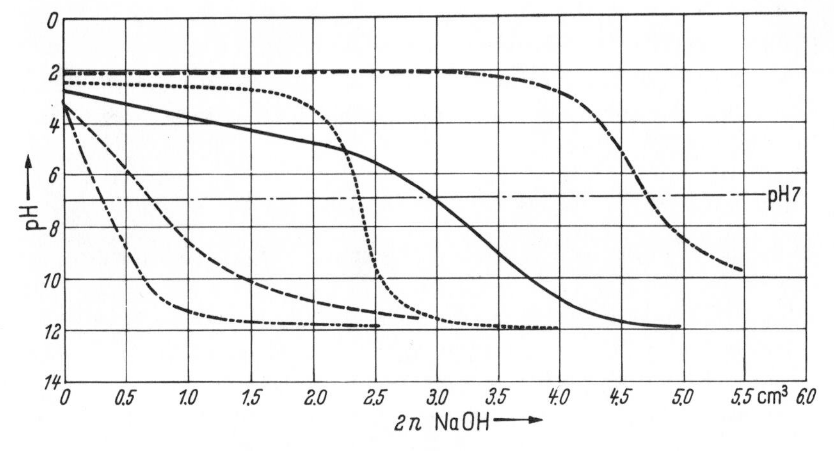

Key	Designation	Active Group
- - - - - -	K-resin	—OH
- - - - - - -	A-resin	$-CH_2SO_3H$
————	C-resin	—COOH
—··—··—	R-resin	$-SO_3H$
— — — —	Glauconite, dealkalized for comparison, test series for 20 g.	

Figure 25 pH-titration curves of some cation exchangers according to Griessbach [236].

the case of alkali ions, this means that these ions will be taken up with different readiness by cation exchanger and that they can mutually displace each other. The exchange affinity of alkali ions increases in the following sequence:

$$Li^+ < Na^+ < K^+ < Rb^+ < Cs^+$$

Thus, lithium is taken up less easily than sodium, sodium less easily than potassium, etc.

Stoichiometric and reversible ion exchange exhibits all characteristics of an equilibrium and follows the law of mass action. Thermodynamic studies have shown that ion exchange leads toward an equilibrium in which the quotient of the concentration ratios of the two exchangeable ions on the exchanger and in the surrounding solution becomes independent of concentration. This state in which the two ions assume a different distribution between the exchanger and the surrounding solution is characterized by a constant known as the selectivity coefficient K_s.

The selectivity coefficient is the ratio of the molar fractions of the exchanging ion pair A and B in the two phases:

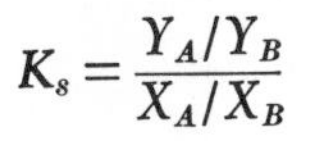

$$K_s = \frac{Y_A/Y_B}{X_A/X_B}$$

where Y is the concentration in the resin and X is the concentration in the solution.

For the sake of clarity, it will be assumed that an exchanger is in the Na-form and is placed into a KCl solution. The following reaction will then take place:

$$Na^+ (\text{in resin}) + KCl(\text{solution}) \rightleftarrows K^+ (\text{in resin}) + NaCl(\text{solution})$$

Its equilibrium is characterized by the constant K_s, which results from the above relation.

Since the ion activity in the resin phase cannot be determined, K_s is not a thermodynamically defined equilibrium constant but only a coefficient defined according to practical requirements.

It must be kept in mind that the selectivity coefficient is not a constant but is influenced by various factors, including: (1) the exchangeable ions (size and charge); (2) properties of the exchanger, *i.e.*, particle size, degree of cross-linking, capacity, and type of functional groups; (3) nature, *i.e.*, total concentration as well as concentration ratio of the existing ions both capable and incapable of exchange as well as the type and quantity of other substances in the solution; and (4) the reaction period.

Various theories have been proposed in an attempt to obtain greater insight into the conditions which determine selectivity [120, 172, 234, 235, 254, 398, 412, 432, 494]. Since none of these theories furnishes quantitative information, only a few rules for practical use, which have been found empirically and by which the selectivity of an exchange process can be estimated, will be presented.

At low concentrations of aqueous solutions and at room temperature, all exchangers give preference to polyvalent ions over the monovalent species in accordance with the selectivity sequence of:

$$Na^+ < Ca^{++} < La^{+++} < Th^{++++}$$

The preferential uptake of polyvalent ions is concentration-dependent, however, and decreases with an increasing concentration of the solution.

In the case of ions of identical valence, the selectivity at low concentrations and ordinary temperature increases with increasing atomic number according to the following selectivity sequences:

$$Li < Na < K < Rb < Cs$$
$$Mg < Ca < Sr < Ba$$
$$F < Cl < Br < J$$

The selectivity for H^+- and OH^--ions depends on the strength of the acids or bases formed from the functional group of the exchanger and the H^+- and OH^--ions. Weakly acid and basic ion exchangers are selective for H^+- and OH^--ions, since the solid acids and bases that are forming have undergone little dissociation and swelling.

The higher the degree of cross-linking, the higher will be the selectivity, *i.e.*, the difference of the selectivity coefficients of different ions. It can be observed, however, that the selectivity decreases again at high cross-linking (starting with 15%). Some theorists ascribe so much significance to this selectivity reversal that they consider the phenomenon equivalent to that of selectivity itself [492].

Table 5 offers an insight into the selectivity of a sulfonic acid exchanger having different degrees of cross-linking with respect to different cations.

Table 5
Relative selectivity coefficients of a sulfonic acid exchanger with different degrees of cross-linking

	Cross-linking		
Cation	*4%**	*8%**	*10%**
Li	1.00	1.00	1.00
H	1.30	1.26	1.45
Na	1.49	1.88	2.23
NH_4	1.75	2.22	3.07
K	2.09	2.63	4.15
Rb	2.22	2.89	4.19
Cs	2.37	2.91	4.15
Ag	4.00	7.36	19.4
Te	5.20	9.66	22.2

*Per cent of divinylbenzene as cross-linking agent

With regard to ionic size and selectivity, the rule is accepted that larger, particularly organic ions are taken up preferentially. Finally, anionic metal complexes show a similar behavior in being exchanged with particular selectivity by anion exchangers.

To determine the selectivity coefficient, the distribution of an ionic species and its concentration in the exchanger phase and the surrounding medium is determined after equilibration. The values found are plotted as molar fractions or per cent equivalents in the exchanger and the solution. Figure 26 shows some representations of this sort.

Using this method of quadratic equilibrium representation, K_s can be easily determined. According to the above equation, we have:

$$K_s = \frac{\dfrac{Y_A}{Y_B}}{\dfrac{X_A}{X_B}} = \frac{Y_A}{Y_B} \cdot \frac{X_B}{X_A} = \frac{\text{area I}}{\text{area II}}$$

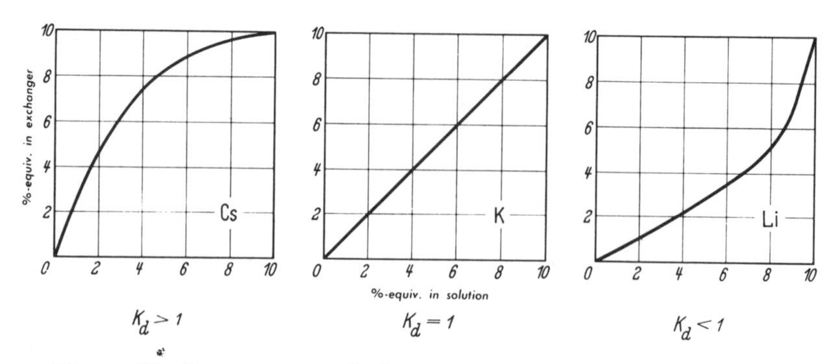

Figure 26 Determination of the selectivity coefficient by the quadratic equilibrium representation (NH_4^+ = reference ion, schematic).

of Figure 27, where the determination of K_s according to the method of Kressman and Kitchener [357] is shown for the examples of thallium and sodium.

2.1.1.8 Stability and attrition

In laboratory experiments as well as in the industrial application of exchanger materials, the stability of the resins which are usually used in cyclic processes plays a decisive role: it influences not only an individual process but the entire life and thus the total costs of a process carried out with ion exchangers in industry. Depending on the nature of the processes taking place in ion exchange, we must distinguish between physical and chemical stability; because of the interest of ion exchangers in nuclear reactor engineering, their radiation resistance must also be considered. Beyond this, two other phenomena, *i.e.*, irreversible adsorption and precipitation, can inactivate an ion exchanger. These cases do not involve a true stability problem. If the functional

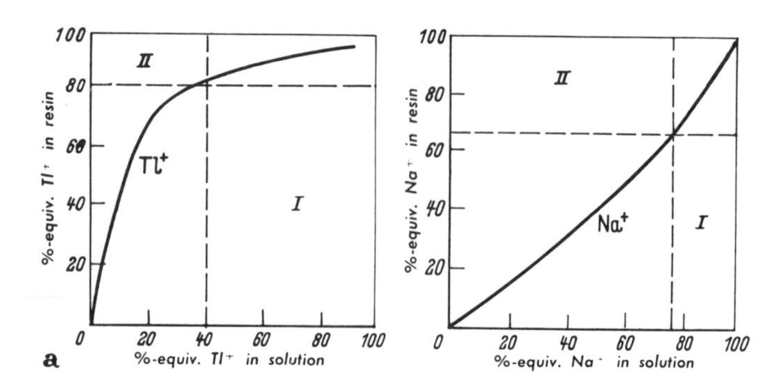

Figure 27 Calculation of the selectivity coefficient K_s by the quadratic equilibrium representation.

groups of an exchange resin are poisoned by irreversible adsorption or if the further activity of the resin is suppressed by precipitates, the operating conditions need to be changed or if the situation cannot be remedied in any way, other separating processes must be used. The question of stability, in contrast, concerns the physical and chemical properties of the resin.

The stability of an exchanger depends on the production method and thus is a major problem of ion exchanger manufacturers. The ion exchanger types available today have been refined to such an extent by many years of research that when the operating conditions recommended by the manufacturer are observed, many years of use can be expected in most cases.

Physical stability. The physical stability of an ion exchanger primarily resides in its strength. Good resins are characterized as being uniformly spherical, containing no internal cracks, being resistant to mechanical compression, and having low brittleness. In practical application the ion exchanger beads may not fracture; otherwise, they will be flushed from the column or the exchanger tank. They must be resistant to osmotic influences, *i.e.*, have a sufficient bursting and disintegration resistance under given loading and regeneration conditions. If this is not the case, too large a proportion of monomer may be the reason for instability.

The thermal stability, at least of anion exchangers, depends on their degree of cross-linking; in weak basic amine types of exchangers, thermal stability decreases with an increasing divinylbenzene concentration. Further studies of the thermal stability of ion exchange resins have demonstrated [250] that strong acid cation exchangers in all forms are completely stable in pure water up to 120°C. The H^+-form becomes unstable at 150°C, where the Li-form, however, is still stable for 30 days. At 180°C the H-form is already highly sensitive, and at 200°C it exhibits pronounced degradation phenomena. Comparatively speaking, however, degradation still remains slight in the Li-form at 180°C and even 200°C and also in the Na- and K-forms at the same temperatures. The thermal stability increases with decreasing divinylbenzene concentration. Weak basic anion exchangers in the OH^-- and the salt form are stable for more than 30 days at 180°C. In contrast, strong basic exchangers in the OH-form are already unstable at 150°C, while corresponding salt forms are still relatively stable. At higher temperatures, the resin is rapidly degraded. The strong basic resins exhibit an increase of stability with decreasing cross-linking. This influence of temperature is still considered a physical stability since the test was performed in pure water. The ratings of stable or unstable, however, refer to a decrease of capacity and to degradation processes. The water-soluble degradation products detected consisted of sulfite and sulfate in the case of cation exchangers, and of trimethylamine and methanol in anion exchangers.

If the physical stability of exchanger particles is to be tested [501] they can be subjected to alternate wetting and drying. Although such cyclic conditions do not occur in practice, they lead to valuable information. Figure 28 is an illustration of the behavior of a recently manufactured exchanger of very high quality and an older one of lower quality before and after a series of drying cycles.

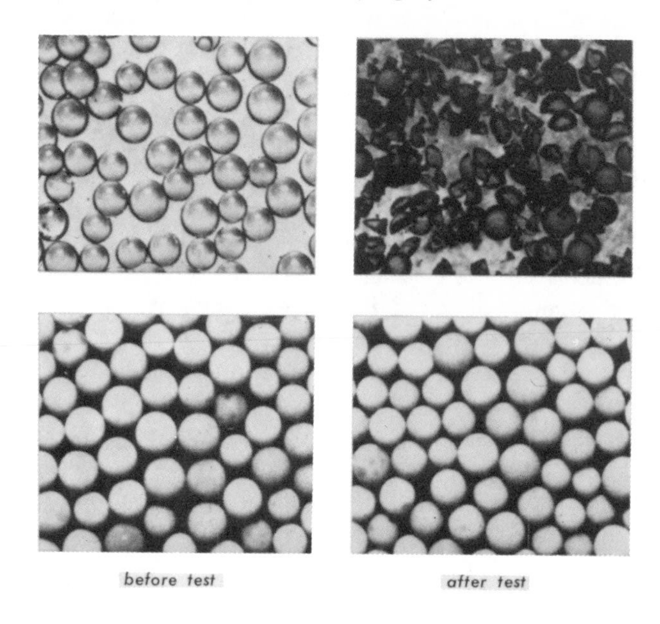

Figure 28 Drying test of two exchanger types of different physical stability [372].

Chemical stability. The chemical stability of an ion exchanger is manifested by the chemical resistance of the functional groups and by its resistance to oxidation. While the sulfonic acid groups of strong acid cation exchangers of the styrene type hardly undergo a change under the influence of pH and temperature, this is no longer fully true for the same exchanger of the phenol type. We can even find a difference between the Na- and the H-forms to the extent that the latter is still more unstable. In the strong basic anion exchangers, the OH-form in particular tends toward irreversible degradation and a quality loss in the form of an aging process. The quaternary ammonium groups converge into tertiary amines and finally into groups without basic properties. Consequently, it is not advisable to allow such a resin in the OH-form to stand in a strong alkaline solution for an extended period of time. These influences on the chemical stability of functional groups become evident in a loss of capacity and consequently in a deterioration of the treated solution.

If an exchanger is degraded in its polymer skeleton because of insufficient oxidation resistance, this leads to a decrease in cross-linking and thus increased swelling. These oxidative attacks are enhanced by the presence of iron or copper. Exchangers of the styrene type with layer degrees of cross-linking are more susceptible to oxidation than those with higher cross-linking. Free chlorine which is present in ordinary city water also has an unfavorable influence on oxidation [60, 635].

Data on the thermal stability of strong basic anion exchangers in the nitrate form are informative and important for the production of nuclear fuels [198]. A critical temperature exists here above which the resins undergo spontaneous thermal decomposition and can even lead to accidents.

A simple laboratory experiment can serve as a test of oxidation resistance. The samples to be investigated are subjected to treatment with 3% hydrogen peroxide solution at 45°C for 72 hours. After this pretreatment, the moisture regain of these samples is compared with that of untreated exchanger samples. Table 6 shows the results of such

Table 6
Oxidation test of different exchangers

	Water content, %		
	A	*B*	*C*
Untreated exchanger	47	45	45
H_2O_2 treated exchanger	54	81	75

tests for three different exchangers. Exchanger A exhibits the lowest increase in water content and thus is most resistant to oxidation.

Radiation stability. Because of the use of ion exchangers in nuclear reactors and for decontamination purposes, their radiation stability is also of special interest. Basically, radiation damage in cation exchangers of the sulfonic acid type is caused mainly by the rupture of cross-linkages, and in anion exchangers by the degradation of functional groups. According to Wiley and Devenuto [633], a degradation induced by CO^{60}-γ-radiation in sulfonated styrene resins cross-linked with 4 and 8 mole% of *m*- or *p*-divinylbenzene isomers or their mixtures, including commercial divinylbenzene, leads to a capacity loss of 5–12.8% with radition intensities of about 290,000 r/h and total doses of $0.91–1.90 \cdot 10^8$ rad in the presence of water. This capacity loss is smaller (2–6%) in resins with up to 8% cross-linking and with irradiation in the dry state. Resins cross-linked with commercial divinylbenzene exhibit a somewhat greater capacity loss (11.2–11.9%) than others (10.2–10.7%) probably because of their alkylphenyl content. The moisture uptake also increases with irradiation, *i.e.*, by 13–16% for 8% resins and by 2.7–3% for 4% resins exposed to $1.90 \cdot 10^8$ rad. The meta-DVB cross-linked resin exhibits a somewhat smaller increase in moisture content (1.5% for 8% cross-linking and 0.8% for 4%), indicating a more stable network. These phenomena can be explained by desulfonation, hydroxylation and cleavage. According to corresponding studies of Hall and Streat [248, 249] with strong basic anion exchangers in the OH-form, a pronounced degradation is produced by Co-γ-irradiation in doses of up to 500 Mrad with the release of water-soluble aliphatic amines. In addition to tertiary, secondary, and primary amines, traces of formaldehyde were also detected. The quaternary ammonium groups form weak basic amine groups, so that an increase in the weak basic capacity occurs.

Generally, cation exchangers have a higher stability than anion exchangers and highly cross-linked exchangers are more stable than those with a low degree of cross-linking.

These influences on the stability of the exchangers on the whole lead to their attrition during practical ion exchange processes. The resins are subject to a physical and chemical deterioration. Their life can be extended if oxidative effects are reduced to a minimum. The life of an exchanger is measured either in m^3 solution/m^3 resin under certain conditions or by the number of loading and regeneration cycles. The former method is recommended if chemical stability is to be determined, and the latter when the physical influences on the life are to be determined.

Martinola developed a method for a rapid routine control of anion exchangers during operation [414]. The various capacity values obtained offer an indication of the existing number of active sites and their base strength. Moreover, information is obtained concerning contamination of the resins by acid compounds and, with this method, measures to regenerate the exchangers in the laboratory can be tested. In strong basic anion exchangers, a change in maximum capacity permits one to derive the change of effective volume capacity attainable under standard conditions. In weak basic exchangers, this is not possible since other influences also play a role here.

2.1.1.9 Behavior in nonaqueous solvents

In some cases nonaqueous solvents are treated with ion exchangers, particularly for purification purposes. Moreover, exchangers in non-aqueous solvents serve as catalysts. These applications have directed interest toward the behavior of ion exchangers in nonaqueous phases and have led to a number of basic research studies.

The investigations of Bodamer and Kunin [70] showed that the behavior in nonaqueous solvents depends highly on the nature of the carbon chain of the exchange resin as well as on its porosity and cross-linking. In principle, ion exchange in nonaqueous solvents proceeds just as in water. However, since mass transfer is slower, the flow rates in the columns must remain lower than in aqueous solutions. Ion exchange takes place more rapidly from polar solvents similar to water than from nonpolar ones. Provided the pores of the resin are sufficiently large, so that exchange can take place at all, no measurable swelling of the resin needs to take place in nonaqueous solvents.

In tests with nonaqueous solvents, the exchanger is transformed into the acid and base or salt form by treatment with 10% hydrochloric acid or 4% sodium hydroxide solution and is washed to a neutral reaction. Cation exchangers in the Na-form are dried for 8 hours at 100°C, while the other types are spread and air-dried for 48 hours at room temperature. However, this method does not completely eliminate the water from the exchanger, and objections can be raised to the results obtained with it. However, in practice it is advisable to know the behavior of such pretreated exchangers in nonaqueous solvents, since the large quantities of ion exchange resin needed in industrial plants cannot be subjected to prolonged and costly drying processes.

To evaluate the applicability of exchangers in nonaqueous media, it is important to determine swelling, capacity, and selectivity.

For the determination of swelling in nonaqueous solvents, 30–40 ml of the dried ion exchanger is placed in a 100-ml graduated cylinder; subsequent shaking must be avoided. The volume is read with an accuracy of ± 0.5 ml. The exchanger is then covered with nonaqueous solvent up to the 100-ml mark. After 120 hours, the bed height of the exchanger is again read. The degree of swelling in per cent is calculated from the final and the initial volumes.

Table 7 and 8 show data on the swelling of some Amberlite types in nonaqueous solvents.

It can be easily seen that polar solvents produce more extensive swelling than nonpolar media. In hydrocarbons anion exchange resins swell more than cationic types.

Table 7
Swelling of different Amberlite cation exchangers in nonaqueous solvents

Cation Exchanger	*% Swelling in*						
	Water	*Ethanol*	*Glycerin*	*Acetone*	*Glacial Acetic Acid*	*Benzene*	*Petroleum Ether*
IR-120 (H)	43	38	24	18	8	0	0
IR-120 Na	73	0	5	0	0	0	0
IR-112 Na	264	0		0		7	7
IR-105 (H)	107	100	120	73	55	0	
IR-105 Na	99	5	5	8	7	5	2
IRC-50 (H)	48	98		0	0	0	0
IRC-50 Na	202	0	6	1		3	1

Table 8
Swelling of different Amberlite anion exchangers in nonaqueous solvents

Anion Exchanger	*% Swelling in*					
	Water	*Ethanol*	*Acetone*	*Pyridine*	*Benzene*	*Petroleum Ether*
IRA-400 OH	37	63	25	20	18	5
IRA-400 Cl	45	63	20	28	11	5
XE-75 OH	220	140	40	20	7	5
XE-75 OH	200	130	23	32	3	3
IRA-401 OH	25	55	38		50	10
IR-4B OH	23	18	0	3	3	3
IR-4B Cl	73	5	0	3	3	0
IR-45 OH	31	52	40	50	35	15
IR-45 Cl	45	30	10	25	0	0
XE-76 OH	35	78	45		55	20

Tables 9 and 10 give information on changes of the capacity in nonaqueous solvents compared to aqueous media. Strontium petronate, the strontium salt of a sulfonated aliphatic hydrocarbon, was used to study purely cationic exchange (Table 9). The values do not indicate maximum capacities since no excess of strontium petronate could be used because of its limited solubility; they represent the fraction of the added material that was exchanged. The acetic acid regain is of interest since it is bound to a greater extent than other substances, especially from benzene. It has been postulated that several of the polar acetic acid molecules are arranged around a polar functional group without forming a true ionic bond.

Table 9
Strontium-hydrogen exchange in nonaqueous solvents

Solvent	*Added mval*	*Exchanged meq/g resin*		
		IR-105	*IR-120*	*IRC-50*
Ethanol	1.69	1.08	1.61	0.00
Acetone	0.59	0.27	0.53	0.00
Dioxane	1.85	0.00	1.26	0.00
Benzene	2.22	0.04	1.49	0.008

The selectivity of exchangers also differs in nonaqueous solvents and water. As an illustration, we may note the studies of Gable and Strobel [200], who investigated ion exchange equilibria with pure methanol. Higher selectivity coefficients were found than with water. It is assumed that ion solvation and ion pair productions are the causes.

Since many organic compounds are not sufficiently soluble in water, a change to nonaqueous solvents is sometimes of advantage for analytical purposes. In addition to a number of other investigators, Inczédy [288, 289] has dealt with these questions. The author studied the swelling behavior of strong acid cation exchangers and strong basic anion exchangers in ethylene glycol, methanol, ethanol, dioxane, dimethylformamide, pyridine, and glacial acetic acid and applied these for the determination of some salts of organic acid and bases. Similar studies were also extended to macroporous ion exchangers and their applicability was demonstrated with the example of chromatographic separation of benzoic acid and anthranilic acid. Phipps [474] shows dimethylsulfoxide as the nonaqueous solvent for inorganic anions in a study which offers much additional information concerning ion exchange from nonaqueous solvents.

Table 10
Acetic acid sorption by different anion exchangers

Solvent	*Added mval*	*Sorbed meq/g resin*				
		IR-4B	*IR-45*	*XE-76*	*IRA-400*	*XE-75*
Water	7.35	3.87	3.81	4.91	2.70	2.50
Ethanol	7.43	2.70	3.13	4.75	2.69	2.58
Acetone	7.50	0.33	3.08	4.10	3.01	2.52
Dioxane	7.71	0.49	4.06	4.06	2.12	1.94
Benzene	9.65	7.94	7.05	8.58	5.30	5.13
Total capacity		10.00	5.20	6.00	2.80	3.00

Table 11. Amberlite
CE = cation exchanger

Designation	*Character*	*Active Group*	*Delivered form*	*Density*	
				Apparent g/cm³	*Real g/cm³*
Resins with acrylic matrix					
IRC-50/75	CE, weak acid	$-COO^-$	H^+	0.69	1.25
IRC-84	CE, weak acid	$-COO^-$	H^+	0.74	1.19
IRC-72	CE, weak acid	$-COO^-$	Na^+		1.16
IRA-68	AE, weak base	$-N(R)_2$	free base	0.74	1.06
XE-258	AE, strong base	Type I	Cl^-		
Polystyrene resins with gel structure					
XE-100/IR-118	CE, strong acid	$-SO_3^-$	Na^+	0.81	1.21
IR-120/121	CE, strong acid	$-SO_3^-$	Na^+ or H^+	0.84	1.27
IR-122/123	CE, strong acid	$-SO_3^-$	Na^+	0.82	1.32
IR-124	CE, strong acid	$-SO_3^-$	Na^+	0.84	1.34
IR-45	AE, weak base	$-N(R)_2$ $-NH(R)_2$ $-NH_2$	free base	0.67	1.15

REMARKS:

(1) pK-value 6.1; effectively separates alkaline salts of polyvalent cations; application for antibiotics, Cu and Ni recovery; IRC-75 for sweet water treatment.

(2) pK value 5.3; separates the alkaline salts of mono- and polyvalent cations; application in water treatment.

(3) Particularly suited for the isolation of antibiotics.

(4) High capacity for large organic molecules; application for deacidifying and desalting of sugar syrups.

ion exchangers
AE = anion exchanger

Bulk density g/l	*Effective particle size mm*	*Moisture content %*	*pH range*	*Maximum operating temperature °C*	*Total exchange capacity meq/ml*	*Swelling with complete conversion %*	*Remarks*
690	0.33–0.50	43–53	5–14	120	3.50	$H^+ \rightarrow Na^+$ 100 $H^+ \rightarrow Ca^+$ 36	(1)
740	0.38–0.46	42–50	4–14	120	3.50	$H^+ \rightarrow Na^+$ 65 $H^+ \rightarrow Ca^+$ 25	(2)
735–780	0.30–0.85	27–29	4–14	120		$H^+ \rightarrow Na^+$ 100	(3)
735	0.35–0.45	57–63	0– 9	80	1.60	FB → Cl 17	(4)
690–750	0.30–1.20	59–62	0–14	40 (OH) 75 (Cl)	1.25		(5)
785	0.42–0.57	56–60	0–14	120	1.50	$Na^+ \rightarrow H^+$ 8.5	(6)
850	0.45–0.60	44–48	0–14	120	1.9	$Na^+ \rightarrow H^+$ 7.0	(7)
865	0.45–0.60	39–43	0–14	120	2.1	$Na^+ \rightarrow H^+$ 6.5	(8)
865	0.42–0.57	37–41	0–14	120	2.2	$Na^+ \rightarrow H^+$ 6.0	(9)
670	0.36–0.46	40–45	0– 9	100	1.90	FB → Cl^- 10	(10)

(5) Highly resistant to contamination; suited for decolorizing resin, adsorbent, and for full desalination.

(6) Open-structure resin; application for the recovery of high molecular weight cations.

(7) 8% cross-linking, mainly for water treatment.

(8) 10% cross-linking, water treatment.

(9) 12% cross-linking; application in electroplating.

(10) Together with a strong acid CE, removes all anions of strong acids.

Table 11

Designation	Character	Active Group	Delivered form	Density: Apparent g/cm³	Density: Real g/cm³
IRA-400	AE, strong base Type I	$-N-(CH_3)_3^+$	Cl^-	0.70	1.11
IRA-400C	AE, strong base Type I	$-N-(CH_3)_3^+$	Cl^-	0.70	1.11
IRA-401	AE, strong base Type I	$-N-(CH_3)_3^+$	Cl^-	0.69	1.06
IRA-401S	AE, strong base Type I	$-N-(CH_3)_3^+$	Cl^-	0.69	1.06
IRA-402	AE, strong base Type I	$-N-(CH_3)_3^+$	Cl^-	0.68	1.07
IRA-410	AE, strong base Type II	$-N \parallel \begin{matrix} CH_2-CH_2OH \\ (CH_3)_2^+ \end{matrix}$	Cl^-	0.70	1.12
IRA-425	AE, strong base Type I	$-N-(CH_3)_3^+$	Cl^-		
Polystyrene resins of macroreticular structure					
200	CA, strong acid	$-SO_3^-$	Na^+	0.80	1.26
252	CA, strong acid	$-SO_3^-$	Na^+	0.80	
IRA-93	AE, weak base	$-N(R)_2$	free base	0.64	1.04
IRA-900	AE, strong base	$-N-(CH_3)_3^+$	Cl^-	0.67	1.07

REMARKS:

(11) Standard cross-linking; treatment of not highly polluted water, uranium recovery.

(12) Special granulometry.

(13) Lower cross-linking than IRA-400.

(14) High decolorizing capacity; for sugar industry.

(15) Lower cross-linking than IRA-400; high capacity.

(continued)

Bulk density g/l	*Effective particle size mm*	*Moisture content %*	*pH range*	*Maximum operating temperature °C*	*Total exchange capacity meq/ml*	*Swelling with complete conversion %*	*Remarks*
705	0.38–0.45	42–48	0–14	60 (OH) 77 (Cl)	1.40	$Cl^- \rightarrow OH^-$ 19	(11)
705	0.45–0.65	42–48	0–14	60 (OH) 77 (Cl)	1.40	$Cl^- \rightarrow OH^-$ 19	(12)
690	0.40–0.50	53–58	0–14	60 (OH) 77 (Cl)	1.0	$Cl^- \rightarrow OH^-$ 27	(13)
690	0.40–0.50	59–65	0–14	60 (OH) 77 (Cl)	0.8	$Cl^- \rightarrow OH^-$ 27	(14)
720	0.39–0.46	50–57	0–14	60 (OH) 77 (Cl)	1.25	$Cl^- \rightarrow OH^-$ 22	(15)
705	0.38–0.45	40–45	0–14	40 (OH) 77 (Cl)	1.40	$Cl^- \rightarrow OH^-$ 13	(16)
720	1.05	52–55	0–14	60 (OH) 77 (Cl)	4.5 per g		(17)
800	0.40–0.50	46–51	0–14	145	1.75	$Na^+ \rightarrow H^+$ 3	(18)
800	0.30–1.10	45–50	0–14	135	1.25	$Na^+ \rightarrow H^+$ <5	(19)
610	0.40–0.50	46–54	0– 9	100	1.4	$FB \rightarrow Cl^-$ 23	(20)
670	0.43–0.52	60–64	0–14	60 (OH) 77 (Cl)	1.0	$Cl^- \rightarrow OH^-$ 20	(21)

(16) Improved regenerant utilization; lower silicate uptake.
(17) For uranium recovery in RIP process.
(18) Low attrition; all applications.
(19) Favorable for condensate treatment.
(20) Good oxidation resistance; long life.
(21) Particularly suited for desilication.

Table 11

Designation	Character	Active Group	Delivered form	Density: Apparent g/cm³	Density: Real g/cm³
IRA-904	AE, strong base	$-N-(CH_3)_3^+$	Cl^-	0.67	1.08
IRA-910	AE, strong base	$-N \Vert \begin{matrix}(CH_3)_2^+ \\ CH_2-CH_2OH\end{matrix}$	Cl^-	0.67	1.09
IRA-911	AE, strong base	$-N \Vert \begin{matrix}(CH_3)_2^+ \\ CH_2-CH_2OH\end{matrix}$	Cl^-	0.68	1.11
IRA-938	AE, strong base	$-N-(CH_3)_3^+$	Cl^-	0.55	1.20
XE-243	AE		OH^-	0.64	
Polycondensation resins					
IR-4B	AE, weak base	polyamine	OH^-	0.56	
XE-265	AE, weak base	polyamine	OH^-	0.70	
Mixed-beds					
MB-1	Mixture of strong acid and strong base	$-SO_3^- -H^+$ $-N-(CH_3)_3^+OH^-$	H^+ and OH^-	–	–
MB-3	Mixture of strong acid and strong base Colored indicator	$-SO_3^- -H^+$ $-N \Vert \begin{matrix}(CH_3)_2^+OH^- \\ CH_2-CH_2OH\end{matrix}$	H^+ and OH^-	–	–

REMARKS:

(22) Effective adsorbent in Cl form.
(23) Improved regenerating effect over IRA-900.
(24) Good capillary properties, high strength.
(25) Adsorbs colloidal silicate.
(26) Exchanger selective for boron.
(27) Granulate; application for deacidification.

(**continued**)

Bulk density g/l	Effective particle size mm	Moisture content %	pH range	Maximum operating temperature °C	Total exchange capacity meq/ml	Swelling with complete conversion %	Remarks
670	0.40–0.50	56–62	0–14	60 (OH) 77 (Cl)	0.7	$Cl^- \rightarrow OH^-$ 5	(22)
670	0.40–0.50	55–60	0–14	40 (OH) 77 (Cl)	1.1	$Cl^- \rightarrow OH^-$ 21	(23)
690	0.41–0.48	42–44	0–14	40 (OH) 77 (Cl)	0.9	$Cl^- \rightarrow OH^-$ 3	(24)
640	0.30–0.90	72–78	0–14	60 (OH) 77 (Cl)	0.53		(25)
		56–60			2.5		(26)
550	0.40–0.55	40–45	0– 7	43	1.14	25	(27)
690	0.30–1.20	62	0– 9	90	2.4	FB → Cl 5–10	(28)
690	0.38–0.60	40–60	0–14	60	(*)	–	(29)
690	0.38–0.60	50–60	0–14	40	(†)	–	(30)

*Column capacity, 0.46 (min).
†Column capacity, 0.55 (min).

(28) Epiamine condensation resin; application for deacidification.
(29) To obtain fully desalinated water of high quality.
(30) Higher volume capacity; higher residual silicic acid.

2.1.1.10 Behavior in mixed-aqueous systems

The behavior of ion exchangers in mixed solvents in several respects resembles that in nonaqueous media. As demonstrated by Bonner and Moorefield [74] in studies of silver–hydrogen exchange on Dowex 50 in ethanol–water and dioxane–water, the resin selectivity increases with the addition of an organic solvent to water, while the swelling volume of a resin in the H-form remains relatively constant in the mixed solvent.

However, the behavior of many ions in strong acid and strong basic exchangers differs considerably in mixed organic-aqueous systems compared to aqueous media alone. The large number of data concerning the exchange behavior of cations and anions from aqueous and inorganic acid media were supplemented by an extended series of studies on ion exchange in mixed organic-aqueous systems. The aqueous component can even consist primarly of organic acids of different normality which contain additions of various miscible organic solvents. The differences become apparent by the fact that the exchange capacity of a given metal ion increases with an increasing solvent concentration at constant acidity and in the presence of one and the same organic solvent. This is true for cation as well as for anion exchangers and differs as a function of the inorganic acids and organic solvent present, although in many cases a dependence on the dielectric constant exists also. Pronounced differences in the exchange behavior of elements compared with pure aqueous media have been observed particularly in mixtures with high concentrations of acetone or tetrahydrofuran and hydrochloric acid.

A large number of data exist on the exchange behavior of ions in systems with methanol, ethanol, *n*-propanol, isopropanol, *n*-butanol, isobutanol, isomeric amyl alcohols, acetone, tetrahydrofuran, dioxane, ethyl glycol, methylethyl ketone, methyl-*n*-propyl ketone, diethyl ketone, ether, diisopropyl ether, acetic acid, propionic acid, mono-, di-, and trichloroacetic acid; a good number of these data were published by Korkisch *et al.* [332, 333, 335, 336, 338] and have been used mainly for chromatographic separations. However, the distribution coefficients determined on this basis probably can also be used for other purposes, *e.g.*, in metallurgy or purification.

If the concentration of organic solvent is high enough, we arrive in a range in which the advantages of ion exchange and solvent extraction become effective at the same time, as demonstrated by Korkisch [330, 331]. This can be used as the basis of a new procedure which combines the advantages of both techniques. Korkisch assumes as a theoretical explanation that the organic solvent, such as tetrahydrofuran, forms an oxonium salt with hydrochloric acid:

$$\begin{array}{l} CH_2-CH_2\diagdown \\ | \qquad\qquad |\overline{O}| \\ CH_2-CH_2\diagup \end{array} + H^+Cl^- \rightleftharpoons \left[\begin{array}{l} CH_2-CH_2\diagdown \\ | \qquad\qquad |O\rightarrow H \\ CH_2-CH_2\diagup \end{array}\right]^+ Cl^-$$

which becomes active as a liquid ion exchanger and will undergo an ion exchange reaction of the following type, for example, with the anionic Fe(III) chloride complex:

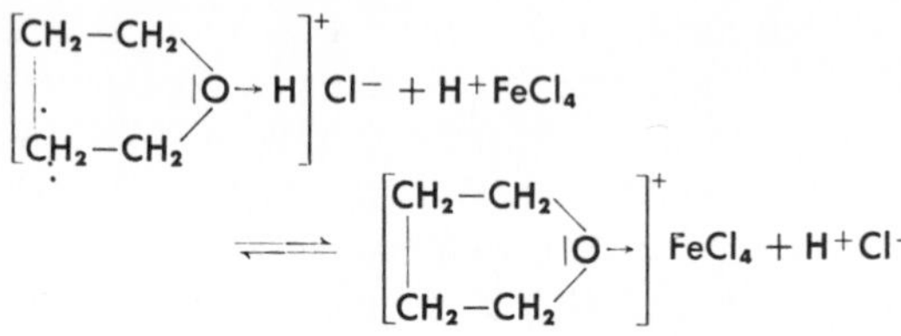

$$\left[\begin{array}{l} CH_2-CH_2\diagdown \\ | \qquad\qquad |O\rightarrow H \\ CH_2-CH_2\diagup \end{array}\right]^+ Cl^- + H^+FeCl_4$$

$$\rightleftharpoons \left[\begin{array}{l} CH_2-CH_2\diagdown \\ | \qquad\qquad |O\rightarrow \\ CH_2-CH_2\diagup \end{array}\right]^+ FeCl_4 + H^+Cl^-$$

where an ion association is formed with very little dissociation. The liquid anion exchanger competes with the solid cation or anion exchanger and, depending on the experimental conditions, either the liquid or the solid exchanger will preferentially sorb the metal ions. It can be assumed that this principle can be employed to solve many analytical and radiochemical problems by the use of other organic solvents or solvent mixtures and different inorganic or organic acids and bases as well as of chelating complexants (Korkisch).

2.1.2 *The Amberlite Ion Exchanger Program*

One of the cited production programs of commercial ion exchangers is the line of Amberlites of Rohm and Haas Company, Philadelphia, which we will describe more fully as an example of synthetic ion exchange resins for scientific and industrial use. The diversity of this line will offer an idea of available types. A suitable exchanger can be selected from this abundance of choices for each application and can be employed in accordance with its particular properties.

Amberlite ion exchangers are available with an acrylic matrix as well as in the form of polystyrene resins with a gel or macroreticular structure, and as polycondensation resins. Moreover, one can obtain resins with a special granulometry, mixed-bed resins, powdered exchangers, and special types for application in nonaqueous media, all of which are sold under the tradename of Amberlyst ion exchangers. Amberlite exchangers have been marketed for other special purposes: "pharmaceutical grade," "nuclear grade," and "analytically pure Amberlite ion exchangers for chromatography."

Table 11 offers a summary of the various types which are primarily for industrial use and their properties. The pharmaceutical-grade resins have been summarized in a table in Chapter 6 (6.1) for reasons of organization. The nuclear grades are of highest purity because of the special requirements of water-cooled nuclear reactors, and in this aspect they differ in particular from products used ordinarily for conventional water treatment. To prevent corrosion, these resins are supplied in different ionic forms, so that a pH can be adjusted at which an electrical potential for the attack of metal surfaces practically no longer exists. Table 12 lists the designations of these grades. The Amberlite exchangers of analytical purity differ from the industrial grades by their high purity with regard to heavy metal and iron contents, particle size distribution, and moisture content. The available analytically pure resins are: Amberlite IR-120-AR, IRC-50-AR, IR-45-AR, IR-4B-AR, IRA-400-AR, IRA-401-AR, MB-1-AR, and MB-3-AR.* The special types for chromatography have a strictly controlled particle size distribution, *i.e.*, type I of 100–200 mesh or 0.16–0.08 mm, type II

Table 12
Amberlite ion exchangers of nuclear grade

Amberlite Type	*Characteristics and supplied form*
Cation exchangers	
IRN-77	strong acid sulfonated cation resin delivered in H form
IRN-218	strong acid cation resin delivered in Li form
IRN-163	strong acid cation resin delivered in Li form
IRN-169	strong acid sulfonated cation resin delivered in NH_3 form
Anion exchangers	
IRN-78	strong base anion exchanger, Type I, delivered in OH form
Mixed resins	
IRN-217	mixture of Amberlite IRN-218 and Amberlite IRN-78
IRN-150	mixture of Amberlite IRN-77 and Amberlite IRN-78
IRN-154	mixture of Amberlite IRN-163 and Amberlite IRN-78
IRN-170	mixture of Amberlite IRN-169 and Amberlite IRN-78

of 200–400 mesh or 0.08–0.04 mm, and type III of 400–600 mesh or 0.04–0.027 mm. Of the resins listed in Table 13, Amberlite CG-120 and CG-50 are available in all three types, while Amberlite CG-4B and CG-400 can be obtained only as types I and II. The Amberlyst ion exchangers which are compiled in Table 13a have an unusually high porosity and specific surface, so that a larger number of their functional groups is accessible in nonaqueous medium. For this reason, they are

*AR is the registered trademark of Mallinckrodt, USA.

particularly suited for ion exchange, catalysis, and adsorption in non-aqueous and mixed-aqueous media.

Table 13
Amberlite ion exchangers for chromatography
(CE = cation exchanger, AE = anion exchanger)

Designation	*Type*	*Characteristic*	*Application*
CG-50	CE	weak acid	column chromatography for separation
CG-120	CE	strong acid	of related substances; inorganic separa-
CG-4B	AE	weak base	tions, rare earths, organic bases, amino
CG-400	AE	strong base	acids, B-vitamins, antibiotics, etc.

As noted above, this broad supply permits the choice of a suitable type for every possible application. In the individual sections devoted to special fields of application, an appropriate Amberlite ion exchanger will always be found again. If a decision has been made in favor of a particular type, additional data can be obtained from the pertinent special sales literature concerning minimum bed heights, backwash rate at a given expansion and temperature, regenerant concentration, specific load during regeneration, regenerant consumption, specific load during washing, wash water consumption, specific load during loading, and pressure drop for a given industrial application. Some types, such as Amberlite IR-120 L, 200 C, IRA 400 C and IRA 900 C, are resins with special granulometries. An indication of their special advantages is offered by Figure 29, where the pressure drop in mixed beds of standard and special granulometries has been plotted as a function of the flow rate.

2.1.3 *The Lewatit-Merck Ion Exchangers*

The use of ion exchangers in analytical chemistry has resulted in a number of special requirements for exchanger products. These are satisfied only partly by commercial brands which are mainly intended for industrial application. To respond to these needs, the Lewatit-Merck ion exchangers have been developed as a special and closed ion exchanger production program for those working in analysis and ion exchange to save them time-consuming and costly preliminary procedures. These products represent the special Lewatit types produced under license of Farbenfabriken Bayer AG, which are intended for analytical and other laboratory applications and are manufactured and sold by E. Merck, Darmstadt.

The entire production program is listed in Table 14. As indicated, it covers strong and weak acid cation exchangers as well as strong and

Table 13a
Amberlyst ion exchangers (CE = cation exchanger, AE = anion exchanger)

Designation	*Character*	*Active group*	*Form*	*Maximum permissable operating temperature* °*C*	*Moisture content as delivered* %	*Application*
15	strong acid CE	$-SO_3^{-}H^{+}$	beads	120	0.5	catalysis of organic reactions in nonaqueous solutions; cation removal from such solvents
A-21	weak base AE	$-N(CH_3)_2$	beads	100	45	deacidification of nonaqueous solutions
A-26	Type I strong base AE	$-N(CH_3)_3^{+}Cl^{-}$	beads	60 (OH) 77 (Cl)	61	removal of anions from nonaqueous solutions
A-29	Type II strong base AE	$-N \parallel (CH_3)_2^{+}Cl^{-}$ CH_2-CH_2OH	beads	48 (OH) 66 (Cl)	45	removal of anions from nonaqueous solutions

weak base anion exchangers with the characteristic functional groups and in standard forms. With the exception of Lewatit-Merck CP 3050, cross-linking amounts to 8% in terms of divinylbenzene used, which can be considered a certain optimum. The available particle sizes are

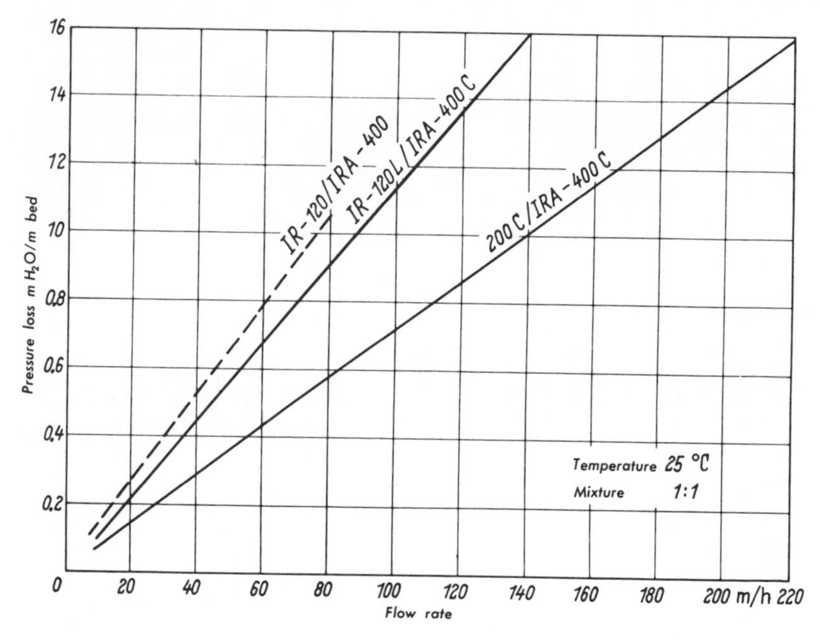

Figure 29 Pressure loss in mixed-beds with the use of exchangers of normal and special granulometry.

suited for general analytical purposes and can be recommended for ion exchange chromatography, since these granulometries are frequently used to determine distribution coefficients. The incorporation of color indicators permits a purely visual distinction between the nearly colorless H-form and the red metal form of Lewatit-Merck S 1080 G 1 and between the blue OH-form and the purple-red form which is loaded with a different anion in Lewatit-Merck M 5080 G 3. Lewatit-Merck M 5080, an anion exchanger, is a strong base resin of type I. The requirements for a more defined porosity are satisfied by macroporous (macroreticular) types of which one strong and weak acid and one strong and weak base exchanger are also available for analytical purposes.

As a result of special production techniques, the Lewatit-Merck ion exchangers have a purity and heavy metal concentration which permit their immediate use in analysis. The fraction of intrinsic resin components which can be eluted is also strictly controlled by the manufactur-

Table 14
Lewatit-Merck ion exchangers (CE = cation exchanger, AE = anion exchanger)

Lewatit-Merck	*Type*	*Active group*	*Form*	*Cross-linking % DVB*	*Particle size US mesh*	*Capacity meq/g*	*Maximum operating temperature °C*	*Remarks*
S 1080	CE strong acid	$-SO_3^-$	Na^+	8	150–200 70–150	4.3	120	
S 1081 G1	CE strong acid	$-SO_3^-$	Na^+	8	70–150	3.9	120	with color indicator
SP 1080	CE weak acid	$-SO_3^-$	Na^+	8	150–200 70–150	4.1	120	macroporous resin
CP 3050	CE strong acid	$-COO^-$	H^+	5	70–150	8.5	80	macroporous resin
M 5080	AE strong base	$-N(CH_3)_3^+$	Cl^-	8	150–200 70–150	4.7	70	Type I
M 5080 G3	AE strong base	$-N(CH_3)_3^+$	Cl^-	8	70–150	4.0	70	with color indicator
MP 5080	AE strong base	$-N(CH_3)_3^+$	Cl^-	8	150–200 70–150	3.9	70	macroporous resin
MP 7060	AE	$-NH(R_2)^+$	Cl^-	8	70–150	4.2	100	macroporous resin

NOTE: While the Lewatit-Merck ion exchangers listed here are to satisfy more demanding analytical conditions, those listed in Table I in the Appendix have been used for a long time for general laboratory applications. The reader is referred to the manufacturer for the state of the art and further expansions in the line of Lewatit-Merck in ion exchangers.

ing process. Therefore, for immediate laboratory use it is necessary only to provide for preswelling, which has been described elsewhere; with the use of columnar processes, transformation into the desired loading form is the only step required after packing the column.

Since the Lewatit-Merck ion exchanger program also covers products of the gel type and those of truly porous structure, the choice of a suitable type to study a given problem can also take this aspect into consideration for research and even routine analyses. This is a particularly welcome addition for operations in nonaqueous and mixed solvent systems, since the exchanger properties of cations on cation exchangers as well as after complexing on anion exchangers can be notably influenced by the use of precisely such systems. We need not go beyond these general aspects. Special applications of ion exchangers in analytical chemistry and elsewhere, which go beyond ordinary laboratory uses —in which the individual types of the Lewatit-Merck ion exchanger production program can be used simply on the basis of the special properties—will be described by many examples in the appropriate sections of this monograph.

2.2 Inorganic Ion Exchangers

Historically, inorganic ion exchangers were the first on which ion exchange was studied [207]. They consist mainly of aluminosilicates, among which the zeolites (chabazite, analcime, natrolite, etc.) have become most important as cation exchangers. Other representatives of this group are green sands (glauconite) and ferroaluminum silicates with a certain potassium content. Moreover, clay minerals such as montmorillonite also have ion exchange capacities because of their lamellar structure and their resulting interlaminar moisture regain capacity [278].

The principle of ion exchange is the same for the mineral exchangers as for the synthetic ion exchange resins [265]. The skeleton carries an excess charge which is compensated by mobile counterions. In the case of the zeolites, this is known to be a consequence of the fact that a part of the Si^{4+} building blocks is replaced by Al^{3+} in the silicate lattice. The lacking positive charge is replaced by alkali or alkaline earth ions, which are present with free mobility in the mineral skeleton in the form of counterions, according to ion exchange nomenclature. As an illustration of these facts, the structure of chabazite is shown in Figure 30.

Synthetic inorganic ion exchangers were first prepared by Gans on the basis of findings obtained with mineral exchangers [206]. Depending on the method of preparation, hydrothermal zeolites or gel exchangers were obtained by complete synthesis; among these, the zeo-

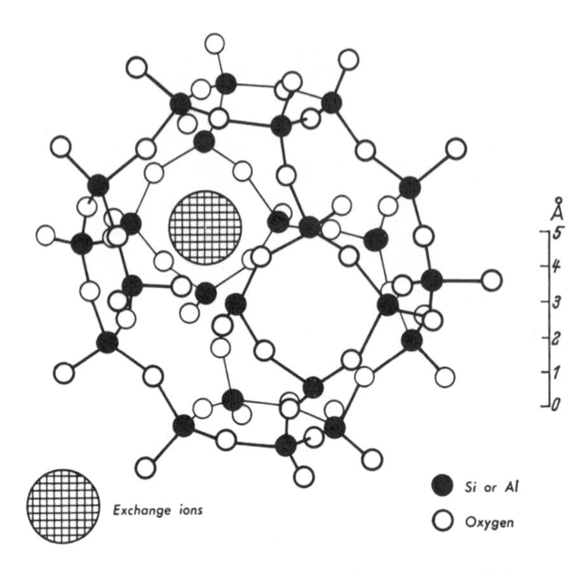

Figure 30 Stereostructure of chabazite [458].

lites were the first ion exchangers which found industrial application for water conditioning under the tradename of Permutit.

The natural and the older synthetic inorganic ion exchangers are no longer of industrial significance. Nevertheless, driven by the wish to obtain ion exchange materials which are stable at higher temperatures and radiation-resistant, more intensive research and development work has resumed in this field in recent years. To the extent to which data are available, this involves the production or modification as well as the characterization and application of insoluble salts, heteropolyacids, new zeolites, and clays.

Insoluble salts. Insoluble salts are the phosphates, tungstates, molybdates, arsenates, antimonates, vanadates, chromates, silicates, sulfides, oxalates, carbonates, and oxides of zirconium, thorium, titanium, cerium, aluminum, tin, bismuth, chromium, and tantalum. The resulting possible combinations permit the production of a number of inorganic ion exchangers which represent microcrystalline aggregates with exchange capacities similar to those of ion exchange resins and with an extremely high selectivity for certain ions together with high stability. These inorganic ion exchangers are resistant to radiation, and exchange reactions are possible even in nonaqueous media.

The insoluble salt which has been most fully investigated is *zirconium phosphate*. Depending on the production conditions [349], it can be obtained in fine microcrystalline form or as a gel cake which can be transformed by drying into a granular product similar to silica gel.

The production conditions also influence the ratio of phosphate to zirconium, which amounts to about 2:1 [7, 31, 588]. In accordance with its structure (an insoluble zirconium oxide matrix with phosphate as the active groups), it is a cation exchanger with pH-dependent capacities of 1 meq/g at pH 1.5 and 4.5 meq/g at pH 9.5. Its stability in strong acids is good up to pH 13, except in hydrofluoric acid, sulfuric acid, and oxalic acid. Ion exchange takes place through the hydrogen atoms of the acid phosphate groups according to well-known reactions represented by the law of mass action [111, 496]. Because of its different selectivity toward alkalis, alkaline earths, rare earths, and other metal ions, zirconium phosphate can be used for separation and purification procedures [32, 122]. Moreover, it has ionic sieve properties.

Of the remaining possible combinations of insoluble salts cited above, *zirconium tungstate* and *zirconium molybdate* have also been more fully investigated. Above pH 6, these products undergo hydrolysis at room temperature, so that their use is more limited. Other *arsenates* and *phosphates* (Ti, Zr, Ce, Th, Sn, Pb) have recently been described according to their stability to hydrolysis, titration curves, distribution coefficients, heat stability, and the hysteresis produced by them [8].

Probably *hydrous zirconium oxide,* which can be prepared as a microcrystalline ion exchanger [14, 15], is the most important of the simple insoluble salts. As an amphoteric compound, it acts as a cation exchanger in alkaline solutions and as an anion exchanger in acid solutions. The exchange capacity amounts to about 1 meq/g. Aqueous titanium dioxide, as an inorganic anion exchanger, has a capacity of about 1 meq/g. Of the remaining tetravalent oxides we should mention tin oxide, which exhibits cation exchange properties only at low pH values but otherwise acts as an anion exchanger [428]. The investigation of aqueous oxides from the ion exchange aspect offers additional information on the nature of these compounds beyond their possible applications [457].

In this connection, it is of interest to note transition phenomena between ion exchange and adsorption on oxides and similar materials, especially aluminum oxide (270). It has been demonstrated (Umland [593–595]) that these processes can be described by a formal exchange theory, so that the adsorption of dissociated salts may be considered as a simultaneous exchange reaction of hydrogen for metal cations and hydroxyl groups for acid rest ions. The presence of hydrogen and hydroxyl ions in Al_2O_3 may be deduced from various observations. On the basis of such investigations, difficulties in terminology result for the term "adsorption," so that it has been proposed by some to consider adsorption as any binding of a solute on a surface. Accordingly, ion exchange would represent a binding of ions on adsorbent sites of opposite charge. Frequently, ion exchangers are also considered exchange adsorbents. However, "adsorption" as it is used in ion exchange should not be confused with its more common interpretation.

In the adsorption of radioactive metals on *silica gel,* which also has ion exchange properties, it was found that uptake by gels which are not saturated with H^+-ions depends on the technique of gel preparation and that the same exchange isotherms are obtained for different gels when these are saturated with H^+-ions. Titration curves can be constructed which are identical to those of weak acid cation exchangers. In the adsorption of metal ions, the quantity of released H-ions is equivalent to the quantity of adsorbed metal ions, a process which takes place for monovalent cations according to the following reaction:

$$\text{Si—OH} + \text{Me}^+ \rightleftharpoons \text{—Si—OMe} + \text{H}^+$$

The degree of metal ion adsorption increases extensively with increasing pH of the solution, so that a capacity of 1.2–1.4 meq/g of gel can be obtained at relatively high pH-values. The pH-dependence of the selectivity as well as general cation exchange kinetics on silica gel then represent the basis for the possibility of metal separations as a logical consequence of silica gel properties [6].

Heteropolyacids. Heteropolyacids are compounds with the general formula $H_3[XY_{12}O_{40}]\cdot nH_2O$ where X is phosphorus, arsenic, and silicon, and Y is molybdenum, tin, and vanadium.

As the first one of this group, *ammonium-12-molybdophosphate* $(NH_4)_3[PMo_{12}O_{40}]$ was again investigated by Buchwald and Thistlewhaite [105] after it had been observed many years earlier that the NH_4-ion in this compound can be reversibly replaced by sodium [43]. Because of its crystalline structure it acts as a cation exchanger with an exchange capacity of 1.2 meq/g. In strong acids it is stable up to pH 6. The most striking property is its pronounced selectivity for alkali ions which are separated in columns where the heteropolyacid must be deposited on a support [559]. The larger alkali ions can be exchanged selectively from solutions with an excess of the smaller ions. The resulting difficulties of eluting cesium were overcome by dissolving the exchanger in sodium hydroxide solution. Ammonium-12-molybdophosphate is superior to other salt forms of the same acid even in the separation of polyvalent ions from a mixture of monovalent ones [91, 92].

Other heteropolyacids which have been investigated for their ion exchange properties are mainly *ammonium-12-molybdoarsenate* and *ammonium-12-tungstophosphate* [560]. The comparative determinations of distribution coefficients made with the objective of chromatographic application show a behavior like that of the organic ion exchange resins, although the high selectivity for silver and thallium and the low degree of exchange of polyvalent ions from acid solutions are remarkable. The exchange of cesium and rubidium under different conditions offers information concerning the ion exchange mechanism [361, 362].

Complex salts based on ferrocyanide. In the search for other new inorganic ion exchangers, complex salts based on ferrocyanide were developed. Potassium hexacyanocobalt(II)ferrate has been made available as a cation exchanger in the form of granulate [483], which has a high selectivity for cesium. The material can be used in large exchange columns for the separation and concentration of Cs^{137} from nuclear fuel wastes. Vanadium hexacyanoferrate(II) also has a useful selectivity for rubidium and cesium, so that this exchanger has made it possible to isolate Cs, split Cs and Rb, separate Rb and Cs from Sr, Ce, Ru, and Zr fission products, and enrich chemically pure Cs^{137} [360]. Titanium hexacyanoferrate (II) [397] served for the determination of the distribution coefficients of a number of metals to permit their separation.

New zeolites. The detailed studies of Barrer in particular aroused interest in the new zeolites. He indicated not only new methods for the production of synthetic inorganic crystals and investigated their crystal structure, but also provided the basis for a field which must be considered to be related to ion exchange, *i.e.*, that of molecular sieves [82, 83, 491].

To the extent to which new zeolites are considered and used as ion exchangers, they primarily consist of the materials prepared by Barrer with hydrothermal methods, *i.e.*, analcite, mordenite, faujasite, and chabazite as well as leucite and basic sodalite [38, 40]. These products have an ion exchange selectivity, for example, chabazite for potassium and sodalite for silver, which can be used in separations. The counter-ion present in the crystal is also largely responsible for their properties as molecular sieves [37]. Dyer and Gettins [170] in particular were concerned with the ion exchange properties of zeolites. Although zeolites are rarely used industrially as ion exchangers because of their low acid resistance, they are of interest in basic research on ion exchange because of their uniform pore and crystal structure. Considerable differences were detected in the exchange behavior of sodium for calcium, strontium, and barium on zeolites A, ZK4, X, and Y in water as well as in anhydrous ethanol and methanol and in mixed solvents. Sherry [548] found that about 25% of the K-ions cannot be exchanged with the new zeolite Linde T in aqueous media below 300°, and he explained this by structural aspects of the unit cell.

The consequences of this research might be significant for separation processes with low concentrations of monovalent and high concentrations of polyvalent ions.

Clays. Clays were investigated for their applicability as ion exchangers particularly because of their specificity and low cost as well as resistance to radiation and heat. The production of suitable products in this case consists not of synthesis but of a suitable preparation of natural raw materials. Significant data have been collected concerning the relationships between the efficiency of clays as cation and then as anion exchangers; such data are important for industrial applications because frequent errors can thus be eliminated.

The studies of Thomas *et al.* [178, 179, 203, 427] demonstrated with reproducible clay samples that clays do allow ion exchange processes when the natural minerals in a given ionic form are transformed into another ionic form by well-known columnar processes, where a reversible exchange can take place between monovalent ions in operating cycles and equilibration of competing ions can be demonstrated. Among the most important groups of clay minerals, a cation exchange capacity predominates. Weiss [624–626] developed the bases for the analysis of ion exchange problems in clays and investigated the most important of these for their exchange properties.

The decisive relationship for the ion exchange capacity is based on structure. In minerals of the mica, vermiculite, and montmorillonite group, the exchangeable cations are located on the base planes of the crystals. As a result of intracrystalline swelling, all base planes of vermiculite and montmorillonite are accessible so that the cation exchange capacity is particularly high and far-reaching in these clay minerals regardless of particle size. In micas, the quantity of exchangeable cations depends on the crystal thickness. Only when the potassium ions located on the internal base planes are exchanged for large alkylammonium ions are these regions accessible for cation exchange with any cations by swelling. In contrast, the crystal edges contain hydroxyl ions exchangeable for fluorite, hydrogen molecules exchangeable for ammonia or amines, and groups which are active for alkali alcoholates. The differences in the ion exchange properties of these three clay minerals probably reside only in the different charge of the silicate layers and the different crystal size. An intermediate position in these results is formed in the studies of Amphlett and McDonald [12, 13], according to whom the exchange properties of green sand are attributable to the clay content of montmorillonite, illite, attapulgite, and kaolinite, which consequently are more readily accessible than in pure clay minerals.

Cation exchange in kaolinite, according to Weiss, is the result of special structural conditions on the hydroxyl-free base plane, while the OH-ions of the hydroxyl-containing base planes and the OH-groups located on the crystal edges can be exchanged for fluorine ions. The quantity of exchangeable cations therefore depends only on the thickness, and that of OH-ions on the thickness and particle size. At pH-values < 10, polyvalent cations can react according to the equation $[\text{kaolinite}]^{-}-Me^{+} + Me^{2+} + X \rightleftarrows [\text{kaolinite}]^{-}Me^{2+}X^{-} + Me^{+}$. The exchange capacity of kaolinite and illite could also be determined by radiochemical analyses [71], in which an extremely high selectivity of kaolinite for Y^{3+} ions was observed.

Because of the complex structural and isomorphic conditions, ion exchange in clays is not stoichiometric and the capacity differs highly in products from different deposits. The cation exchange capacity decreases during drying with an irreversible release of water and finally leads to a hysteresis in which the ion exchange reaction has become irreversible [577]. Anion exchange in particular is still quite obscure. In kaolin, pyrophyllite, halloysite, and montmorillonite, phosphate and sulfate are preferentially exchanged against chloride [519], again with evidence of a relation with the total surface, *i.e.*, with the structure of the investigated product.

The ion exchange behavior of bentonites is of interest for pharmaceutical purposes for the production of retard drugs, and its form with large organic molecules has been investigated [545]. A cation exchange mechanism is also assumed to take place in the adsorption of polymeric diallyldimethylammonium chloride–SO_2 on bentonite, although this reaction is irreversible [592].

Production of inorganic ion exchangers. The production of inorganic ion exchangers naturally is only of subordinate interest since these products are offered in specified forms by custom manufacturers. Beyond this, however, Table 15 shows a compilation of the most important types with a brief description of their production together with literature references in which the detailed methods of preparation can be found.

2.3 Cellulose Ion Exchangers

As a result of a small number of carboxyl groups in its constitution, natural cellulose has ion exchange properties. Numerous reaction products can be obtained by the oxidation of cellulose. These products include hydroxycelluloses with 15% COOH-groups, which in powdered form must theoretically be considered ion exchangers although they find little practical use because of their water insolubility. Powdered

Table 15
Inorganic ion exchangers (CE = cation exchanger, AE = anion exchanger)

	Type	*Production*	*References*
Insoluble salts			
Zirconium phosphate	CE	from $ZrOCl_2 \cdot 6HFO$ and H_3PO_4 in HCl	[16, 7]
Zirconium tungstate	CE	from $ZrOCl_2 \cdot 8H_2O$ and $Na_2WO_4 \cdot 2H_2O$ in HCl	[133, 496]
Zirconium molybdate	CE	from $ZROCl_2 \cdot 8H_2O$ and $(NH_4)_6(Mo_7O_{24}) \cdot 4H_2O$ in HCl of pH 1.2	[345, 496]
Hydrous zirconium oxide	CE and AE	from $ZrO(NO_3)_2$ in HNO_3 by precipitation with alkali	[15]
Hydrous titanium oxide	AE	from $TiOSO_4$ in 1N H_2SO_4 by precipitation with alkali	[15]
Hydrous tin oxide	CE	by dissolving granulated tin in 40% HNO_3 at 80°	[428]
Heteropolyacids			
Ammonium-12-molybdophosphate	CE	from NH_4NO_3, $NH_4H_2PO_4$ and $(NH_4)_2MoO_4$ at 80° in water	[92, 105]
Ammonium-12-phosphotungstate	CE	commercial 12-heteropoly acid is precipitated in 0.1N HCl with NH_4Cl	[362]
Potassium hexacyano-cobalt(II) ferrate	CE	from 1 volume 0.5M $K_4Fe(CN)_6$ and 2.4 volumes 0.3M $Co(NO_3)_2$ in 30 minutes at room temperature	[483]
Vanadium hexacyano-ferrate(II)	CE	from $H_4Fe(CN)_6$ and $NaVO_3$ (V:Fe = 3:1) plus 1–2M HCl	[350]
Molybdenum hexacyanoferrate	CE	from Na-molybdate in HCl and $H_4[Fe(CN)_6]$ at H/Mo > 2 and Mo/Fe > 2	[33]
Zinc ferrocyanide	CE	from 0.1M $Na_4[Fe(CN)_6]$ and 0.1M Zn_2No_3 and heating for 1–2 hours on a water bath	[342]

New zeolites			
Analcite, mordenite harmatome, faujasite chabasite, sodalite	CE	hydrothermal synthesis between 60 and 250° of aqueous gels with the composition Na_2O, Al_2O_3 and SiO_2 with an excess of NaOH	[40, 41, 37]
Zeolite Type A	CE	by crystallization of Na-Al-silicate gels	[82]
Clays			
Montmorillonite	CE	by treatment of suitable deposits	[179, 624–626, 39, 13, 239]
Montmorillonite, kaolin, pyrophyllite, haloysite	AE	by treatment of suitable deposits	[519]
Green sand	CE		[12]

cellulose ion exchangers which are now used in columnar processes have become very important as a result of the studies of Peterson and Sober on protein separation [470, 471].

Cellulose ion exchangers have ion exchanging properties just as do the synthetic ion exchange resins, but they differ from the latter on a few important points. For example, the cellulose matrix is hydrophilic in contrast to the hydrophobic synthetic resin matrix. As a result of the fiber properties of cellulose, *i.e.*, its loose network cross-linked by hydrogen bridges, the majority of active exchange groups located at distances of about 50 Å can become active for the exchange of large molecules which normally cannot penetrate the pores of a synthetic ion exchange resin. This leads to rapid exchange and to a higher capacity for large molecules in spite of the smaller absolute exchange capacity of cellulose exchangers compared with synthetic ion exchange resins.

According to its structure, the cellulose molecule has hydroxyl groups at the carbon atoms 2, 3, and 6, to which the active exchange groups are bound. Nothing can be said about their position, but most recent research findings in cellulose chemistry indicate that C atoms 6 and 2 are the most reactive. Because of the heterogeneous nature of the reaction, it must naturally be kept in mind that unmodified cellulose molecules are also present in the exchangers, while others are mono-, di-, or trisubstituted.

In appearance the most common cellulose ion exchangers do not differ from the starting materials. Their particle size of 15 to 20 μ depends on the irregular shape of cellulose and is difficult to define, but this is not decisive for the properties. The development of more granular types with a modified matrix and a somewhat different distribution of charged groups offers a variant in physical shape, which also becomes evident in a more rapid equilibration and higher capacity [628]. On repeated use, cellulose ion exchangers do not lose in quality, and they can be stored for months in aqueous solutions of 0.5 M NaH_2PO_4 to 1N NaOH. Contact with solutions of higher acidity than pH 4 should be avoided if possible, even though a brief contact with N HCl does not have a degrading effect. Regeneration is carried out with 0.5 N NaOH or with acid buffers or dilute HCl. The capacity of the described cellulose ion exchangers ranges between 0.25 and 1.0 meq/g referred to dry weight. As noted above, these values of acid-base capacity should not be underestimated, since the main advantage of these exchangers is that their capacity is sufficiently high for large molecules, such as proteins, nucleic acids, enzymes, and the like. Their low buffer action and low capacity for the exchange of small electrolytes is an additional advantage for the separation of proteins.

Columns of up to 40 cm length (ratio of 1:10) are used in work with cellulose ion exchangers. The exchanger powder is prepared in a separate graduated cylinder by suspending 10 g in 500 ml of the buffer or solvents to be used, and after about 1 hour discarding the supernatant solution which has become turbid due to fine exchanger dust; the remaining slurry is quickly slurried into the column. The bed which forms by settling is compacted further by the application of a weak vacuum. Such columns do not drain dry spontaneously. The pressure of the liquid layer over the bed is sufficient to produce satisfactory flow rates, which should generally range between 5 and 25 ml/hour.

In the preparation of cellulose ion exchangers (for instructions see the original literature [295, 296, 430, 471, 480, 485, 561]), wood celluloses are more suitable as raw materials than linters celluloses. Cellulose powder No. 123 and the linters cellulose powder No. 124 of Schleicher and Schüll can be recommended. The reaction always consists of the conversion of an alkali cellulose with a chlorine compound. It is important to avoid reaction conditions which dissolve or disperse the material, for although a reprecipitation is possible, this leads to gel-like products with too high a flow resistance to aqueous solutions. A reduced swelling with simultaneously increased exchange capacity can be obtained by cross-linking of the cellulose chains with di-(2-ethylamine) sulfate, 1,4-butanedisulfate, 1,3-dichloro-2-propanol, divinyl sulfone, and formaldehyde prior to introduction of the active groups. A cross-linking agent which simultaneously introduces exchange groups is dichloroacetic acid [244]. Quaternary cellulose ion exchangers can be produced from commercial DEAE cellulose by conversion with methyl iodide or ethyl bromide in absolute methanol [51]. By the reaction of cellulose with ethyleneimino-N-ethylacetate, β-ethyleneimino-N-methylpropionate and β-ethyleneimino-N-ethyldiethylphosphonate, it was also possible to prepare amphoteric cellulose ion exchangers [407].

The main field of application for cellulose ion exchangers is the separation and purification of proteins, nucleic acids, oligonucleotides, polysaccharides, lipids, nucleosides, purines, and pyrimidines, to name just a few of the almost confusing number. However, they can also be used for enzyme stabilization, as specific adsorbents, and in thin-layer chromatography. Their large-scale use is only a question of the development of suitable methods [583].

Serva-cellulose ion exchangers. The Serva-cellulose ion exchangers were the first to be manufactured and sold in large quantities, and they consequently contributed to the wide acceptance and successful use of these exchanger types. The large production program for acid and basic types, with their most important properties and data, is listed in Table 16.

These products are manufactured fr~~n special wood celluloses

Table 16
Serva-Cellulose ion exchangers (CE = cation exchanger, AE = anion exchanger)

Name	*Ion exchanging group*	*Properties*	*Capacity* *meq/g*	*Prevailing particle size* μ
CM-Cellulose	$-OCH_2COOH$	CE, weak acid	0.62±1	50–200
P-Cellulose	$-OPO_3H_2$	CE, medium acid	0.8 –0.9	50–200
SE-Cellulose	$-OC_2H_4SO_3H$	CE, strong acid	0.2 –0.3	50–200
DEAE-Cellulose	$-OC_2H_4N(C_2H_5)_2$	AE, strong base	0.4 –0.55	50–200
TEAE-Cellulose	$-OC_2H_5N^+Br^-$	AE, medium base	0.55–0.75	50–200
PAB-Cellulose	$-OCH_2C_6H_4NH_2$	AE, weak base	0.15–0.2	50–200
ECTEOLA-Cellulose	unknown	AE, weak base	0.3 –0.4	50–200
AE-Cellulose	$-OC_2H_4NH_2$	AE, weak base	0.33±0.1	50–200
BD-Cellulose	$-OC_2H_4N(C_2H_5)_2$	AE, medium base	0.8 ±0.05	50–200
GE-Cellulose	$-OC_2H_4NH \cdot C{=}NH_2NH_2{}^+Cl^-$	AE, strong base	0.2 –0.3	50–200
BND-Cellulose	$-OC_2H_4N(C_2H_5)_2$	AE, medium base	0.8 ±0.05	50–200

which are cross-linked with hydrophilic diepoxides prior to the introduction of active groups to increase mechanical and chemical resistance. They can be considered to be of analytical purity, so that they can be used without pretreatment. As a result of their more granular structure, they have favorable packing and flow properties. The irregular shape of cellulose fragments makes their particle size definition arbitrary and of less significance. A direct microscopic particle size determination leads to distribution curves such as those shown in Figure 31.

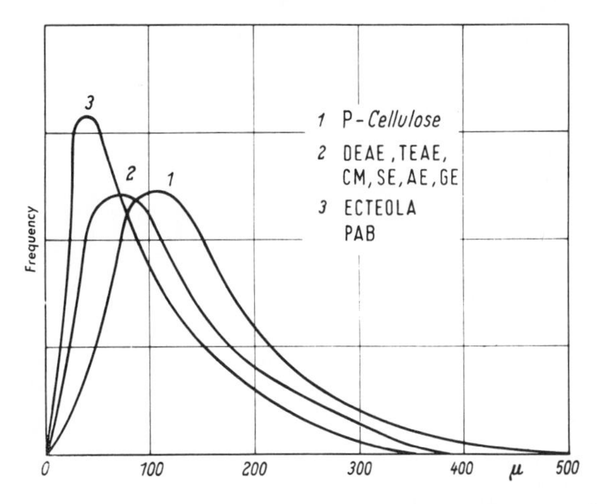

Figure 31 Typical particle size distribution of Serva-Cellulose ion exchangers. Particle diameter: 15–20 μ. Water content: 10–15%. Density: 0.1–0.2 g/ml.

For a capacity determination of Serva-cellulose ion exchangers, the products are dried overnight at 50°C, 2 g of sample is stirred for 10 minutes with 50 ml 0.5 N NaOH for the basic exchangers and with 50 ml 0.5 N HCl for the acid products, and the products are washed to neutrality with deionized water on a Buchner funnel. The exchanger is then slurred into a titration vessel with a small amount of water, treated with 25 ml N NaCl, and titrated with 0.1 N HCl to pH 3.0 or 0.1 N NaOH to pH 10 with continuous stirring and with the use of a pH meter and a glass electrode. This method leads to titration curves such as those shown in Figures 32 and 33, in which the inflections indicate the approximate pK-values and the pH-value at which 50% of the ionic groups have been dissociated.

A few points need to be kept in mind during columnar operation with Serva-cellulose ion exchangers. During transformation into the wet state, 1 g of exchanger corresponds to about 6–8 ml of the wet column volume. About one-half of the liquid is firmly bound to the exchanger, while the other half represents column liquid. The flow rate can be tested by loading a suspension of the cellulose material into a 70-cm column with a porous glass filter support in such a way

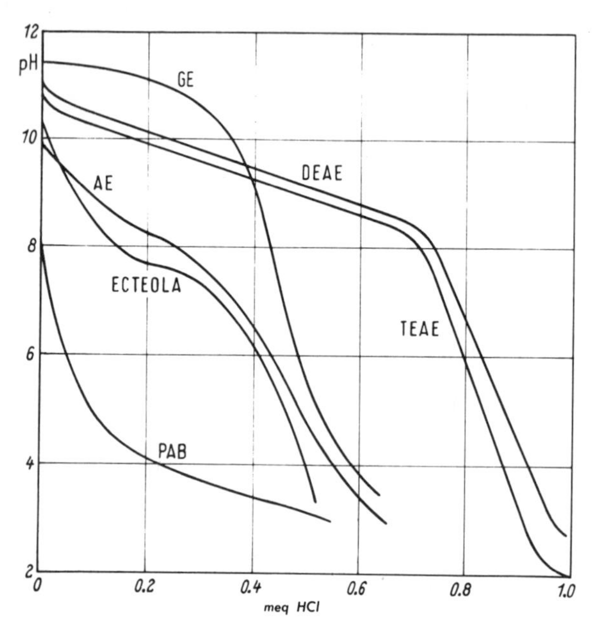

Figure 32 Typical titration curves of base cellulose ion exchangers.

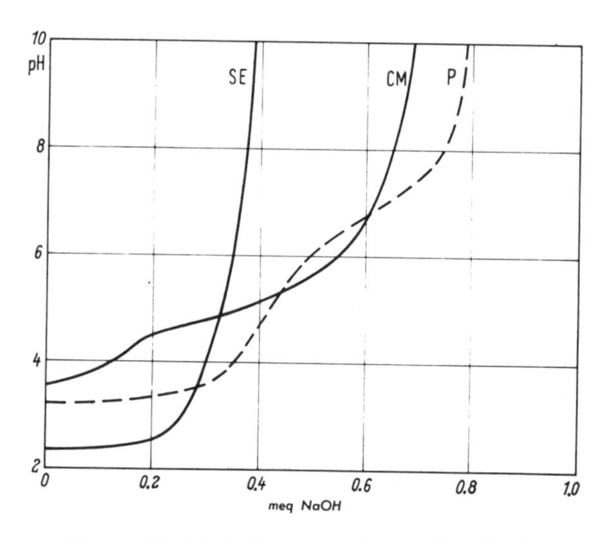

Figure 33 Typical titration curves of acid cellulose ion exchangers.

that the suspension fills 10 cm of the column after settling. Subsequently, 50 cm of water is added on the exchanger bed and the distance by which the liquid level drops in 10 minutes is used as a measure of the flow rate. Values of 15–25 cm are desirable. The development of artifacts must be kept in mind. Possibly, polysaccharides may also bleed out. This can be detected by a green color in the reaction with a 0.2% anthrone solution in sulfuric acid.

For application in column chromatography, 10 g of the exchanger is stirred with 500 ml of the buffer first used, allowed to settle for 1–2 minutes in a high, graduated cylinder to decant very coarse components, and is then allowed to stand for 45 minutes. The somewhat turbid supernatant solution is then discarded, and the exchanger in a charge of about 5% is poured in one step into the chromatography column while the drain cock is kept open. The flow rate should be 5–25 $ml \cdot cm^{-2} \cdot h^{-1}$, which adjusts by gravity. Since the capacity by weight of cellulose ion exchangers is very high for high molecular weight materials, 100 mg of substance to be exchanged can be calculated per gram of exchanger. For reuse with proteins, a treatment with four column volumes of 0.1 N NaOH and four volumes of water can be recommended as sufficient. Anion exchangers should be washed subsequently with two volumes of 0.1 N HCl to remove carbon dioxide which leads to artifacts.

As a concluding brief indication of the selection of experimental conditions with the use of Serva-cellulose ion exchangers, it might be noted that for proteins as well as other materials, DEAE-cellulose is used in 60% of the cases, CM-cellulose in about 20%, and TEAE-cellulose in about 5%. ECTEOLA-cellulose is used almost exclusively for nucleic acid, nucleoproteins, and nucleotides, while SE-, GE- and P-celluloses serve for substances of lower molecular weight. Certain rules apply to the choice of eluant: (1) cationic buffers (Tris-HCl, piperazine-HCl) are used with anion exchangers, and anionic buffers (phosphate, acetate) with cation exchangers; (2) a decreasing pH-gradient must be used for anion exchangers and an increasing one for cation exchangers; (3) pH-values near the pK-values of the exchangers should be avoided. For further treatment of the eluates, particularly freeze-drying, a choice of volatile eluants naturally is of advantage.

2.4 Dextran Ion Exchangers (Sephadex)

The polysaccharide dextran (poly-α-1, 6-glucan), which is produced from sucrose by microbiological methods, consists of fiber molecules which can be converted with epichlorohydrin and as a result of this cross-linking can be transformed into polymers with a three-dimensional structure; according to the studies of Porath and Flodin [481], these can be used for the fractionation of water-soluble substances. These dextrans, which are obtained as beads from the suspension polymerization process have proved to be useful as molecular sieves based on the method of gel filtration or in molecular sieve chromatography [543] for chromatographic separations based on molecular weight. They can be obtained in standardized form under the trade name Sephadex [145, 473].

Since the number of hydroxyl groups is only insignificantly modified by cross-linking, it is possible to prepare esters and ethers by further reactions with suitable reactants and to introduce ionic groups into the dextran molecule, thus arriving at ion exchange materials with polysaccharide as the matrix. In an early study Porath and Lindner [482] demonstrated that oxytocin and vasopressin can be separated on dextran diethylaminoethylether by the ion exclusion process.

Sephadex ion exchangers are obtained when exchanger groups are substituted in neutral Sephadex types G-25 and G-50. The functional groups are bound to the glucose building blocks of the network by ether bridges. Sephadex etherified with carboxymethyl or sulfopropyl groups results in weak or strong acid cation exchangers, CM- and SP-Sephadex, while substitution with diethylaminoethyl or 2-hydroxypropyldiethylaminoethyl groups furnishes weak and strong base anion exchangers, DEAE- or QAE-Sephadex.

Sephadex ion exchangers are insoluble and because of the hydrophilic character of their matrix, they swell easily in water. The degree of swelling is determined by the degree of cross-linking (the types produced from G-50 retain more moisture than those from G-25) and on the ionic strength and pH-value of the surrounding medium. While a higher ionic strength always causes shrinkage, pH changes in the range of 3–10 cause shrinkage in DEAE-Sephadex, and in the range of 3–5 produce greater swelling with increasing pH in CM-Sephadex. The strong acid or base exchangers, SP- and QAE-Sephadex, in contrast, have a nearly constant swelling volume in the entire pH range.

Sephadex ion exchangers are stable to alkalies. However, they are attacked by strong inorganic acids below pH 2, since these hydrolize the glycoside bonds of the matrix, particularly at high temperatures. Inorganic acids of up to 0.5 N can be used for regeneration. Strong glycol-splitting oxidants damage the exchangers.

Sephadex ion exchangers are comparable to corresponding cellulose ion exchangers and in many cases the same methods can be used. In terms of exchange capacity, however, they resemble synthetic exchange resins since their total capacity is 5–10 times higher than that of cellulose exchangers. The beadlike particles offer favorable flow characteristics in chromatography and good filtering properties in the batch process. While the individual production lots of cellulose exchangers depend highly on the raw material [274], Sephadex exchangers can be synthesized with a constant lot quality.

Sephadex ion exchangers are produced in more highely cross-linked types carrying the additional designation of A-25 and C-25 as well as with lower degrees of cross-linking in types A-50 and C-50, where A represents anion exchangers and C the cation exchangers. For separations in the molecular weight range of up to 30,000 for proteins, the types designated by 25 have their full capacity. They are preferable in this range since they also have the greater mechanical stability. For higher molecular weights of up to about 200,000, the types designated by 50 should be used since the mesh width of their matrix permits large molecules to gain access to the internal functional groups. For example, the hemoglobin binding capacity of SP-Sephadex C-50 (at pH 5 and $I = 0.01$) is equal to 7 g per g of exchanger (dry), while with the more highly cross-linked SP-Sephadex C-25 this capacity is only 0.2 g/g because the large hemoglobin molecules cannot penetrate the internal network of the exchanger. For separations of even larger molecules, preference should be given to Types 25 since only the functional groups on the surface of the exchanger bead can become effective. It must be kept in mind that a degree of cross-linking alone limits the useful exchange capacity when the molecular weight is too high and does not influence the separation characteristics with regard to the ion exchange process. In contrast, for neutral molecules the Sephadex ion exchangers also act as molecular sieves.

Sephadex ion exchangers are particularly suited for the separation of sensitive and high molecular weight biogenics such as proteins, nucleic acids, and polysaccharides.* Their nonspecific adsorption is minimal, and protein and biologically active materials generally are not denatured.

Before use, the exchangers should be allowed to swell in water or buffer and be equilibrated with the separation medium or the starting buffer. For this purpose, it is advisable to digest and drain them repeatedly with concentration up to 2 M buffer. This removes any interfering counterions of Na^+ or Cl^- in the delivered form of the product and at the same time adjusts a correct pH value. Subsequently, the exchanger is digested with dilute buffer until the composition of the medium coincides with the starting buffer. For regeneration, the same procedure is used, and more highly contaminated exchangers can be previously digested in 0.5 N sodium hydroxide solution, par-

---

*See [473] for literature references.

ticularly with protein contaminations, or in 0.5 N hydrochloric acid. Lipids can be removed with alcohol or neutral detergents. Cycling with hydrochloric acid/sodium hydroxide solution, which is customarily employed before the use of synthetic resin as well as cellulose exchangers, is superfluous in Sephadex ion exchangers.

The separation conditions should be selected so that the ionic strength of the starting buffer if possible amounts to 0.1 since the shrinking effect of the exchanger is only minimal at ionic strengths above 0.1. During separation of amphoteric molecules with anion exchangers, the pH value should be adjusted to about 1 unit above the isoelectric point, and in cation exchangers to about 1 unit below this point to bind the substance to the exchanger. A positive ionic strength gradient, applied in steps or continuously, or a pH gradient can be used for elution. The pH gradient must be adapted to the isoelectric point. Since QAE-Sephadex, as a quaternary ammonium base, does not change its charge during pH variations, its state of swelling is not influenced by a pH gradient. Consequently, serum proteins can be separated in closed systems by pH gradients and the exchanger can be directly regenerated [303]; this is important for industrial processes.

Table 17 lists the available Sephadex ion exchangers with their properties.

2.5 Coal Ion Exchangers and Other Materials with Ion Exchanging Properties

If wood, peat, firewood, soft coal, or anthracite is treated with oxidizing reagents or concentrated sulfuric acid, it is possible to introduce additional groups with an exchange capacity besides the existing OH- and COOH-groups with their exchange capacities. Heating with caustic soda under anhydrous conditions permits the production of coal with ion exchange properties [210]. Another group, the ammoniated coals, are also anion exchangers based on coal. The most important coal ion exchangers which are also commercially available are the sulfonated coals. These products contain a mixture of polyfunctional sulfone, carboxyl, and phenol groups of very different acid strengths. Depending on these different acid strengths, only phenol and carboxyl groups or, with a sufficient excess, also the sulfonic acid groups are transformed into the H^+-form during regeneration. If sulfonic acid groups are also present in the H-form, sulfonated coals are capable of neutral exchange according to the equations:

$$\text{Coal}^-\text{H}^+ + \text{Na}^+\text{Cl}^- \rightleftharpoons \text{Coal}^-\text{Na}^+ + \text{H}^+\text{Cl}^-$$
$$2\ \text{Coal}^-\text{H}^+ + \text{Ca}^{2+}\text{Cl}_2^- \rightleftharpoons \text{Coal}^{2-}\text{Ca}^{2+} + 2\text{H}^+\text{Cl}^-$$

Table 17
Sephadex ion exchangers (CE = cation exchanger, AE = anion exchanger)

Type	*Properties*	*Total capacity meq/g*	*Hemo-globin capacity g/g*	*Most favorable operating range pH*
DEAE-Sephadex A-25	AE weak base	3–4	0.5	2– 9
DEAE-Sephadex A-50	AE weak base	3–4	5	2– 9
QAE-Sephadex A-25	AE strong base	2.6–3.4	0.3	2–10
QAE-Sephadex A-50	AE strong base	2.6–3.4	6	2–10
CM-Sephadex C-25	CE weak acid	4–5	0.4	6–10
CM-Sephadex C-50	CE weak acid	4–5	9	6–10
SP-Sephadex C-25	CE strong acid	2–2.6	0.2	2–10
SP-Sephadex C-50	CE strong acid	2–2.6	7	2–10

NOTE: The earlier SE-Sephadex (sulfoethyl groups) has nearly the same properties as SP-Sephadex.

If about 100% of theory is used for regeneration, only a conversion with the alkaline salts takes place according to:

$$2\,\text{Coal}^{-}\text{H}^{+} + \text{Ca}^{2+}(\text{HCO}_3^{-})_2 \rightleftharpoons \text{Coal}^{2-}\text{Ca}^{2+} + 2\text{H}_2\text{O} + \text{CO}_2$$

a reaction which is very economically used in the "starvation process" for partial deionization.

Among commercial coal ion exchangers, the products Dusarit S and Imac C 19 of Imacti Industrielle Maatschappij Activit N.V., Amsterdam, should be cited. Dusarit S is a polyfunctional cation exchanger in the form of granulate (0.3–1.2 mm) based on sulfonated coal of high physical stability and with the properties of activated carbon. It is delivered in the Na-form and serves for the treatment of water and purification of glucose juices in the Na- and H-cycle. Imac C 19 is a polyfunctional strong and weak granular cation exchanger (0.3–1.2 mm) delivered in the H-form. It is used for the decarbonation of carbonate-rich water with a high regenerant effect. Table 18 lists the other properties of these coal ion exchangers together with other commercial products. Occasionally, coal ion exchangers produced

Table 18
Coal ion exchangers*

Designation	*Capacity*	*Heat stability °C*	*Permissible range pH*
Dusarit S	20 or 10 g CaO/l	80	0–14
Imac C 19	28 g CaO/l	80	4–14
Permutit S 53	10–15 g CaO/l	40	0–11
Permutit HI 53	1–28 g CaO/l	40	0–10
Zerolit Na	1.8 meq/g		
Zeo-Carb HI		30	8– 8

*Cation exchangers.

commercially or in the laboratory are also used for ion exchange chromatography.

A number of other materials [143] can also be converted into ion exchangers by chemical treatment. Tar, paper, cotton, pectin, tannin, and lignin have become well-known examples. Under the influence of sulfuric acid, lignin is converted into lignic acid, with ion exchange properties which were recognized by Freudenberg at the time when the first synthetic ion exchange resins were produced. By treating lignin-containing substances, particularly sawdust, with sodium sulfite solutions, very inexpensive ion exchangers which can be used for the adsorption of radionuclides are obtained [30].

Alginic acid [437, 438] has always been of some interest as a cation exchanger. Occasionally, it was also used for separations by ion exchange chromatography [563]. The possible therapeutic applications of alginates are of particular interest because of their Sr-Ca-ion exchange reactions [257].

2.6 Ion Exchange Membranes

Ion exchange membranes in the simplest forms are synthetic ion exchange resins in the form of sheets and films. Such materials have been known for many years; however, because of their low stability and high electrical resistance, no industrial applications could be found for the first membranes, so that research in the field initially remained limited to studies of their physicochemical properties [524–526, 562]. Today, methods are known for the production of ion exchange membranes with acceptable mechanical, chemical, and electrical properties.

On the basis of their structure, the membranes can be permeable for water as well as for electrolytes, thus inhibiting only convection.

However, they may also allow only the permeation of water and not of the electrolyte, thus producing a separation effect. Such membranes are known as semipermeable. Among these semipermeable membranes, there are again those which allow only the permeation of anions or of cations. These are known as permselective membranes.

Three types of ion exchange membranes are distinguished:

1. heterogeneous membranes,
2. interpolymer membranes, and
3. homogeneous membranes.

The heterogeneous membranes are produced from finely milled ion exchange granulate and are formed into sheets by compression with an inert elastic binder. The structure of such products, for which other manufacturing processes have also become known, is apparent in Figure 34. Swelling in water results in the formation of water-filled interstices into which electrolyte particles can penetrate. High molecular weight colloidal particles can also penetrate the channels and can cause undesirable contamination and precipitation phenomena. To attain a sufficiently low resistivity, 30 wt.% of binder is generally used for production, although this leads only to low mechanical stability.

a

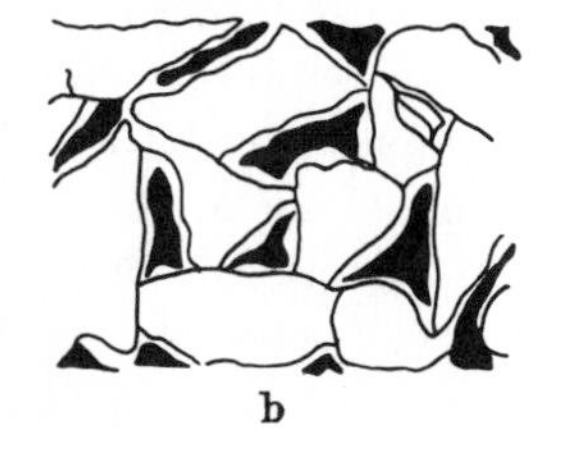
b

Figure 34 Structure of heterogeneous ion exchange membranes (a) before and (b) after swelling; binder in black.

Interpolymeric membranes are obtained by casting a film from a homogeneous solution of two polymers, one of which represents the polyelectrolyte and the other the water-soluble filmogenic material. Thus, for example, membranes have been produced from polystyrenesulfonic acid and polyacrylonitrile in dimethylformamide. The polyelectrolyte is built so firmly into the chains of the matrix polymer that even long immersion in water does not elute the polyelectrolyte. This property admittedly is obtained only under certain conditions; thus, the polyelectrolyte must have relatively little chain branching and its molecular weight must range between 30,000 and 160,000.

In the homogeneous membranes, which in principle are the best, the film-forming matrix itself carries the ionic groups. Such materials can be produced from methacrylic acid, phenolsulfonic acid [382], or styrenesulfonic acid [233, 500] by copolymerization or polycondensation in analogy to synthetic ion exchange resins. The method of producing a cross-linked styrene-divinylbenzene skeleton followed by sulfonation in even greater analogy with synthetic exchange resins is very difficult and was successfully used for the first time by Schwab *et al.* [652]. After many years of intense research, industrial processes were developed in which copolymers of styrene and divinylbenzene were manufactured in sheets into which active groups were introduced for the formation of a cation or anion exchange resin in membrane form. By a suitable choice of monomer and cross-linking agent, ion exchange membranes with a broad range of ion selectivity, resistivity, and good mechanical stability can be produced by these and similar processes.

Ion exchange membranes, also known as heterogeneous types but manufactured by binding the active polymers to fiber materials serving as the matrix or by later chemical or photochemical treatment of already existing film materials, are more difficult to classify.

Ion exchange membranes can carry cationic as well as anionic groups and thus can be considered cation or anion exchangers. According to the membrane theory of Gibbs and Donnan, they exclude mobile ions of equal charge by the high concentration of fixed ionic groups in the membrane. As a result of this property of ion exclusion, a flow of electrical current through a cation exchange membrane will be carried in its entirety by the mobile cations. In the same manner, an anion exchange membrane is impermeable to cations and permits only the transport of anions. Thus, ion exchange membranes are capable of selective ion transport (permselectivity).

It has also become possible to construct composite ion exchange membranes leading to structures which simultaneously contain cation and anion exchange groups. This can be accomplished by mixing cationic and anionic regions which can no longer be distinguished so that an amphoteric membrane is formed. When the components are organized and combined in a distinguishing way, "mosaic" membranes can be formed by parallel arrangements or "sandwich" membranes by series arrangements, as shown in Figure 35 for both types. These membranes have a number of interesting properties. For example, the mosaic membranes have an unusually high salt permeability [340, 341, 637] and the sandwich structures behave like rectifiers [34, 329]. Their theoretical and practical investigation is of interest because such structural principles can also be found in biological membranes.

Characterization of ion exchange membranes. To characterize ion exchange membranes to serve as a basis for the discussion of research findings or as a specification for commercial products, some data are necessary which are obtained from the membrane properties. To a great extent these data are identical to those of other ion exchange materials as far as a matrix, cross-linking, functional groups, etc. is concerned. However, some additional properties must be added for a more complete description.

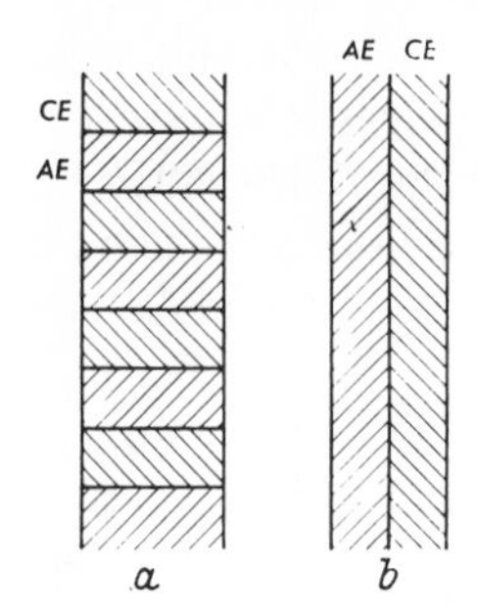

Figure 35 Composite membranes: (a) mosaic; (b) sandwich structures. CE = cation exchanger, AE = anion exchanger.

The fixed ion concentration or capacity is determined by stirring the cation exchange membrane in the H^+-form with 20 ml 0.02 N NaOH containing 29 g NaCl/l and, after equilibration, the excess of alkali is determined by titration with 0.02 N HCl. It is also possible to add the superficially dry membrane in H-form to 50 ml of a 2 N NaCl solution and titrating with 0.1 N NaOH with the use of Bromcresol Green.

The water content of ion exchange membranes is of interest, since in contact with aqueous media the membranes adsorb water in a quantity which varies highly in different types; but at the same time the water content determines membrane efficiency by the possible dissociation of functional groups. A determination is made either by drying the membrane surface with filter paper, weighing it in the wet state, subsequently drying it on phosphorus pentoxide in vacuum, weighing back, and reporting the weight loss in per cent of water content or by shaking the membranes with distilled water under standard conditions until no acid reaction is obtained with Bromcresol Green. The samples in the surface-dried state are then placed in a closed vessel at 100% relative humidity and 25° C up to constant weight, followed by drying at 110° C for 16 hours in an oven, and weighing back.

The flexibility of ion exchange membranes gives information on their mechanical properties. Rosenblum, Tombalakian, and Graydon [500] developed a method for their measurement according to which the number of flexes up to failure, produced with an aluminum rod on clamped membrane specimens, serves as a measure of flexibility.

The strength of the membranes is important for their practical application. As a criterion the tensile strength is used, determined with a tensile tester and a standard method—for example, ASTM-D 638.

The electrical resistivity depends on the exchange capacity, the nature of the membrane (heterogeneous or homogeneous), and its thickness. In industrial application its value influences the process costs, so that minimum resistivities, for which homogeneous membranes offer the best conditions, are desirable. The electrical resistivity is measured by the method of Kohlrausch in 0.1 M KCl solution in which the ohmic resistance is determined at 25°. Other sources cite the resistivity of the membrane in equilibrium with Cl^--ions in 0.5 seawater, expressed in ohm/cm^2.

The cation or anion transfer coefficients serve for a more precise description of permselectivity and offer a criterion of the capacity of ion exchange membranes to transport any ionic species selectively. Customarily, two methods are used to determine the transfer coefficients and the results of both are usually cited. In one method, the membrane is placed between two solutions of identical concentration with Ag- and AgCl-electrodes and a current of known voltage is allowed to flow through the membrane for a given time. Subsequently, the two solutions are removed, the concentrations are determined by analysis, and the transfer coefficients are calculated from the quantity of electrolyte which migrated through the membrane according to [409, 586, 644]:

$$\bar{t}_i = \frac{FJ_i}{I}$$

In the second method, the membrane is placed between two solutions of different concentration with Ag- and AgCl-electrodes and the transfer coefficient is determined from the emf according to the equation [247]:

$$\bar{t}_i = \frac{E}{2\,E_{\max}} + 0.5$$

The two methods differ by the fact that water transport is neglected in the second.

The entire complex of physicochemical problems connected with the structure and use of ion exchange membranes and giving information on transport phenomena as a general concept and on diffusion phenomena—electroosmosis, electrodialysis, etc.—as special consequences has been investigated; the literature is very extensive and offers exhaustive information on many problems [291, 381, 383, 523, 528]. Ion exchange membranes are used for water desalination. The principle is shown in Figure 36. In this three-cell unit, a center cell is separated from the cathode and anode compartment by a membrane permeable only for cations on one side and only for anions on the other. The figure clearly shows that a current flow must result in an

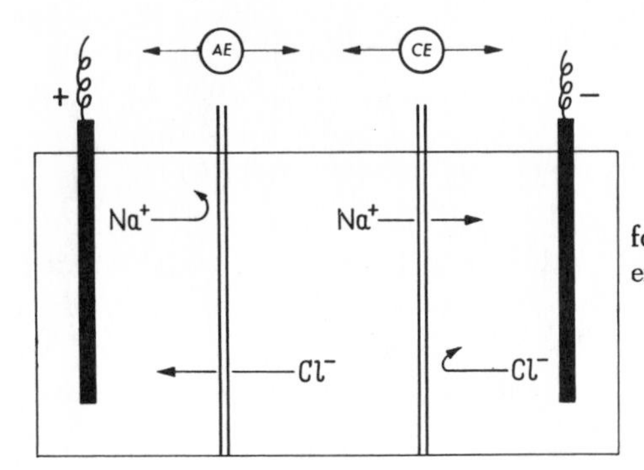

Figure 36 Three-cell system for the removal of salts by ion exchange membranes.

electrolyte depletion in the center cell. In practice, multicell units are used [177, 406], as shown schematically in Figure 37. Industrial applications of ion exchange membranes are found in the field of concentration and desalination, including the treatment of brackish water [117], recovery of valuable chemicals, separation of organic electro-

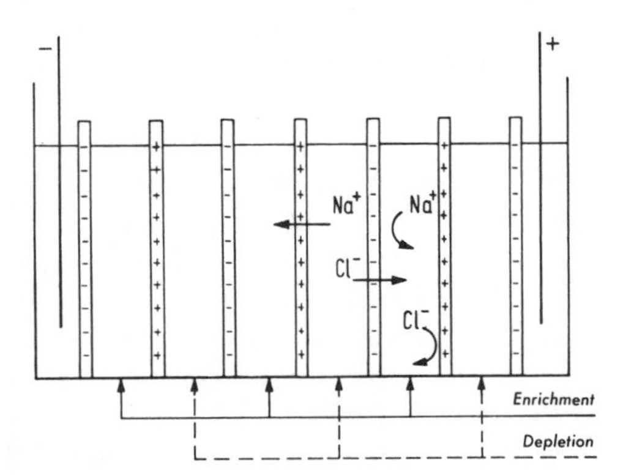

Figure 37 Multicell system with cation exchange and anion exchange membranes.

lytes such as amino acids, and sucrose and glucose refining. It is reported that electrolysis with ion exchange membranes permits the production of caustic soda and sulfuric acid, reduction of uranyl chloride, production of acetic acid from sodium acetate, and the recovery of sulfuric acid from salt baths. In addition to other applications in chemical reactions, the analytical usefulness of ion exchange membranes has also been investigated and demonstrated in some cases [59, 173, 287, 299].

Various types of ion exchange membranes are commercially available. Rohm and Haas Company is delivering the membranes Amberplex C-1 and Amberplex A-1, which have a high chemical stability and can be used up to temperatures of 95° C. A large production program for industrial membranes has been developed by the Japanese Asahi Chemical Company which manufactures homogeneous cation exchange membranes CK-1 and DK-1 and homogeneous anion membranes CA-1, DA-1, CA-2, and DA-2. Additional products are marketed in the form of Permaplex C20 and Permaplex A20 by Permutit Company, London, and as Nepton AR 111 A of Ionics, Inc., USA. The use of ion exchange membranes that are available in sizes larger than 1 m^2 must follow instructions of the manufacturer, which usually need to be observed very carefully and should be procured from the suppliers depending on need.

2.7 Liquid Ion Exchangers

Various high molecular weight basic and acid organic liquids can be used as liquid extractants for the recovery, purification, and concentration of different anions and cations. Such substances are water-insoluble electrolytes. Their mechanism of action can be equated to the general ion exchange processes, and as a result they have aroused increasing interest as "liquid ion exchangers" [378, 521]. In this case, the ion exchange process takes place between the two immiscible liquid phases. They are particularly suited for liquid countercurrent processes.

Quite generally, aliphatic and aromatic primary, secondary, and tertiary amines as well as quaternary ammonium bases which are readily soluble in water-immiscible organic solvents but insoluble in water are suited as liquid anion exchangers. The most common products in the case of the amines are tri-*n*-octylamine (TnOA) [95], triisooctylamine (TiOA) [95], tri-2-ethylhexylamine [127], tribenzylamine (TBA) [439], didecylamine [9], methyldioctylamine [439], methyldidecylamine [72], tridodecylamine [95] and tributylamine [650] as well as the commercial amine mixtures—*i.e.*, trialkylmethylamine (homologous mixture with 18–24 C-atoms each, Primene JM), N-dodecenyl-N-trialkylmethylamine (mixture with 24–27 C-atoms each, Amberlite LA-1), N-lauryl-N-trialkylmethylamine (Amberlite LA-2), N,N-didodecenyl-N-*n*-butylamine (Amberlite XE-204) and tricaprylamine (Alamine 336). Aliquat 336 can be cited as a quaternary ammonium compound. Liquid cation exchangers primarily are the acid esters of phosphoric acid, among which monododecylphosphoric acid (DDPA), monoheptadecylphosphoric acid (HDPA), and di-(2-ethylhexyl) phosphoric acid (D2EHPA), have become important and commercially available as weak acid cation exchangers, while dinonylnaphthalenesulfonic acid (DNNS) has been investigated as a strong acid cation exchanger [630].

The analogy of liquid ion exchangers with ion exchange resins is readily apparent from the equation:

$$2R_3NHCl + H_2SO_4 \rightleftharpoons (R_3NH)_2SO_4 + 2HCl$$

for anion exchange and from

$$2(RO)_2PO(OH) + Me_2SO_4 \rightleftharpoons [(RO)_2POO]_2Me_2 + H_2SO_4$$

for cation exchange. Discussions as to whether it is appropriate to call such reactions ion exchange can be answered affirmatively here; liquid ion exchangers should be included as a special type of exchanger in this description. The process forming the basis of ion exchange with a liquid exchanger, and accomplished by thorough mixing and subsequent separation, consists of the transfer of an electrolyte from the organic phase of the liquid exchanger into the aqueous phase and vice versa. This process takes place on the basis of separation and recombination of electrostatic bonds and consequently is an ion exchange. When amine extraction is considered as a liquid ion exchange, this is certainly more complicated than indicated by the above equation. A more detailed discussion than is possible here is merited particularly by questions of complex chemistry dealing with the extraction of metals with liquid anion exchangers. However, some studies exist on this mechanism according to which trilaurylamine in the extraction of hexavalent uranium is present in an equilibrium between a monomeric and a dimeric form, both of which participate in the extraction by their own equilibrium constants. The formation of higher amine aggregates cannot be ruled out and the dependence on the solvent utilized must be investigated separately in each case [400].

The required properties of liquid ion exchangers consist of water insolubility, miscibility with inexpensive solvents as carriers, selectivity, and stability to common reagents. Moreover, they may not have an emulsifying action and it should be possible to strip and regenerate them with common chemicals.

For the development of a liquid ion exchange procedure in the laboratory, calibrated separatory funnels are used. The liquid ion exchanger is prepared in a concentration of 5 vol.% in 2.5% isodecanol and 92.5% kerosene, and the metal-containing aqueous solution is treated with this in a ratio of 1:4. These extraction experiments furnish the basic data on phase separation, emulsifying behavior, distribution of the metal, and contaminations in the two phases under different extraction conditions. Equilibrium curves can be plotted with the concentration in the organic phase and the aqueous phase as coordinates which can be interpreted similar to the selectivity isotherms of solid ion exchangers. The same simple systems can then be used to investigate the influence of the organic phase composition on the liquid ion exchangers, additive and carriers, influences of temperature, contact time, and contamination. The recovery of an exchanged particle differs from extraction since by nature it cannot be obtained by distillation in procedures with liquid ion exchangers. Therefore, in analogy to elution in solid ion exchangers, a reextraction must be performed with an aqueous phase (stripping) to separate the exchanged particle. Additional technological steps can also be developed in the laboratory. The principles for a later industrial installation, which basically represents only a scale-up, can be developed in extraction columns of different design, in centrifugal extractors, and particularly in mixer-settler systems.

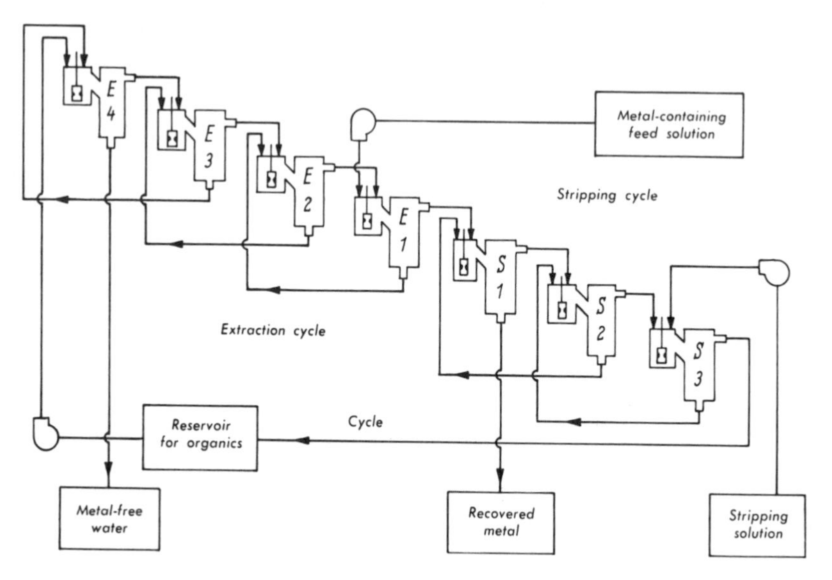

Figure 38 Flow sheet of a laboratory cycle for the recovery of uranium by liquid ion exchange.

Figure 38 shows the flowsheet of a laboratory cycle according to Lewis [392] with a commercial mixer-settler system in which data for the design of an industrial plant can be obtained after an operating time of about 100 hours without a loss of extractant, emulsification, or other malfunctions. The optimum flow rates for the metal solution, organic exchanger phase and aqueous stripping phase, the necessary optimum exchange time, the surface requirement for good separation, the optimum residence time of the organic phase in the settlers, and the additional purely technical data must be determined in such a system for the design of the industrial installation.

Similar to ion exchange resins, which attained their present importance by their use in water treatment, liquid ion exchangers were developed in connection with the recovery of metals, particularly uranium, and separation of fission products. As the technology of liquid ion exchangers is becoming more sophisticated, the occasionally noticeable stagnation in this field will be overcome again and new areas of application will be found in hydrometallurgy, waste treatment, and inorganic synthesis (amine process in analogy to the Solvay process).

Chapter 3

Ion Exchangers as Preparative Agents

3.1 Ion Interchange

Ion interchange (also known as metathesis) is the simplest preparative application of ion exchangers. Benefits are derived from the general ion exchanger property of exchanging their counterions for other ions. Cation exchangers in the H-form and anion exchangers in the OH-form exchange the hydrogen ion or hydroxyl group for cations or anions, respectively:

$$\text{Cation exchanger–}SO_3H + MeX \longrightarrow \text{Cation exchanger–}SO_3Me + HX$$

$$\text{Anion exchanger–}N(R_3)OH + MeX \longrightarrow \text{Anion exchanger–}N(R_3)X + MeOH$$

Thus, a preparative possibility is afforded to obtain free acids and bases by a simple method. The known procedure of recovering inorganic and organic acids by converting their barium salts with sulfuric acid or their lead, silver, or other heavy metal salts by conversion with H_2S is considerably simplified by the use of ion exchangers. Salts that can be obtained only through the free acids or from the alkali salts of the corresponding acids are also easily obtainable in this manner.

Generally, strong acid and strong base ion exchangers are used for this purpose. In some cases, weak base exchanger types are also applied.

Ion exchange processes are usually characterized by their simplicity and offer four additional advantages for preparative work: high rapidity, high purity of the products, suppression of insoluble precipitates, and low consumption of reagents.

3.1.1 *Preparation of Acids*

The preparation of acids by ion interchange was first described by Klement [314]. He prepared thiocyanic acid, hypophosphorous acid, phosphoric monamide, triphosphoric acid, and polyphosphoric acid [316]. As a practical example, the procedure for the preparation of thiocyanic acid is described below.

Preparation of Thiocyanic Acid. Sulfonic acid exchange resin (70 g) in an exchange column is transformed into the H-form with 2 *N* HCl and washed with distilled water until the eluate remains clear after the addition of silver nitrate solution. Subsequently, a solution of 10 g ammonium rhodanide in 50 ml water is charged on the column, allowed to stand for 10 to 15 minutes, then allowed to drain. After draining, it is washed with 50 ml water. An approximately 8% solution of the free thiocyanic acid is obtained in the eluate.

The method described by this example is generally applicable.

In this connection the research of Hein and Lilie [259], who developed the preparation of different complex acids on the basis of the mild conditions afforded by ion exchange, is of interest: when 1.5 g $K_3Co(CN)_6$ in 50 ml water is reacted with Wofatit P and subsequently dried in a vacuum desiccator, 0.69 g of the free acid $H_3Co(CN)_6 \cdot aq$ was recovered.

3.1.2 *Preparation of Bases and Salts*

Just as acids can be prepared from salts on cation exchangers, free bases can also be recovered with the use of anion exchangers. Thus, d'Ans *et al.* [135] were able to obtain the bases forming various complex salts. As an example, the conversion of hexamminocobalt(III)-chloride is described below; this can serve as a general working method.

Preparation of Hexamminocobalt(III)-Hydroxide. Permutit ES, Dowex 2 or Amberlite IRA-400 (50 g) is transformed into the OH-form in 1 liter *N* NaOH in a column and washed until the eluate is neutral. The yellow-brown solution of the cobalt complex in the form of its chloride is then charged on the column and allowed to percolate through the column at a flow rate of 10 ml/minute. Subsequently, it is washed with about 300 ml water. The filtrate contains the free base $Co(NH_3)_6(OH)_3$, identified by titration with 0.1 *N* HCl.

If the solution of free complex base obtained by this method is treated with ammonium oxalate, the sparingly soluble oxalate of the hexamminocobalt(III) ion precipitates. This route has been used to convert the chloride into the oxalate on ion exchangers.

That the preparation of bases is not limited to inorganic chemistry is demonstrated by a report of Rebek and Semlitsch [487] on the preparation of a dye base with ion exchangers.

A special method which must be mentioned in connection with the preparation of salts is represented by so-called ligand exchange developed by Helfferich [262, 263]. In this procedure the complex water molecules of a metallic ion bound to a cation exchanger of the carboxylic acid type are exchanged for ammonia. The method can be extended, the ammonia being then exchanged for diamine. In contrast to ion exchange in general, this method in accordance with complex chemistry leads to an exchange of ligands on the counterion while the latter is retained in the resin phase. Carboxyl resins are used for this purpose because of their higher selectivity compared to sulfonic acid exchangers.

Additional possibilities in the preparative use of ion exchangers which go beyond their common application for purification and separation can be expected in reactions leading to the formation of salts from acids analogous to the Solvay process. In this application of weak base anion exchangers, the exchange process depends on the relative selectivity of the anion exchanger for chlorides and bicarbonates [371]. The production of inorganic brines such as SiO_2, $Fe(OH)_3$, $Al(OH)_3$, thorium, and zirconium brines and the formation of hydrosols [42] are also examples for the preparation of bases and salts.

3.1.3 *Preparation of Normal Solutions*

Another example of the use of ion exchangers as a preparative medium is the preparation of normal solutions containing practically no carbonate or silicate.

A strong base ion exchanger such as Amberlite IRA-400 is treated with dilute hydrochloric acid in a column, washed, and subsequently transformed into the OH-form with carbonate-free 2 *N* NaOH. The exchanger is then washed free from alkali with boiled distilled water. When such an exchanger is loaded with a carbonate-free common salt solution, a quantitative ion interchange leads to a carbonate-sodium hydroxide solution with the equivalent quantity of NaOH. In practice, the method of Steinbach and Freiser [568] can be used. A solution of 2.922 g NaCl (p.a.) in 50–100 ml boiled distilled water is loaded on 40 g Amberlite IRA-400 at a flow rate of 4 ml/minute. The column is subsequently washed with freshly boiled distilled water. The eluate is collected in a 500-ml graduated cylinder in the absence of air and finally is brought to the mark. The 0.1 *N* NaOH obtained is practically free from carbonate. For the determination of silicate concentrations of 0.1–0.01 ppm in water by colorimetry, the blank value of an ordinary *N* NaOH solution (p.a.) is too high. However, with the above method an ultrapure silicate-free potassium hydroxide solution can be prepared by loading an equivalent solution of potassium sulfate on a strong base anion exchanger in the OH-form as described by Fisher and Kunin [181].

These examples offer an insight into the broad field of application of ion exchangers. A few other examples will only be enumerated: silicic acid can be prepared from sodium silicate on a strong acid cation exchanger; a strong base cation exchanger in the K-form is used for the recovery of the potassium salt of penicillin; and sodium citrate is obtained on a strong base anion exchanger in the citrate form.

3.2 Purification of Solutions and Substrates

Ion exchangers have found an important and rapidly growing application for the purification of solutions and substrates; thus, ionic components of solutions can be completely removed by cation or anion exchange resins. These processes, which have become known by the terms desalination, deionization, or demineralization, play an important role in the application of ion exchangers as preparative media as well as in technology.

Laboratory applications of this type include: (1) extraction of acids and bases from neutral salts, organic materials and solvents; (2) extraction of metal salts from acids and bases; and (3) desalting of organic or biological substrates and nonaqueous solvents.

In the first two methods, cation exchangers in the H-form or anion exchangers in the OH-form are used. A mixed-bed process can also be applied for desalting. This refers to a bed containing a cation as well as an anion exchanger which simultaneously removes cations and anions from the solution and replaces the salt by an equivalent quantity of water.

The practical application of such purification methods will be demonstrated by a few examples.

Example 1: Extraction of copper ions from a weakly sulfurous cupric sulfate solution on a strong acid cation exchanger. In this procedure, a copper-free solution can easily be obtained. The purification effect is clearly manifest because of the inherent color of copper ions.

A preswollen, strong acid cation exchanger (5 g) is slurried into a column of 15×1.5 cm and washed once with distilled water. Subsequently, 200 ml 2 N sulfuric acid is percolated through the column to transform the exchanger into the H-form. The column is now washed with distilled water until the eluate is neutral.

After this preparation of the column, the weakly sulfurous, blue cupric sulfate solution is loaded on the column and percolated through it at a flow rate of 2 ml/minute. The eluate, which is collected in a beaker, is colorless. The exchanger bed assumes a greenish opalescence as soon as it is partially transformed into the copper form. Loading of the cupric sulfate solution can be continued until breakthrough for copper has been attained; this can be easily recognized by the appearance of the deep-blue copper complex in an ammonia-treated receiver.

For the regeneration of the exchanger column, 300 ml 2 *N* sulfuric acid is charged on it and percolated through the column at 2 ml/minute. After brief washing with distilled water, the column is ready for a further purification cycle.

Example 2: Extraction of small amounts of iron from technical-grade hydrochloric acid on a strong base anion exchanger. This example will demonstrate that metals can be bound not only on cation exchangers but also on anion exchangers. The mechanism of action involves the conversion of the metal into an anionic complex so that it can be sorbed by an anion exchanger.

In concentrated hydrochloric acid solution, iron is present in the form of the anion $FeCl_4^-$. Therefore, if a concentrated hydrochloric acid solution containing iron(III)-chloride as an impurity is loaded on a column of strong base anion exchanger, the eluate is free from iron.

Since the iron complex decomposes with decreasing acid concentration,

$$FeCl_4^- \rightleftharpoons FeCl_3 + Cl^-$$

the iron can be removed from the column in the form of $FeCl_3$ by simple washing with water and the column can thus be easily and inexpensively regenerated.

In the practical procedure, a column of 110 × 2.5 cm packed with anion exchangers in the OH— or Cl-form is used. To test the activity of the exchanger, the column is first loaded with about 25% hydrochloric acid containing 50–500 ppm iron(III)-chloride. Subsequently, 2 liters water is loaded on the column to elute the iron and regenerate the exchanger. The ion exchanger is then ready for further acid purification cycles.

Since technical-grade hydrochloric acid, which is produced by various processes, always contains traces of iron, this method is of industrial importance.

Another example for the removal of metals is the purification of quaternary ammonium bases from alkali metals, which leads to final impurity concentrations of less than 1 ppm Na^+ with a yield of 80% [455].

Example 3: Extraction of small quantities of ammonia from water in a mixed-bed process. Ammonia-free water can be produced from distilled water by ion exchange in a mixed-bed process. For this purpose, it is best to use an exchanger mixture of 2 volume parts of Amberlite IRA-400 and 1 volume part of Amberlite IR-120.

The mixed exchange resins are used in the column process. When loaded with distilled water, for example, the nitrogen content will be reduced from 0.860 mg N_2/50 ml to 0.006 mg N_2/50 ml. The ammonia concentration in the eluate is determined by photometry with Nessler. In such cases it is more convenient in laboratory practice to discard the exhausted exchanger than to regenerate it.

The degree to which the removal of ammonia from substrates can be extended is demonstrated by the example of blood purification according to Moretti *et al.* [442].

Blood contains unstable ammonia substances which easily decompose with the formation of NH_4-ions. When the strong acid cation exchanger, Dowex 50X8, was used in the Na^+-form in this procedure, the cation equilibrium in the blood shifted in favor of sodium at the expense of potassium and calcium until the exchanger was used in such a ratio of loading with sodium, potassium, calcium and magnesium that no coagulation or its consequences occurred; this demonstrates another interesting possibility of ion exchange. If such a column is installed into a heart-lung machine for extracorporeal circulation and blood is pumped through it for the removal of NH_4^+-ions, a clinical treatment of hepatic coma may become feasible. From the purely preparative standpoint, blood—a valuable substrate—has been freed from the undesirable impurity of NH_4^+ by ion exchange.

Example 4: Extraction of pyrosulfite from noradrenaline solutions. Noradrenaline solutions for medical purposes are generally stabilized with pyrosulfite. For a quantitative determination of noradrenaline in dilute solutions (up to 0.001%), the fluorescence method of Erne and Canbäck [174] can be applied; however, pyrosulfite represents an interference substance and must be removed from the solutions. An anion exchanger can be used advantageously for this purpose.

The anion exchange resin (*e.g.*, Amberlite IRA-400 or Dowex 2, particle size 0.4–0.8 mm) is washed with 20% sodium chloride solutions in a 100 × 10-mm column and the excess of chloride is removed with water. Thus prepared, the column is loaded with the noradrenaline solution and washed with water. The noradrenaline determination can be made in the filtrate.

The value of such an ion exchange process can be seen from the test data of Table 19 (according to the cited authors) showing the fluorescence values before and after ion exchange.

Example 5: Extraction of formic acid from formaldehyde solutions. In the production of formaldehyde from methanol by air oxidation according to the equation:

$$2\ CH_3OH + O_2 \longrightarrow 2\ HCHO + 2\ H_2O$$

a small quantity of formic acid also forms according to:

$$2\ CH_3OH + 2\ O_2 \longrightarrow HCOOH$$

Formic acid interferes with the production of plastics. Although it is present in aqueous formaldehyde solutions in concentrations of only about 0.1%, it can nevertheless lead to complications. An ion exchanger can be used for its extraction.

Table 19

Fluorescence values of noradrenaline analysis with and without ion exchange

Noradrenaline	*Fluorescence values* — *without ion exchange*	*Fluorescence values* — *after ion exchange*
10 mg	78, 79, 80, 78	77, 79, 78, 80
10 mg with 10 mg pyrosulfite	35, 40	79, 79, 78, 79

Amberlite IR-45 is transformed into the OH-form with sodium hydroxide solution and is carefully washed to neutrality. The formaldehyde solution, in which the formic acid content had been previously determined, is loaded on the column and percolated at a low flow rate. By varying the operating conditions, elevating the column temperature if necessary, the process can be controlled so that formic acid is extracted from 35–50% formaldehyde solutions to residual concentrations of 0.001%.

Example 6: Purification of a nonaqueous solvent; extraction of mercaptans from hydrocarbons. Genuinely porous or macroreticular ion exchangers can be very advantageously used in nonaqueous systems. The procedure is first carried out in water in order to transform the exchanger into the desired loading form and for regeneration. If the nonaqueous solvent is immiscible with water, an intermediate treatment is necessary with a solvent which is immiscible with water but miscible with the solvent to be treated subsequently. Methanol and ethanol are particularly suited for this purpose when used at a flow rate of 0.05–0.1 ml/minute. After this step, the column is treated with one to two column volumes of the pure solvent and is thus ready for the purification step. When the latter has finally been completed and the exchanger is exhausted, two column volumes of alcohol are again charged through the column and displaced with water in the transition to regeneration. A practical example of a purification process which is frequently used in a similar form is the extraction of mercaptans from hydrocarbons.

Amberlyst XN-1002 (25 ml) is regenerated in a suitable column with three column volumes of aqueous N sodium hydroxide solution at a flow rate of 0.1 ml/ml/minute, washed with seven column volumes of water and treated with four column volumes alcohol. A solution of 500 ppm octylmercaptan in isooctane is passed through the column at the same flow rate. The eluate is analyzed by iodometry. Under these conditions, 150 column volumes can be percolated through the column until breakthrough is observed at 5 ppm.

3.3 Concentration of Solutions

Ion exchange has also proved useful for the concentration, isolation and recovery of ionic components from highly dilute solutions. Usually, the H^+-ions of a cation exchanger or the OH^--ions of an anion exchanger are exchanged for the ions contained in a solution. In this manner, traces of ions can be concentrated on an ion exchanger. Its high capacity permits the ions contained in a relatively large volume to be stacked on the small volume of the ion exchanger and to be recovered in concentrated solution by elution with a small quantity of concentrated eluant.

The attainable effective enrichment depends on the capacity of the ion exchanger utilized and the position of the equilibrium. It is always possible to find conditions under which the total sorbed ionic fraction can be recovered with a minimum volume of eluant.

For example, if a cation exchanger has a capacity of 1 meq/ml, 1 ml can sorb the cations from 1000 ml 0.001 *N* solution. Under favorable equilibrium conditions, it may be possible to perform the elution with 5–10 ml of a 1 *N* solution.

Some practical examples will illustrate the resulting possibilities.

Example 1: Recovery of copper from fiber spinning dope. During spinning of cuprammonium rayon, "blue liquor" or cuprammonium solution containing ⅓ of the charged copper and a "spinning acid" containing ⅔ of the charged copper are usually obtained. Copper recovery is of technological importance and is carried out with the aid of ion exchangers [218]. This can be performed in the laboratory for demonstration purposes as described below.

Blue liquor (2 liters) is prepared with 0.08 g Cu/liter and 0.75 g NH_3/liter, while 500 ml spinning acid is prepared from 18 g Cu/liter and 65 g H_2SO_4/liter. A simple ion exchange column is prepared with Lewatit S 100 cation exchanger in the H-form. The copper content in the cuprammonium solution and in the spinning acid is determined by iodometry.

For the concentration of copper, the cuprammonium solution is percolated through the column, leading to a copper enrichment according to the equation:

$$2\ IA^{-} \cdot H^{+} + Cu(NH_3)_4^{++} \rightleftharpoons IA^{--} \cdot Cu(NH_3)_4 + 2\ H^{+}$$

Cuprammonium solution is charged up to breakthrough. The column is then briefly washed with water. The regeneration and recovery of copper might now be carried out with acid. However, it is of industrial importance that one can also use spinning acid which already contains a part of the initially charged copper. By eluting copper with the spinning acid, the copper from the cuprammonium solution and from the acid is practically recombined in the eluate. Another copper analysis of the eluate can give information on the attained enrichment.

Example 2: Recovery of molybdenum. Metals present only in small quantities in ores or other products should be concentrated prior to their recovery. The advantages offered by ion exchange will be demonstrated with the example of molybdenum [575].

This example as well as the following are described as preparative laboratory experiments. It is self-evident that these methods are similarly applicable to large-scale processes.

Molybdenum is extracted from an ore (0.015% MoS_2), a copper concentrate (1.05% MoS_2), or a molybdenum concentrate (63% MoS_2) with 3% NaClO solution at 45°C. As a rule, 30 minutes is sufficient to convert the molybdenum into molybdate according to the equation $7\ NaClO + MoS_2 + 4e \rightarrow MoO_4^{--} + S_2O_3^{--} + 7\ NaCl$; for its enrichment, the solution obtained is loaded on a strong base anion exchanger in the Cl-form. Elution should be carried out with 8% NaOH solution. After washing of the column, the exchanger is again brought into the Cl-form.

In the eluate, molybdenum is precipitated as MoO_3 with NHO_3 or as $CaMoO_4$ with $CaCl_2$.

Example 3: Separation and recovery of phenol. Organic ion exchangers sometimes adsorb organic substances with a capacity which is higher than the exchange capacity. This is also true for phenols which, as will be shown in this example, can be concentrated on ion exchangers and recovered [19]. This example is of industrial interest, since phenols and phenolic compounds present a difficult problem in wastewater treatment.

An 0.1 *N* phenol solution (500 ml) is prepared and loaded on a suitable exchange column with 75 ml Dowex 2 in the Cl-form. Breakthrough is controlled in the filtrate and the adsorbed quantity of phenol is calculated from this as a comparison with the exchange capacity. Pure methanol or methanol with 0.8 *N* sodium hydroxide is used for elution.

Example 4: Isolation of alkaloids. Ion exchangers are being used increasingly for the isolation of alkaloids from plant extracts. An example is the recovery of quinine from bark extracts, which can be carried out as in the following model experiment [22].

A 0.003 *M* quinine solution is passed through 100 ml of a cation exchanger in the NH_4-form until breakthrough occurs. The quinine solution contains 1% sulfuric acid. To elute the quinine and simultaneously regenerate the exchanger, ammoniacal alcohol is percolated through the column. After the exchange column has been used for two or three runs, the recovery of quinine should be quantitative with a total column capacity for quinine of 3.5–4 g.

Atropine, morphine, and strychnine can be bound similarly on a strong acid cation exchanger and can be simultaneously purified during elution [77].

Example 5: Enrichment of metal traces from seawater. This possibility has been repeatedly discussed in the literature. Brooks [94] offered an example to demonstrate that some trace elements of seawater can be separated by an ion exchange process. The basis of the method is that the principal elements of seawater do not form anionic chlorocomplexes, while many of the trace elements, such as bismuth and thallium, form complex salts in 0.1 *N* HCl.

To this end, seawater is concentrated to 0.1 *N* HCl with distilled hydrochloric acid and is slowly percolated through a column with strong base anion exchanger (Amberlite IR-400). After loading with a sufficient quantity, the column is drained and the exchanger is subjected to combustion. The trace elements can be detected in the ash by spectroscopy.

The latter example already extends to the application of ion exchangers in analytical chemistry where the concentration of sample components is often of special interest [631]. Trace elements can also be enriched on ion exchangers in other cases. For examples, such a concentration has become important for the detection and identification of trace elements in plant materials. With the use of colorless or lightly colored exchangers, an enrichment can often be detected with the naked eye. Consequently, the analyses are simplified.

3.4 Ion-Exchange Catalysis

Ion exchangers are used as catalysts. Industrial interest first turned to inorganic zeolites whose catalytic activity usually depends on the presence of a metal such as manganese, iron, chromium, and vanadium. Numerous publications exist on the suitability of inorganic ion exchangers for the catalysis of oxidation reactions, hydrogenations, cracking processes, or alkylations. In this monograph, we are interested in the organic synthetic ion exchange resins which Griessbach *et al.* [238,

447, 493, 515] introduced into catalysis. This research resulted in new organic and preparative applications for ion exchange resins.

The catalytic activity is always present in the counterions. Cation exchangers in the H^+-form and anion exchangers in the OH^--form catalyze processes which are accelerated by acids and alkalis, respectively. Exchangers loaded with Hg^{2+}, cyanide or acetate exhibit the catalytic properties of these ions. The exchanger network serves solely as a catalyst support.

Ion exchange resins show a particular affinity for acid and base catalysis in organic reactions, since they can be easily transformed into the H^+- and OH^--form and their porosity and swelling offer a sufficiently large active surface, so that even large organic molecules can penetrate the interior of the resin.

In this connection, special attention should be given to the Amberlyst ion exchangers of Rohm and Haas Co., USA, whose macroreticular structure makes them particularly suited for heterogeneous catalysis in nonaqueous systems. Their unique structure provides them with a sufficient genuine porosity and a strong matrix together with a small volume change in different solutions and solvents. Their large surface leads to high catalytic activity. The functional groups on the entire exchange particle are readily accessible for liquid and gaseous reactants, a condition which is not available to the same degree in conventional ion exchangers.

Ion exchangers offer all the advantages of solid catalysts:

1. The catalyst can be easily separated from the reaction products.
2. The product remains free from impurities.
3. High yields can be obtained since secondary reactions can be suppressed as a result of the short contact time.
4. The catalysis can be controlled to proceed in a given direction as a result of the variable properties of ion exchangers.
5. The ion exchangers can be used repeatedly and in continuous processes.

Only their limited thermal and mechanical stability sets contain limits for the applicability of ion exchange resins in catalysis.

As an example, the description of the esterification of ethylene glycol with acetic acid into glycol diacetate follows. In a three-neck flask with reflux condenser, stirrer, and thermometer, 62 g of ethylene glycol and 120 g of acetic acid are treated with 2 g of a strong acid ion exchanger and heated on a boiling water bath. Every 30 minutes, 1 ml of the reaction mixture is removed and the conversion is determined by titration with 0.5 *N* NaOH. As shown by Figure 39, the esterification of

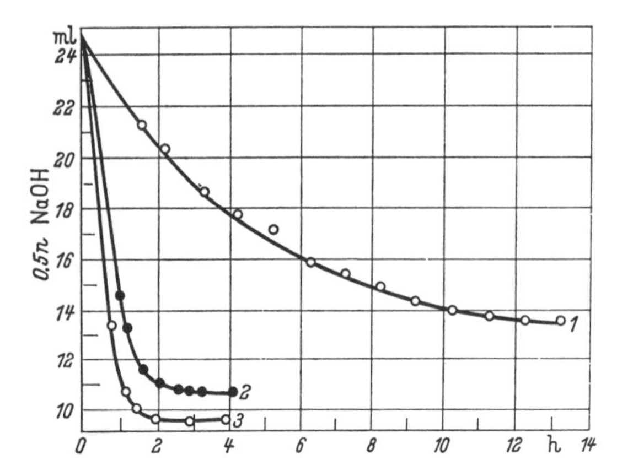

Figure 39. Esterification of ethylene glycol with acetic acid into glycol diacetate with the use of a strong acid cation exchanger. For comparison: curve 1 shows esterification without catalyst; curve 2, esterification with Dowex 50; and curve 3, esterification with H_2SO_4.

ethylene glycol with acetic acid into glycol diacetate takes place just as rapidly with a strong acid cation exchanger as with sulfuric acid.

Catalytic processes with ion exchangers can be performed discontinuously in a batch operation or continuously in a column. Glass equipment, semiworks, or industrial metal equipment is used in batch operation; conventional glass columns or special columns can be employed (Figure 40). With the use of columns the raw materials are loaded on the exchanger and the end products appear in the eluate. In batch operation the starting materials are mixed with a sufficient quantity of ion exchanger and, if necessary, are heated with refluxing. After completion of the reaction the mixture is decanted from the ion exchanger or filtered, and the solution is purified. In both types of operation the ion exchanger can be immediately used for another catalysis.

The full utilization of ion exchangers in industrial catalysis frequently fails because of the equilibrium state of the reactions involved since only such conversions lead to economically satisfactory results in a simple ion exchange process in which the equilibrium has been shifted entirely to one side. By the continuous removal of reaction products the equilibrium can be continuously shifted and the reaction can be completed in the desired direction. To realize this principle in ion exchange catalysis, the exchanger must be packed into the columns in such a configuration that the liquid and gas phases can flow unhindered through the bed. The solution of this problem was fully discussed by Spes [565]. Since molded polyethylene ion exchangers containing up

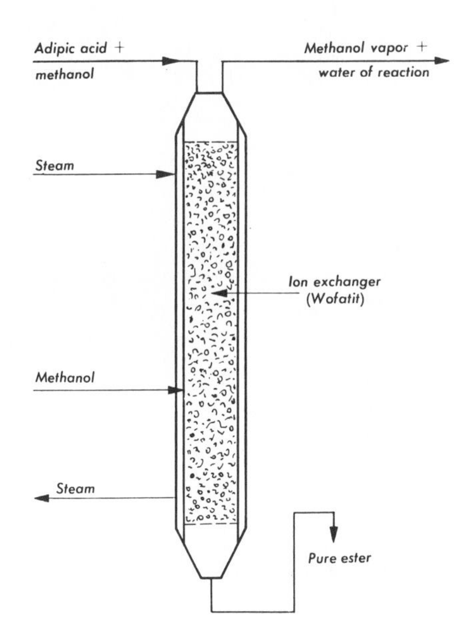

Figure 40. Column assembly for the continuous esterification of adipic acid with methanol catalyzed by ion exchange resin.

to 50% of the exchanger and having a sufficient mechanical stability up to 30° can be used for such biphasic flow conditions, catalytic reactions with ion exchangers in which the equilibrium is shifted have become economical. Their practical performance is demonstrated in Figure 41 by the hydrolysis of methylacetate and the production of methylal from methylacetate–methanol mixtures in specially designed columns.

Only sulfonic-acid–containing exchangers are suited as acid catalysts; according to the studies of Bodamer and Kunin [70], exchangers with carboxyl groups are too weak. For catalysts with a basic reaction, anion exchangers of various base strengths can be used. An interesting variant of this is the simultaneous use of a strong acid cation exchanger in the H^+-form as well as of hydrogen chloride gas as a dehydration catalyst; this improves the catalytic activity in terms of reaction rates and degree of conversion [640].

The reaction rate increases with the quantity of catalyst used, although a limiting value is attained as a rule beyond which the reaction is not further accelerated. Its progress is influenced by the particle size and degree of cross-linking so that the highest reaction rates are attained when these values are low. The findings of Deuel *et al.* [144] are also connected with this rule; according to these authors, low molecular weight esters and saccharides can split completely, while their high molecular forms are split incompletely or not at all by the same resin. Helfferich offered an explanation of the reaction kinetics of

Table 20

Examples of the use of synthetic ion exchange resins as catalysts

Reaction	*Exchanger; loading*	*Process*	*Temperature*	*Literature*
Hydrolysis of different acetates	sulfonated phenol–formaldehyde resin			[452]
Saponification of ethylacetate	Dowex 2; OH^-	batch	20°	[623]
Hydrolysis of proteins	Dowex 50; H^+	batch	100°	[600]
Hydrolysis of nitril	Amberlite IRA-400, OH^-	reflux		[600a]
Esterification of glycerol with ethanol	Zeo-Karb; H^+	reflux	115°	[479]
Esterification of *n*-butyl alcohol and acetic acid	Amberlite IR-100; H^+	reflux	114–116°	[304]
Converse of glucose and fructose	Amberlite IRA-400; OH^-	batch		[241]
Formation of ether				[219]
Formation of acetal	Wofatite			[292]
Aldol condensation and Knoevenagel condensation	different exchangers			[293]

Table 20 (continued)

Reaction	*Exchanger; loading*	*Process*	*Temperature*	*Literature*
Acyloin condensation from benzoin from benzaldehyde	Amberlite IRA-400; CN^-	cycle	80°	[640]
Cyanohydrin synthesis	different exchangers			[211]
Rearrangement	Dowex 50; H^+	reflux		[582]
Hofmann decomposition	Amberlite IRA-400; OH^-	column	18°	[144]
Hydration of olefins	sulfonated coal			[466]
Styrene polymerization				[204]
Polymerization of unsaturated hydrocarbons	sulfonic acid exchanger			[574]
Hydration of propylene	sulfonic acid exchanger	pressure	150°	[391]
Decomposition of diazoacetate	sulfonic acid exchanger, H^+	column	130°	[488]
Production of polyesters	sulfonic acid exchanger, H^+	reflux	140°	[418]
Condensation of methylstyrene with formaldehyde	Kationit KU-2, H^+	sirred reactor		[503]
Esterification of methacrylic acid	Kationit KU-2, H^+	reflux and continuous plant	125°	[28]
Dehydration of hydroxamic acids	Wofatit KPS			[529]
Aromatic nitrogen bases	Amberlyst 15	batch	100°	[27]

Figure 41. Examples of the shape of polyethylene ion exchange packings according to Spes [565]. (A) cubes of 3-cm sides suitable as packings; (B) sheet, 3-cm thickness, post-finished; (C) column base; (D) right parallel-piped, 6 cm × 6 cm × 20 cm; (E) cylinder, 10 cm × 10 cm.

this phenomenon. According to the latter, a direct relation exists between the catalytic activity of a strong acid cation exchanger and the distribution coefficient of the molecules to be converted. Thus, if the reaction takes place in the interior of the exchanger particle, the following relation must be valid:

$$K_{KA} = k_{AT} \cdot \gamma$$

where K_{KA} is the overall conversion of the catalytic process in the presence of the resin, k_{AT} is the conversion of the reaction in the resin, and γ is the distribution coefficient of the substrate between resin and solution. For a comparison of the exchanger-catalyzed reaction with the same reaction catalyzed by free inorganic acid, we furthermore have the relation:

$$q = \frac{K_{KA}}{k_{LS}} = \frac{k_{AT}}{k_{LS}} \cdot \gamma$$

where q is the catalytic activity which permits a comparison of the catalytic reactions conducted with the resin and an equivalent quantity of inorganic acid. Wolf [638] confirmed these relationships. The ester hydrolyses and sugar inversion carried out take place under the influence of molecular adsorption, which can be expressed by corresponding distribution coefficients in the resin and in the solution.

During operation with nonaqueous solvents, the stability of the ion exchanger with respect to the medium must be kept in mind. Before an ion exchanger is used as a catalyst, it must have been completely transformed into the desired loading state; a regeneration is necessary only when secondary ionic components of the reaction mixture cause a charge reversal of the exchanger. The reaction solution must therefore be low in ionic impurities. If necessary, these must be extracted prior to the reaction.

Table 20 lists some reactions which are catalyzed by ion exchangers. Reported first by the former IG-Farbenindustrie AG, such processes were later applied to numerous other purposes.

Finally and incidentally, we might note in connection with the use of ion exchangers as preparative reagents that they have also been used in substitution reactions. Thus, Urata [597] heated benzyl bromide, β-phenylethylbromide and amyliodides with the Cl-form of Amberlite IRA-400 and obtained the corresponding chlorides; he also converted organofluorine compounds by the same route. The preparation of alkyl nitriles from alkyl halides and ion exchangers in the CN-form was described by the same author as well as by Gordon *et al.* [228, 229].

Chapter 4

Ion Exchangers in Industry

4.1 General Technology

Although development of the new synthetic ion exchange resins led to improved and more specific exchanger materials compared to the earlier zeolites, they certainly would not have attained their present importance had they not become industrially accepted in a broad field of applications. When we hear of ion exchangers in industry, our first thought is of water treatment. This is correct; for not only did the development of ion exchange technology originate to meet this need, but 99% of all ion exchange resins manufactured in the world are used for water treatment. Beyond this, however, ion exchange represents an industrial process on a competitive level with conventional distillation and precipitation processes.

The object of an industrial ion exchange process primarily is the removal of low concentrations of impurities from solutions, purification of valuable materials, or their recovery. If the decision is made that an ion exchange process might be suitable for one of these purposes, the ionic character of the exchanging species must be well known or, should this not be possible or not available in the literature, this character must first be determined. In solutions of strong electrolytes, metal ions form complexes of varying charge and strength which may depend on the pH as well as on the nature and concentration of the solution. Thus, as ampholytes, amino acids may react in different ways and little may be known about the ionic character of other compounds. A knowledge of the chemistry of a system which is to be treated is a basic requirement of the ion exchange process and the starting point of all considerations leading to satisfactory application.

The industrial requirements for exchanger materials are higher than in preparative or analytical laboratory work. As a result, research in process technology and improvements of exchanger properties always went hand-in-hand. New ion exchangers consequently will continue to be introduced on the market.

The selection of an ion exchanger type for a given process generally can be made only when the reaction between the exchanger and the reaction solution has been investigated and its suitability as to mechanical and physical properties has been tested. If the commonly available types are not appropriate, a "tailor-made resin" can be produced in certain cases by some modifications and can thus be adapted to special requirements. Opinions differ concerning the use of special types. While one side supports the view that further advances in the industrial application of exchangers can be made by special resins, the other points out that the number of standard products available on the market is sufficient if not already too large. Special types (magnetic ion exchangers, for example) are unfavorable because of their high cost [68, 608]. Even with weak acid cation exchangers in water treatment, the costs are so high in spite of high capacities and high utilization of regenerating agents that their application has remained limited. However, we will consider the industrial application of ion exchangers from the point of view that originally only small impurity concentrations were removed from solutions and the purification of the latter was continued in order to recover valuable materials; the object at this time is to remove large quantities of impurities and even to produce major chemicals with ion exchangers. This evolution has been possible only through the development of ion exchangers of higher capacity and stability and with greater regeneration capacity, and by resins—particularly the macroreticular types—which can be used in nonaqueous media.

The major part of ion exchange costs is absorbed by regeneration since it has become possible to automate the processes to a certain degree. The purchase of exchange resins, in spite of their expense, is not so large a contributing factor in the current costs since relatively small quantities are usually sufficient [536]. However, it is important to decide whether a strong or weak electrolytic exchanger will be used in a given ion exchange process. The weak electrolytic exchangers require much less regenerant than the strong ones and therefore should always be preferred. However, in many cases, the acidity or basicity may be insufficient and the effective pH range of the various ion exchange types in any case is the basis for their application. Weak acid cation exchangers in the H-form are most active in basic media and weak basic exchangers in the OH-form are most active in acid media, although both can be effectively used in the salt form in neutral media. A weak basic anion exchanger, however, can exchange neither carbonic acid nor silicate, so that a strong base exchanger must be used in this case.

The ion exchange material itself is a part of the depreciating costs of an exchange process, so the life of most resins will need to be kept in mind. General data concerning the expected life cannot yet be presented since the exchange material is subject to different degrees of stress depending on the processes and conditions involved and may be physically or chemically degraded or may deteriorate irreversibly. The time, number of operating cycles, or volume of treated solutions can be used to express the exchanger life. Cation exchangers have a longer life than anion resins, with an assumed ratio of 5:1–2 of the treated volumes as a rough estimate.

As noted above, ion exchangers are used industrially to remove ionic components from solutions, to recover ions from otherwise nonionic solvents, and for ion separation. An undesirable or valuable ion is exchanged for a less important one or the different affinity of different ions for a given exchange material is used to isolate a desired component. Ion exchange is also appropriate when a material is to be produced with properties which depend entirely on a particular counterion. Chromatography has not yet been fully developed into an industrial process, but this is a question of the volume to be produced since amino acids and rare earths—the latter on a kilogram scale—are separated in semiworks plants. However, one characteristic must be taken into account. In chromatographic procedures such as those used for analysis, only a small part of the total exchange bed is used for the first exchange while the other is used for separation, so that on an industrial scale large quantities of exchanger with many chemicals will be needed in large plants. As mentioned earlier, a number of applications have become known for ion exchange catalysis, and additional advances can be expected by the use of macroporous resins since their structure offers a considerably improved access to the active groups in the interior of the beads. With a suitable arrangement, however, gases can also be purified by ion exchange if the materials are capable of this process. In the food industry, sugar syrups can be desalted or decolorized, again particularly well with macroporous resins. In the dairy industry, low-Na or -Ca products can be produced with ion exchangers and the shelf-life of powdered milk can be increased by treatment with anion exchangers in the OH-form. It can practically be claimed that there is hardly a sector of industry which is not directly or indirectly connected with some ion exchange process.

If the use of ion exchangers is not to be disappointing, it is necessary to know the efficiency of exchange materials and the applicable processes. In general it can also be said that pretreatment of the process solution, general operating conditions (temperature), properties of the exchanger, and design of the ion exchange installation are all basic requirements for an optimum operation of an exchange process. The selectivity and capacity are two other points which must be recalled in this connection. The chemistry of the exchanger and the process solution together determine the exchange process. Moreover, it is not possible to switch arbitrarily among different exchange processes. Those which are feasible for industrial ion exchange are the batch operation, columnar, or fixed-bed processes, and the continuous processes. Before we turn to the details of these, some comments are indicated in connection with process and equipment design and technological engineering problems.

It has already been pointed out that, initially, only dilute solutions were treated with ion exchangers and that processing of concentrated solutions and the production of major chemicals by ion exchange was only a later development. This includes another distinction in clear process solutions or slurries or suspension. The available equipment possibilities are exhausted by installations which operate discontinuously or continuously and which can process only clear solutions or also slurries and suspensions. The choice of a type of installation is based on technical and economic considerations which must have a suitable foundation in design studies. Thus, a complex of problems evolves, including the following questions according to Stamberg [567]:

1. choice of ion exchanger and of regenerant and eluant
2. choice of a discontinuous or continuous process
3. choice of a type of installation according to design feasibility
4. determination of optimum parameters for the construction and operation of the selected plants
5. determination of process control and automation.

An exact solution to these problems is very difficult since a discontinuous or continuous process operation and individual types of installations are intended to result in an optimally functioning plant whose optimization criterion consists of the profit attainable by its operation.

4.1.1 *Batch Operation*

The industrial units for batch operation were based on customary stirred tanks and were only occasionally modified for special purposes. As shown in Figure 42, for example, the tank is equipped with a type of perforated tray near the bottom to support the exchanger and with a feed system at both ends.

For the performance of an exchange process, the exchange solution is injected through the base valve into the tank (Figure 42b). By a continued gas injection the exchanger is fluidized, and consequently equilibration is accelerated. When equilibrium has been attained, the ion exchange process is completed and the tank is drained through the base valve.

Regeneration takes place by the same cycle, the regenerant taking the place of the exchange solution.

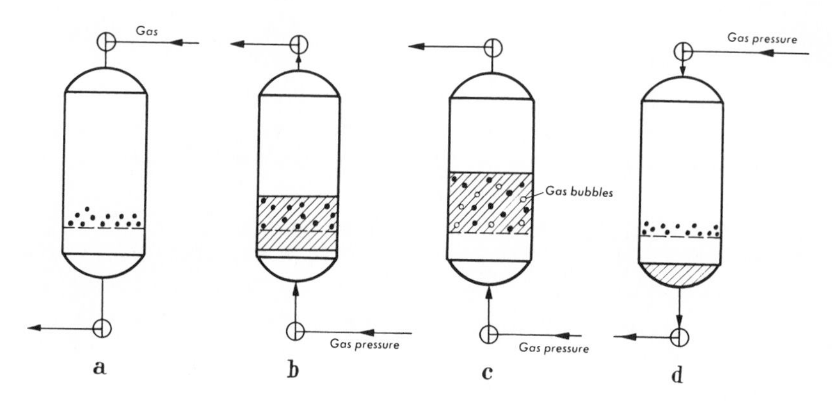

Figure 42. Operating cycle in an exchanger tank in batch operation. (a) empty, (b) charging, (c) equilibration, (d) discharging.

The success of a batch operation is determined by the position of the ion exchange equilibrium. The process can be applied only if the ion to be exchanged has a high affinity for the exchanger compared to the initially present counterion. In the case of strong acid and strong base exchangers, with efficient mixing equilibration usually occurs in minutes. The quantity of solution per bulk volume of exchanger should amount to a minimum of 0.4 and a maximum of 15 volumes.

Finally, we need to mention a variant, the "resin-in-pulp" (RIP) process, for the industrial recovery of uranium. In this case, the ion exchanger is charged into baskets and is moved up and down in the tanks which are filled with crude and regenerating solution. Although developed for uranium recovery, this process appears to be applicable also for other purposes [280]. In the United States, the possibilities of the method with regard to its efficiency and cost position were investigated by Argonne National Laboratory [24].

4.1.2 *Column Process (Fixed-Bed Process)*

4.1.2.1 General description

The column process is the most common and efficient ion exchange method. It is carried out in steps. However, laboratory conditions of working with columns cannot be easily applied to industrial conditions. Only in some rare cases can an industrial plant represent the scale-up of such a laboratory column, *e.g.*, for water treatment such design methods are not feasible since the pumping costs would become too high.

The capacity of the ion exchanger is the most important process engineering parameter and is predetermined in the total exchange capacity as a numerical value. This parameter, which naturally cannot be exceeded, must be converted into an optimum column capacity by the technology employed. To achieve this, an exchange material of high selectivity, small particle size, and high apparent or real porosity and a process control with elevated temperature and low flow rate in a column with a large length/diameter ratio may be favorable. The degree to which these parameters can be realized, however, depends on the hydraulic conditions, on diffusion, and, in the case of high selectivity, on the economically tolerable amount of regenerant. In the columnar process in particular, ions of high selectivity exhibit low diffusion; this is the reason why the leakage of salts in water treatment on a sulfonic acid exchanger at high flow rates is smaller than on a carboxyl exchanger. The economical and feasible optimum column capacity generally amounts to between 50 and 75% of the total exchange capacity. This furnishes the quantity of exchanger necessary for a given performance. However, the nature and control of regeneration essentially determine how the column capacity is attained. This leads to relationships between the capacity and consumption of regenerant, which can be described by curves as a function of regenerant concentration and are practically expressed in per cent of efficiency.

The hydraulic conditions of an ion exchange column are those requiring consideration in the engineering design. Flow of a solution through an ion exchanger must be uniform through the cross section of the bed and the linear flow rate must be identical in all layers. During percolation of the solution through the column, the pressure drop increases and the exchanger bed expands during backwashing. Darcy's law concerning the hydraulic conductivity serves to describe the pressure drop, while expansion of the exchanger bed is treated by the Stokes law. Information on the behavior of an ion exchange column can be obtained in first approximation by using both of these theories.

After the exchanger has been loaded into the container or the column, any fine exchanger particles are removed by backwashing with water in ascending flow through the column and the exchanger is layered in order of decreasing particle size. Subsequently, the regenerant is passed through the column and the bed is washed with water to remove excess and depleted regenerant. The process solution is then loaded on the column until the capacity of the exchanger has been exhausted. Backwashing then takes place again to remove retained fine particles. The final step of this cycle is regeneration or elution, depending on whether the exchanged ions are to be recovered or not.

If an ion exchange process is to be performed by this method, the solutions fed through the exchanger bed must be in uniform flow and must contact each exchanger particle [461]. The exchanger bed must be in suitable configuration and the equipment must be provided with systems for backwashing and for the feed and distribution of the regenerant and wash water.

Figure 43 shows an ion exchange filter with its typical components. It consists of a cylindrical tank containing a layer of porous material (quartz, Anthrafilt, gravel or the like) to permit spreading of the exchanger bed; the ion exchanger is charged on it somewhat above midpoint of the tank.

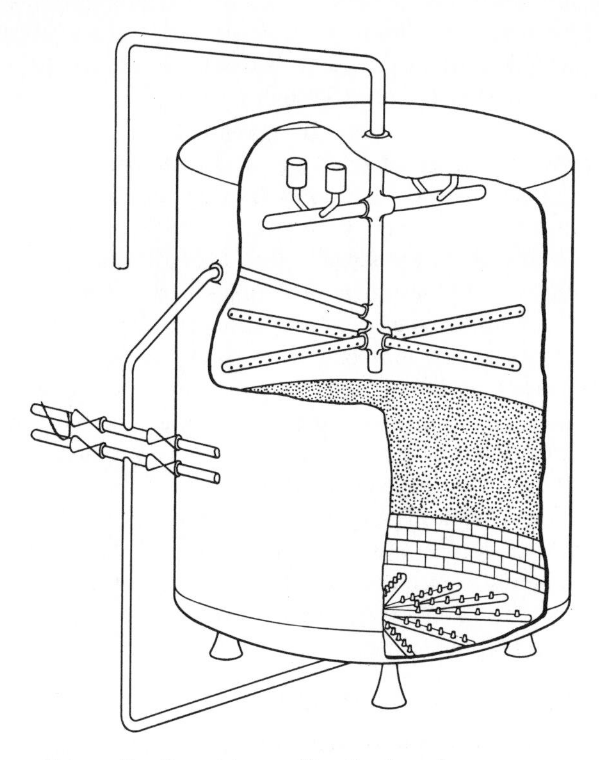

Figure 43. Ion exchange filter for the column process.

The filter tanks often consist of stainless steel. However, other construction materials can also be used depending on the size and application involved. Thus, glass-reinforced plastics are used for small softening plants of up to about 30-cm diameter and ordinary steel serves for larger units. In plants for deionization in which acid corrosion may play a role, soft steels lined with rubber, polyvinyl chloride, Neoprene, or polypropylene are usually employed. Plastic pipes of PVC or polystyrene can be used for the necessary pipelines of up to about 3-cm diameter. For larger sizes, soft steel pipes, rubber-lined on each side of the filters and fully encased in rubber at filter contact, must be considered. Many of the valves commonly used for chemical equipment can be employed, although the possibility of corrosion must be kept in mind.

Because of the different length/diameter ratio in industrial ion exchange filters, it is necessary to give special attention to the distribution and discharge of liquids. The collecting system at the base must provide for a uniform discharge to avoid premature exhaustion of the exchanger in certain levels with preferential flow conditions. Various processes again require different preventive measures of which the most diverse types are used for industrial conditions. In mixed-bed filter tanks, a central distributing system is usually designed in a manner similar to the collecting system. It serves either for the collection of depleted regenerating agent from the anion exchanger bed or for the distribution of acid over the cation exchanger bed. The maximum utilization of capacity is decisively influenced by optimization of the design of this distributor. The regenerating agent distributor or water inlet at the upper end can be designed as a unified or separate system. One design or the other primarily affects the period of changeover from washing to regeneration and back to washing.

The ion exchange filter itself is completed by the tank for the regenerant and any necessary pumps to form an ion exchange installation. Its instrumentation is a special complex of problems which must be treated and solved specifically on the basis of the process involved. Automation has advanced very far in this respect, so that completely automated plants can be constructed. On the whole, the construction of ion exchange units with their ancillary equipment, the pros and cons of which may give rise to long discussions, requires equipment manufacturers and much experience.

4.1.2.2 Mathematical treatment

If the processes in an ion exchange column are to be determined mathematically, these can be based either on theoretical kinetic considerations or on the empirical results of tests. As far as possible, the solution flowing through the exchanger bed is to make exhaustive utilization of the exchanger for technological as well as economic reasons.

The operation of an ion exchange column during one working cycle can be described best on the basis of the curve shown in Figure 44 and with the following notation [97]:

H equals quantity of resin [l], h equals specific amount of resin [m^3/l]

Q equals throughput volume

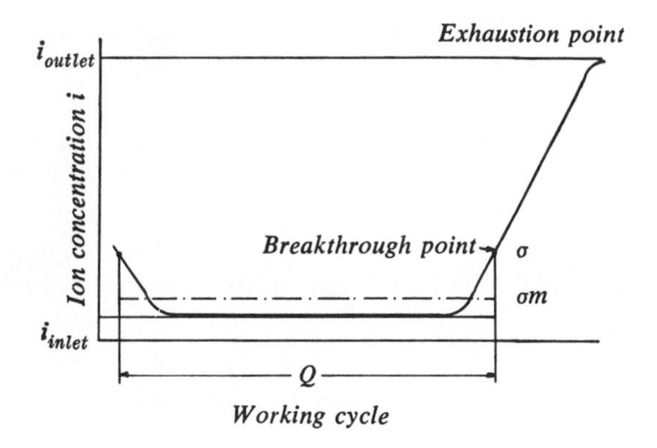

Figure 44. Loading characteristics of a strong acid cation exchanger where i is the cation concentration which is exchanged for hydrogen ions.

i equals ion concentration of solution [meq/l or eq/m^3]
σ equals ionic slip
C equals total capacity [eq]
c equals net volume capacity (NVC) [eq/l]
R equals amount of regenerant [g], r equals specific amount of regenerant [g/l]
b equals specific load [$m^3/h \cdot m^3_{IE}$ or $l/h \cdot l_{IE}$]
s equals bed height [m]
N equals performance [m^3/h]
L equals flow time [h]
W_E equals column water requirement [m^3]
f_Q equals factor for resin swelling
w_A equals specific column water requirement [l/l_{IE}]
W_W equals wash water requirement
Ψ equals regenerant excess
f_Q equals factor for resin swelling
F equals container cross section [m^2]
d equals container diameter [mm]
l equals container length, cylindrical [m]
Δ_p equals pressure drop in resin bed [m H_2O]
δ_p equals specific pressure drop [m H_2O/m bed]
v equals linear flow rate [m/h].

This curve can also be considered an example of partial desalination in which the ion concentration i of the water is plotted as the ordinate with the values i_i equal to the ion concentration at the inlet and i_o equal to that of the eluted water at the outlet as a function of the throughput volume Q as the abscissa. In accordance with the net volume capacity

(NVC) c, in which ion slip is taken into account, the cation exchanger in this column exchanges the ions of the water for H^+-ions until breakthrough of ions which are no longer exchanged occurs after a certain eluted volume indicated on the curve of Figure 44 as the breakthrough point. The breakthrough point which occurs prior to exhaustion of the column is an important parameter. The S-curve of Figure 44, which describes the exchange of cations in water for H^+-ions of the exchanger at the same time, is the breakthrough curve, because the working cycle has ended with breakthrough since here the eluate of the column begins to contain more ions which have not been exchanged. Depending on whether their slope is steep or flat, breakthrough curves can be used to evaluate column efficiency.

The amount of exchanger required for such an exchange process is calculated according to:

$$H = Q\frac{(i_i - i_o)}{c}$$

or, since the ion concentration expected at the outlet is equal to the ion slip σ_m, according to:

$$H = Q\frac{(i_i - \sigma_m)}{c}$$

or for the case where $\sigma_m < 5\%$ of i and can therefore be ignored, it is calculated according to:

$$H = Q\frac{i}{c}$$

The net volume capacity c which has been equated to the breakthrough capacity here will be considered more fully. The various definitions and methods of determination of the ion exchanger capacity were discussed earlier. In industry, the capacity is preferably expressed in eq/l on the basis of the customary commercial form of an exchanger. Since the industrially useful capacity is smaller than the total capacity C and depends on the amount of regenerant r, the specific load b, the available ions i, and the temperature, a net volume capacity or breakthrough capacity c (sometimes also known as the normal operating capacity) is used for calculations of continuous processes. However, c differs from a total operating capacity c_t, which would be obtained if the exchanger were converted completely into a given loading form by a large excess of regenerant and would then be operated with the exchangeable ions up to breakthrough. These two parameters lead to the efficiency of the exchanger:

$$\eta = \frac{c}{c_t}$$

The parameter must be known to design an installation. The sales bulletins of resin manufacturers list data for the mathematical determination of *c*, but assumptions have to be made for certain variables, such as the specific load and specific quantity of regenerant. The parameter can also be determined in the laboratory however. For this purpose, the exchanger is first regenerated with an excess of regenerant; then it is thoroughly washed, and the test liquid is loaded until breakthrough occurs. The column is then regenerated with the intended regenerant; the capacity is determined analytically from the quantity of eluted ions and its value can be used as the operating capacity *c*.

This capacity determination is to be recommended for industrial conditions because the influence of a number of factors on the behavior of the exchanger in the process is also determined with it. This is due to the presence of the exchangeable ions in the same ratio as during the industrial process, as far as species and concentration are concerned. Moreover, a possible impurity content must be taken into account which may reduce the operating capacity. The influence of pH on the capacity has been discussed earlier. The capacity of exchange resins, particularly those which are weakly acidic, increases with increasing temperature, thus leading to a method for increasing the capacity. The residence time also influences the capacity as shown in Figure 45 for a strong and a weak acid cation exchanger according to Becker-Boost [45–47].

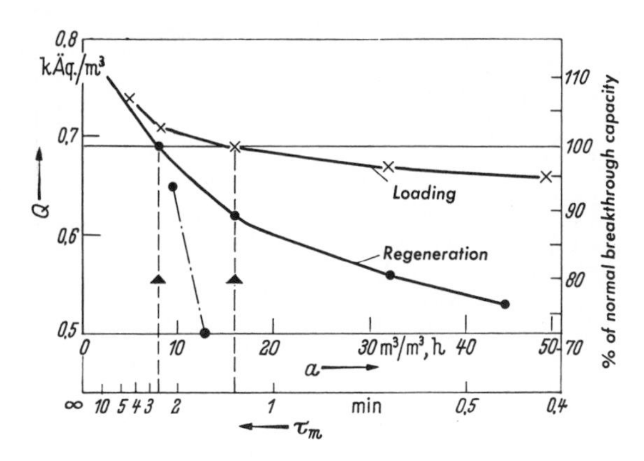

Figure 45. Relation of normal operating capacity Q on the residence time τ_m according to Becker-Boost.

As noted earlier, the regeneration method is an important factor for the technically attainable capacity. Considerable excesses (up to 300% of theory) of the conventional regenerants—sulfuric acid and hydrochloric acid as well as common salt for cation exchangers and sodium carbonate, ammonia, and sodium hydroxide solution for anion exchangers—are necessary for strong acid cation and strong base anion exchangers. For weak acid and weak base exchangers, stoichiometric quantities of regenerant are often sufficient. The concentrations of regenerant used fall between 5 and 7% hydrochloric acid, 2 and 4% sodium hydroxide, and 10 and 15% sodium chloride solution. The operating capacity of a strong acid ion exchanger can be raised by 30% if it is regenerated with 12% instead of 2% sulfuric acid. In Figure 46, curve *a* shows the capacity, expressed as net volume capacity (NVC), as a function of the quantity of regenerant used. This curve applies to the case in which the quantity of exchanger calculated for a given performance is used in a single filter. If this quantity of exchanger is distributed over two series-connected filters and the same quantity of acid is used to regenerate the second filter first while the remaining excess is used for the first filter, curve *b* will represent the second filter and curve *c* the first. This leads to the average shown by curve *d;* in other words, a considerably higher capacity is obtained simply by series-connection and separate regeneration of the same quantity of exchanger with the same consumption of acid.

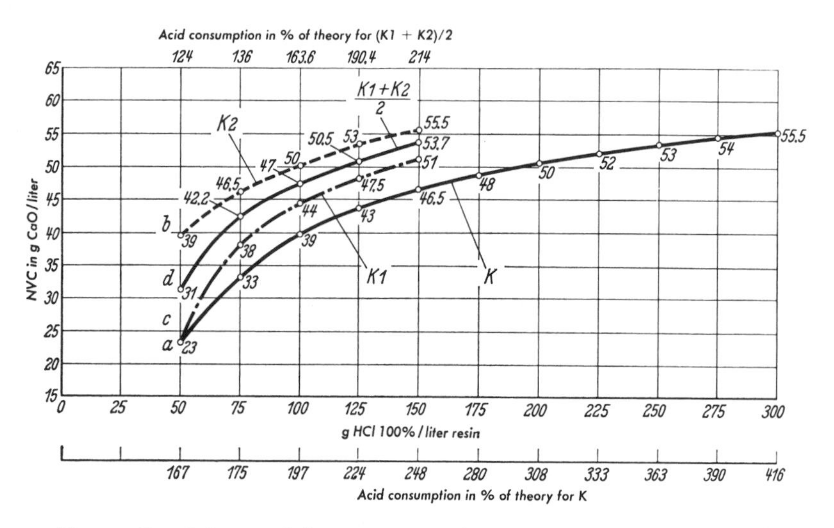

Figure 46. Relation of the capacity to the quantity of regenerant used in a single-unit and double-unit system.

The throughput volume Q is given by the performance N and the flow time L. However, a distinction must be made between the net throughput eluted by the installation and the overall throughput, which includes the column water requirement W_E:

$$Q_{\text{overall}} = Q_{\text{net}} + W_E = NL + W_E$$

Since the overall throughput volume is the determining factor in calculating the quantity of resin (which is why a calculation of multistage plants always starts with the last stage and thus takes the column water requirement of the following stage into account), the specific quantity of resin h:

$$h = \frac{(i - \sigma_m)}{c}$$

or

$$h = \frac{i}{c}$$

for $\sigma \rightarrow O$ is used for calculations with a variable throughput volume. With consideration of a safety margin f_s, we then obtain the following for the quantity of resin:

$$H = Q_{\text{overall}} \cdot h \cdot f_s$$

After calculating the quantity of resin H, the specific load b must be determined according to:

$$b = \frac{N}{H}$$

where the following must apply:

$$b_{\text{actual}} \leqq b_{\text{assumed}} \text{ or } b_{\text{max}}$$

If the calculated specific loading is higher than that assumed originally for the capacity determination, c and H must be corrected until the indicated relation is satisfied.

The water requirement of an installation is composed of the washwater, dilution water and backwash water. For a calculation of the quantity of resin, it is important whether additional loading takes place during backwashing, regenerant dilution, or elution. Thus, the installation is not loaded by the backwashing and dilution water of the cationic stage, since industrial water is used for this purpose; on the other hand, the column water requirement for the following anionic stage as well as the washwater requirement of the cationic stage must be taken into account with the latter. The washwater requirement W_W is calculated from the specific plant water requirement according to:

$$W_W = \omega_A \cdot H$$

The backwash water requirement depends on the necessary bed expansion, the filter cross section and resin density. Usually 70–100% bed expansion and 15–30 minutes of backwashing time are assumed.

The following relations, which as such are self-understood, apply to the quantity of regenerant R:

$$R = r \cdot H$$
$$R = \psi \cdot C$$

with

$$C = c \cdot H$$

and

$$r = \psi \cdot c \cdot a$$

where a is the equivalent weight of the regenerant. A back-calculation of r is necessary only in systems with sequential regeneration, *i.e.*, when the discharging regenerant of the second exchanger is utilized also over the first stage. Then, the following also applies:

$$R_{tot} = R_1 + R_2$$

and

$$r_2 = \frac{R_{tot}}{H_2} \geqq r_{assumed}$$

If we know the quantity H of necessary exchanger from the throughput volume and the ion exchanger capacity and its variation due to the cited factors, the bed height and filter cross-section can be calculated from it. In common resin exchangers, the maximum bed height is 1.2–1.6 m, and should not be less than 0.6 m. With consideration of swelling by the swelling factor f_Q, we have:

$$H_{max} = H \cdot f_Q$$

With consideration of the bed heights s = 1.2–1.6 m, this results in the container cross section of:

$$F = \frac{H_{max}}{s}$$

and the container diameter:

$$d = \sqrt{\frac{4F}{\pi}}$$

as well as the cylindrical container length:

$$l = (1.8 \text{ to } 2.0) \cdot s$$

It is also necessary to consider the pressure drop of the resin bed which is a linear function of the flow rate. Essentially, it is influenced by the kinematic viscosity v and the particle size. In connection with the flow rate:

$$v = \frac{N}{F}$$

the specific pressure drop, which is a function of temperature and of the resin:

$$\delta_p = f(v, T, \text{resin})$$

can be derived for a given operating state from the manufacturer's resin data sheets and the pressure loss Δ_p can be calculated from it according to:

$$\Delta_p = \delta_p \cdot s$$

Too great a pressure drop can be partly compensated by increasing the operating temperature, since the solution viscosity decreases with increasing temperature.

These results, obtained on the basis of empirical relations between the loading capacity and its dependence on the regenerant and exchange properties, and together with the ion transfer coefficient, can offer additional possibilities for the design calculation of ion exchange units [44]. With regard to the geometry of such units and, in the final analysis, the L/D ratio, Beyer and James [56, 130] have found that in columns in which a single counterion participates in the exchange process, these processes are independent of the column dimensions. More precisely, breakthrough is independent of the column. As indicated above, the breakthrough time is an important parameter in ion exchange columns since it is a measure of the possible total exchange until the ions leak through the column without binding.

A more profound treatment of design problems for the column process is an interesting task for the basic research scientist and a desirable prerequisite of efficient planning for the process engineer. However, available theories are only of limited applicability. So far, they have not allowed predictions on the optimum construction and capacity of an ion exchange column from available basic data for equilibria and time processes. Moreover, in part the theoretical considerations contradict each other. Some theories merit emphasis nevertheless. Attempts have been made to use the process curves to draw conclusions on the equilibrium position and exchange kinetics; this was done with the use of an exchange parameter according to Vermeulen [615]. Later, a theoretical analysis of multicomponent ion exchange in fixed beds [312] and a general analytical treatment for the material balance and equilibrium conditions in zones of variable composition of such systems [587] were developed. On the basis of kinetic measurements, Klamer and van Krevelen [311] calculated ion exchange columns and compared the calculated and measured breakthrough curves; for the processes in which film diffusion is rate-determining, the agreement between the two types of curves was good. The other direction in which generally known isotherms and exchange rates as well as other data are used for the theoretical treatment of processes in exchange columns, the findings then serving as the basis for the calculation of industrial columns, again include a number of theories with these common aspects; Helfferich [261] has subjected these to a critical review. Cooper [131] derived a general solution for the character of an ion exchange column under the conditions of particle diffusion and an irreversible equilibrium. This solution includes the time dependence and is not based on the usual assumption of a "constant pattern." A number of diffusion models and the exact solution of diffusion in spheres were investigated. The quadratic approximation was found to offer the best agreement with an exact solution.

In addition to basic formulas for the design of ion exchange columns, their mathematical treatment should also contain relationships for its economy. Although ion exchange processes have proved to be useful in many fields, the question will always arise as to which degree optimum plant and operating costs can be predicted by an economy calculation. More detailed discussions of this subject are beyond our present scope but some indications on it should be mentioned. The costs of water treatment can be determined by capital investments, regenerant costs and labor costs [84]. This immediately results in questions on how depreciation is to be evaluated with different cycle times per day, how the most inexpensive domestic chemicals can be best utilized by improved process control, for example, countercurrent regeneration, and how labor costs can be reduced by extensive automation. Attempts have also been made to calculate annual costs with a single formula in which the main factors are regenerant costs, resin depreciation, plant depreciation, and labor costs [164]. An analysis of such a formula can offer a more reliable decision on whether a two-stage plant with a strong acid cation exchanger and strong base anion exchanger or a three-stage installation with a strong acid cation exchanger, weak base anion exchanger and strong base anion exchanger is the best possibility. The expansion of formulas available for the design of installations into generally applicable equations on which to base economy considerations for the selection of certain circuit diagrams in full desalination plants as well as the most favorable design of resin volumes demonstrates the far-reaching information which can be obtained from a cost calculation based on sufficient design data [98].

4.1.2.3 Special processes

Ion exchange process technology includes a number of special developments, some of which have led to great advances and thus to a more widespread use of ion exchangers. In many cases, the primary interest was oriented toward a better utilization of regenerants and thus a more economic design of the processes.

Countercurrent processes. These represent the next important advance in process engineering of ion exchangers in general. They permit the production of an improved quality of water, for example, with the use of less regenerant [276, 277, 642, 645] and a smaller technological investment. While the parallel flow technique is still most common, *i.e.*, the solutions flow through the exchanger bed from the top, in countercurrent processes regeneration takes place in one direction and loading in the other. In parallel flow processes, the ions are separated during loading and in subsequent regeneration, which is carried out with a deficit of material for economic reasons, the ions at the lower end are no longer separated, leading to the well-known problem of sodium leakage. In countercurrent processes, the opposite direction of loading and regeneration first attacks the ion with the most similar selectivity, requiring a considerably smaller quantity of regenerant [318]. This principle, which has a good theoretical foundation, could be applied relatively easily with the zeolites of higher specific gravity [604], but it involves technical difficulties with the synthetic ion exchange resins of lower density, since the exchange bed expands with ascending flow, so that more regenerant is consumed and resin particles are lost and since more washwater must be used due to dilution at the head of the bed. Several well-defined processes have been developed to find a technical solution of the countercurrent principle; these have been described, usually in the patent literature, as simple countercurrent processing, pulsed countercurrent regeneration, fluidized-bed processes, pressed-bed processes, completely packed filters, etc.

The *simple countercurrent process* can be realized by installing a distributor at the level of the fluidized bed, in accordance with Figure 47, and introducing water together with the regenerating acid which is fed from the bottom, so that expansion of the bed is maintained in tolerable limits. This process can be carried out in filters up to a certain size and with certain limiting flow rates of regenerant.

In *pulsed counterflow regeneration* [186] attempts are made in larger filters to control rearrangement of the exchange bed (which is the result of unsatisfactory distribution) and locally higher velocities by loading the regenerant from the bottom in a pulsating cycle ranging between a supercritical velocity and a following stationary interval,

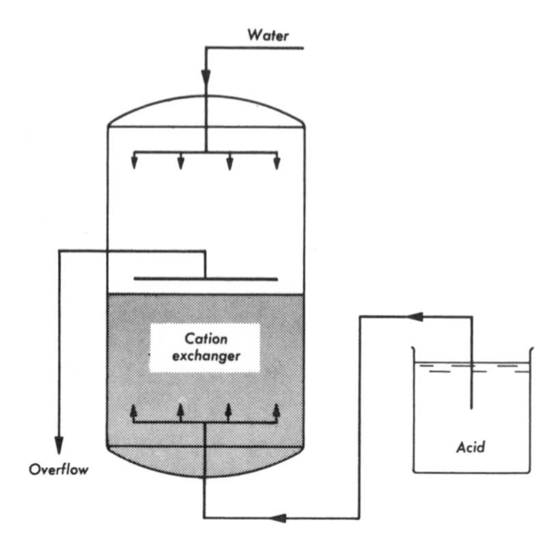

Figure 47. Simple countercurrent regeneration with simultaneous water feed.

so that entrainment of ion exchange particles is suppressed up to a certain point.

The *fluidized-bed process* of Farbenfabriken Bayer AG [168] operates in countercurrent so that regeneration takes place in descending flow and loading in ascending flow. During loading, the water flow rate is adjusted so that 25–75% of the total exchange resin is in the fluidized state. The balance of the exchanger is pressed on the top distributor plate as a compact bed, as shown in Figure 48. According to the patent owners, this process offers a technically and economically particularly favorable solution. With a high capacity of the resins and optimum quality of treated water, the consumption of the theoretically necessary chemicals is hardly exceeded. The filter resistance and wash water consumption are lower than in conventional processes. Conventional filter designs can be used. Since the backwashing space is eliminated, the filters become considerably smaller. Provision must be made for cleaning the resins. Very encouraging findings [212] have been made in the practical realization and application of another countercurrent process which has also been designated as a fluidized bed system.

In a *countercurrent filter with two distributors* regeneration takes place in descending flow, and during ascending loading the exchanger bed is forced on the upper distributor as a closed piston, as shown in Figure 49, thus preventing rearrangement of the bed. To prevent settling of the exchanger below a critical flow rate, a secondary circuit must be superimposed on it. A variant of this is a similar filter in which the upper distributor is located above the exchanger point only at a

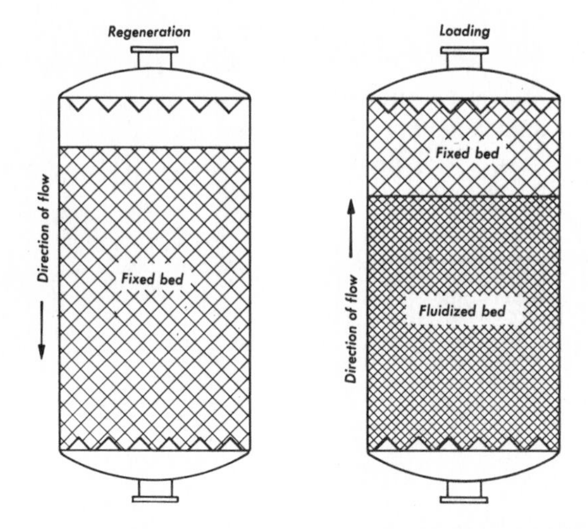

Figure 48. Ion exchange in the fluidized bed process.

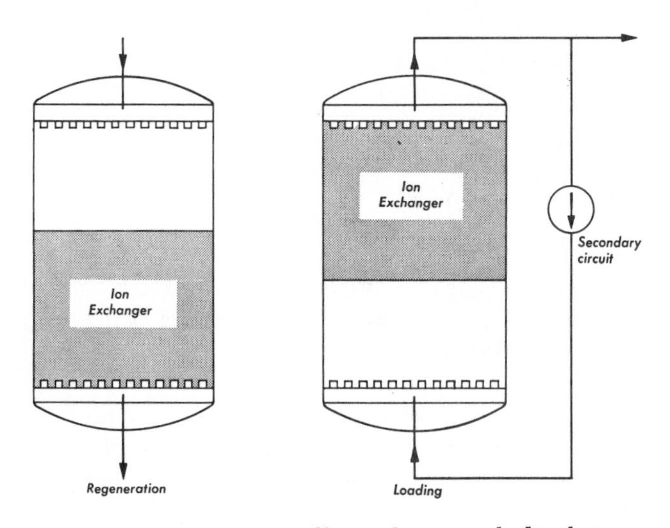

Figure 49. Countercurrent filter with two nozzle distributors.

height corresponding to resin swelling. To wash out contaminants, the resin must be transferred into another container, however.

In the techniques which have become known by the name of *pressed-bed processes,* the interior of the exchanger bed contains a rubber bag which can be filled with water as shown in Figure 50 (British patent [90]). This bag compresses the exchanger bed during ascending regeneration and thus holds it in position. A bed of coarse polyethylene chips is located at the head of the tank above the ion exchanger. The volume of the water-filled bag is as large as the free space over the exchanger bed, thus producing the effect of a "pressed-ion exchanger bed" which eliminates all difficulties during regeneration. A similar design operates with a rubber balloon located axially in the container [35, 36], which has approximately the volume of the backwashing

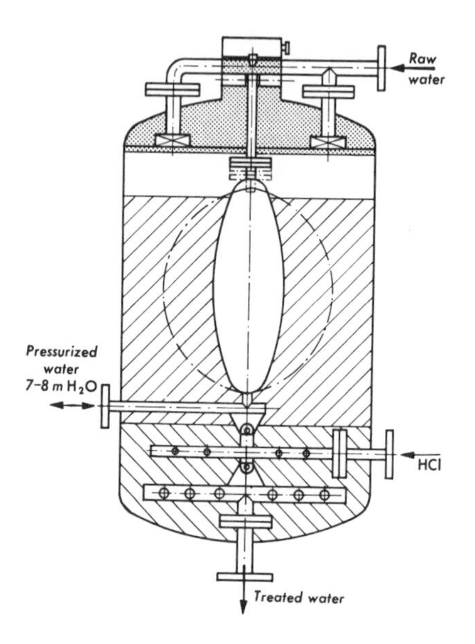

Figure 50. Pressed-bed filter with regeneration in countercurrent.

space in the inflated state and offers no room for rearrangement of the exchanger during normal cycles. Another pressed bed process is obtained [214] by an elastic bellows placed above the exchanger bed which, as can be seen in Figure 51, limits the backwashing space and thus provides for a stable filter bed during ascending regeneration. In this process the filter requires two distributors. In the pressed-bed processes, the chemicals consumption in the case of neutral exchange is reported to amount to 130% and in complete desalination, to 110% of theory in acid and 120% of theory in alkali.

A *countercurrent filter with an inactive blanket* [216] according to Figure 52 is suited for working in counterflow because an air cushion prevents fluidization of the filter bed during regeneration. The filter has a drainage system at mid-height through which regenerant is withdrawn together with a small amount of air. A liquid level forms at the level of the drainage outlets over which wet, dimensionally stable exchanger is still located. This liquid level prevents fluidization of the exchanger bed under the influence of the regenerant flow. Filters operated in parallel flow can be modified for this process. Operating

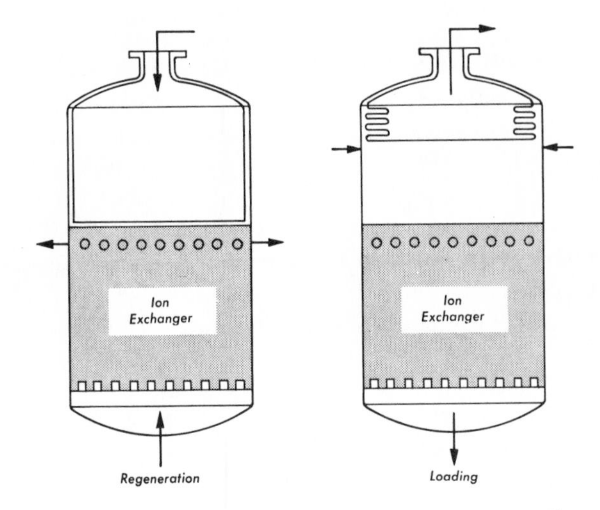

Figure 51. Pressed-bed filter with cylindrical elastic bellows.

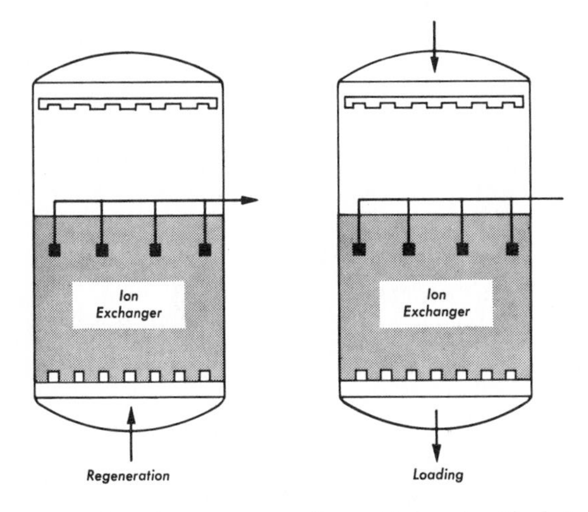

Figure 52. Countercurrent filter with inactive blanket.

data are available for this countercurrent technique in installations of up to 4000-mm diameter and 22,000-liter exchanger.

Other special countercurrent processes which should also be mentioned are those with a buried collector system [87], counterflow filters with adjustable trays, countercurrent filters with a secondary flow [215, 217], and countercurrent filters with axial compartmentalization [651].

Countercurrent processes also result in savings of washwater, which have been claimed to be 2–3 m^3/m^3 of exchanger compared to 7–10

m^3/m^3 exchanger in parallel flow processes. In addition to the cost reductions in water treatment plants, this saving becomes even more important in decontamination installations for radioactive effluents or in closed-cycle wastewater treatment plants of galvanic installations.

Multilayer bed processes. In the search for possibilities to save regenerant, compound regeneration was carried out for a long time, in which considerable improvements were obtained in full desalination plants. Another step was then the series-connection of a weak and a strong electrolyte exchanger in separate tanks. Finally, with a logical extension of the advantages gained, the multilayer bed processes with filters were developed, in which weak acid and strong acid as well as weak basic and strong basic exchangers are arranged above each other in a single filter tank. The high capacities of weak electrolytes compared to the relatively low capacities of strong electrolytes can be used beneficially with this principle. If a multilayer bed in parallel flow is used, the exchangers are loaded from top to bottom in separated layers, are mixed as thoroughly as possible by regeneration in the same direction, and are backwashed before renewed loading; a relayering of the weak and strong exchanger is thus produced at the same time. A multilayer bed is more favorable and still simpler in a countercurrent process, however. In this case, a large quantity of regenerant first flows through the strong acid or base layers, and then a small volume traverses the weak acid or base layers which consume less regenerant. When all technological details are observed, such as sharp separation of exchanger layers by selection of exchangers of suitable specific gravity and particle size and in sufficient bed heights, the regenerant consumption may be reduced to 110% of theory. Certain limiting values for the flow rates of the regenerants are also important; in filters of up to 600-mm diameter, 1.2 m/h for strong acid cation exchangers and 1.1 m/h for strong base anion exchangers of Type II do not lead to turbulence. Figures 53 and 54 are impressive proof for the regenerant consumption in different cycles [632].

The Amberlite ion exchangers IRA-93 and IRA-402 have been tested from the above aspects as a multilayer bed with anion exchangers. Figure 55 shows the arrangement of these two exchangers, Amberlite IRA-93 at the top and IRA-402 at the bottom, after one and then several operating cycles in which an improvement of the layer separation can even be observed.

Powdex process. The Powdex process of Graver Water Conditioning Co. [20, 384] can be considered a combined filtering and ion exchange process. It operates with exchangers with a mean particle size of 0.05 mm, so that a larger number of active groups becomes accessible because of the larger surface, leading in turn to higher

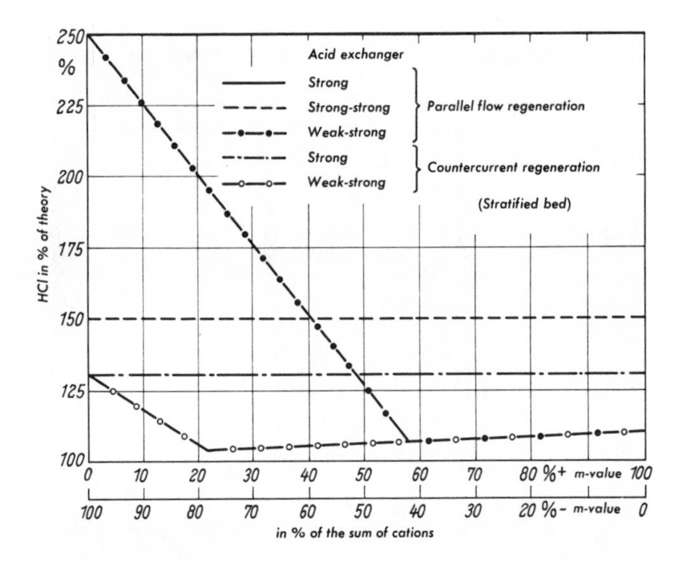

Figure 53. Regenerant consumption of cation exchangers in different flow systems.

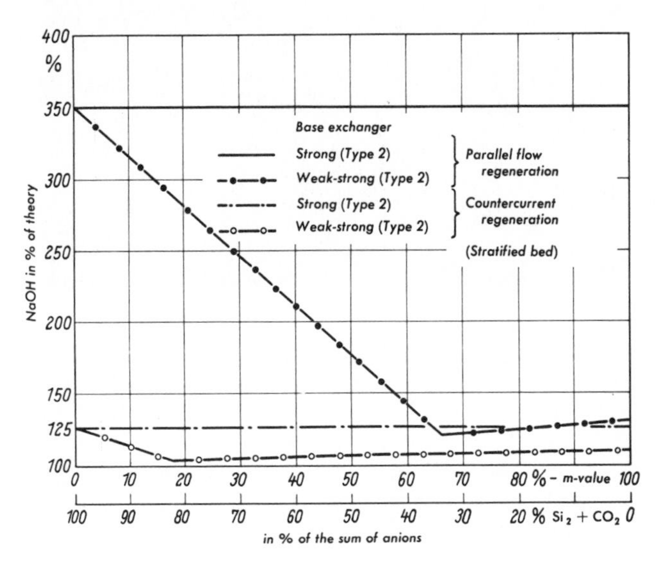

Figure 54. Regenerant consumption of anion exchangers in different flow systems.

exchange rates. The cation and anion exchangers in a given mixing ratio are loaded on filter elements in pressure vessels in layer thicknesses of between 3 and 6 mm. Since powdered ion exchangers exhibit a special electrical interaction, this design also results in a highly effi-

Figure 55. Stratified anion exchange bed from Amberlite IRA-93 and IRA-402 after one (left) and several working cycles (right).

cient filter effect. Thus, colloidal silica is separated by a Powdex plant as efficiently as by membrane filtration. The reaction rates as well as utilization of resin capacity are very favorable in the Powdex process. Moreover, the cation exchanger can be used in the NH_4-form so that the alkalinity of the water is favorably influenced. The operating temperatures in the main Powdex process may amount to up to 120°C.

Until now, the Powdex process has been used industrially in the United States for condensate purification. The particularly favorable investment and operating costs, small space requirement and labor costs, constant efficiency because of the continuously fresh ion exchange material, and suppression of a regenerating station are stressed as advantages [504].

Special regenerating processes. Improvements in the operation of ion exchange installations are expected from the application of special regenerating and cleaning processes. Thus, Kunin and Vassiliou [375] investigated the regeneration of carboxylic cation exchangers with carbon dioxide, which depends on the acid strength of the resin but which subsequently furnishes usable solutions of $NaHCO_3$ for the regeneration of weak base anion exchangers. The regeneration of weak base anion exchangers with 4% aqueous ammonia solutions is favored by Klump [321, 322] because this process causes neither unpleasant odors nor other disadvantages compared to NaOH regeneration. The process is recommended if the exchanger has already been damaged by humins, because an immediately effective remedy is provided by NH_3. In addition, the electrical regeneration of mixed-bed exchangers and regeneration with acid mixtures have been proposed [184, 189], particularly to prevent precipitates on exchangers. According to some patents, attempts are even made to get along without a regenerating solution by shifting the adsorption and desorption equilibrium by varying the pressure or temperature [185, 187, 507]. Sulfite [610] and complexing agents [48] have been used to clean exchange resins with iron oxide deposits and when the general performance dropped, sodium hypochlorite solution [57] has been used to improve the operation.

4.1.3 Continuous Processes

Continuous ion exchange processes are characterized by the simultaneous performance of ion exchange and regeneration in different parts of the equipment. In a columnar process according Figure 4, an exchange zone is provided in which ion exchange takes place at a given time, while the layer above it has already been exhausted and is not used and the layer below it is also not used. Accordingly, an ideal continuous ion exchange would be designed in such a way that the consumed ion exchanger is continuously removed and transported to regeneration and freshly regenerated exchanger is supplied to the process. Such designs have been realized in laboratories.

Two advantages can be obtained by continuous ion exchange. As a result of its removal, less ion exchanger is needed than in a columnar process. This is not the case to such an extent, however, that only the amount of exchanger corresponding to the exchange zone is needed, because regeneration is carried out simultaneously in another part. The actual required amount of ion exchanger nevertheless is far below that of the columnar process. Furthermore, with the separation regeneration in a continuous ion exchange process, the countercurrent principle is applied leading to the advantage of countercurrent regeneration of an increased utilization of regenerant as documented by countercurrent columnar processes. Thus, the second advantage is a smaller consumption of regenerant.

Some difficulties had to be overcome for an industrial realization of continuous ion exchange processes. A uniform flow of exchanger particles in large vessels was one of these so that an undisturbed exchange zone would be attained in continuous operation. This also required a homogeneous liquid flow. The exchanger material must have sufficient mechanical resistance to prevent plugging with the frequently pulsed movements. A more or less sharp exchange zone, particularly with varying composition of the treatment solution, requires special controls by a suitable switching system and careful maintenance.

A number of processes for continuous ion exchange have been described. In some of these, the ion exchanger moves from top to bottom, while moving stationary beds work in one apparatus or in separate units, fluid beds operate in multistage perforated tray columns, and fluidized beds are used in a completely continuous countercurrent system. In all others, the ion exchanger is transported in ascending direction. This can take place in parallel flow with a descending exchanger or in a moving stationary bed with ascending exchanger.

Although interest in the development of continuous ion exchange installations is great because of their undeniable advantages, only a few have been developed to industrial maturity, while others have been operated only as pilot plants.

The design of continuous ion exchange processes is based extensively on theoretical considerations. Generally, equilibrium curves and the necessary kinetic relations for a given process must be determined. An evaluation of these data then leads to a determination of the number and height of theoretical plates and mass transfer coefficients, which then finally permit the partly theoretical and partly practical determination of the column height and efficiency [53, 121, 230, 567].

4.1.3.1 The Higgins process

In the Higgins process, which dates back to the extensive research of I. R. Higgins at Oak Ridge National Laboratory [272, 273], several zones are functionally connected in series with one phase of the entire ion exchange cycle taking place in each zone. The ion exchange resin is transported from one zone to the next. Each zone is optimized in design for the process taking place within it. In the zones where ion exchange proper proceeds, the solution flows toward the resin. As indicated by Figure 56, the system consists of a treatment zone, a backwashing zone, a regenerating zone, and a rinsing zone. Furthermore, a pulsed column is added, which controls the resin cycling speed. During the treatment phase, the water to be treated flows through inlet valve F and is homogeneously distributed over the vessel cross-section by a distributing system in the treatment container. It then flows through the treatment zone to the discharge system and leaves the loading vessel through valve G. At the same time, the regenerant flows through the regenerating valve into the regenerating column, and rinsewater enters the rinse column through the rinse valve. The regenerant and rinsewater flow out through the outlet valve. A control instrument built into the rinse column controls the rinsewater outlet valve so as to maintain the interface between regenerant and rinsewater at a certain level. In the rinse column, the ion exchanger is backwashed for the removal of entrained suspended particles and purification of the resin. Following the loading phase, valves B, C, and D are opened and the resin charge present in the pulsed column is transported further by water flowing in through the pulse valve; after this step, a

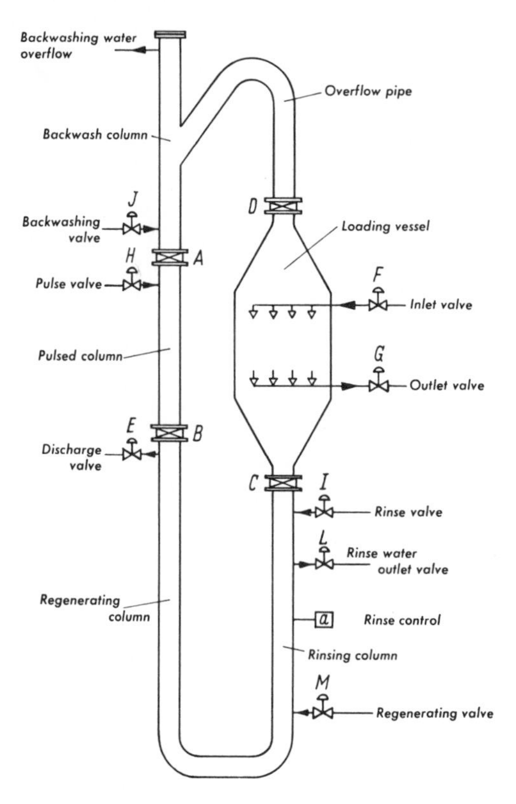

Figure 56. Continuous reactor according to Higgins [272].

treatment phase immediately begins again, while the regenerating rinsing and cleaning phases take place simultaneously in the other zones.

The advantages of the Higgins process reside in the descending direction of flow, the small consumption of treated water, the possibility of treating slurries, and especially in the low regenerant consumption. Only a fraction of the amount of resin required in fixed-bed ion exchange plants is needed in these installations. The space requirement of a Higgins system is only 20–50% of that of a fixed-bed ion exchange installation with a room height of 7–9 meters.

When the Higgins process is used for water deionization, a strong acid cation exchanger is used in a first system and a weak base anion exchanger in a second. A CO_2 vent is installed in between and, if necessary, a strong base anion exchanger is provided at the end as a fixed-bed filter for the removal of silicates. Ammonia in the form of a 3–5% solution can also be used to regenerate the weak base anion exchanger.

4.1.3.2 The Asahi process

The Asahi process, which was developed in Japan by the Asahi Chemical Company for the recovery of copper from a synthetic fiber plant was later expanded by licensing and became available for additional applications of continuous ion exchange. In principle, the process consists of three columns: ion exchange, a regenerating column, and a wash column. As shown by Figure 57, the relatively wide exchange column is located in front of the narrow regenerating column and the wash column beside it.

In the simplest case of application of this continuous plant, the treatment liquid enters the exchange column at 1, flowing through a compact but relatively thin exchanger bed and discharging in purified form at 2. The exchange resin is transported into the regenerating column by the pressure conditions generated during this process and the automatic controls of the system, and is regenerated in countercurrent. Another pressure system and automatic valves transport the regenerated exchanger into the wash column where it is continuously washed in countercurrent and transported through the bottom of the wash column into the reservoir of the exchange column. The continuous countercurrent cycle with the moving exchanger bed has thus been closed.

This simple case can be readily operated for water softening by exchange of sodium for water hardness formers or as a partial desalination. For its application to other problems such as deionization, only two of these systems need to be connected in series and operated

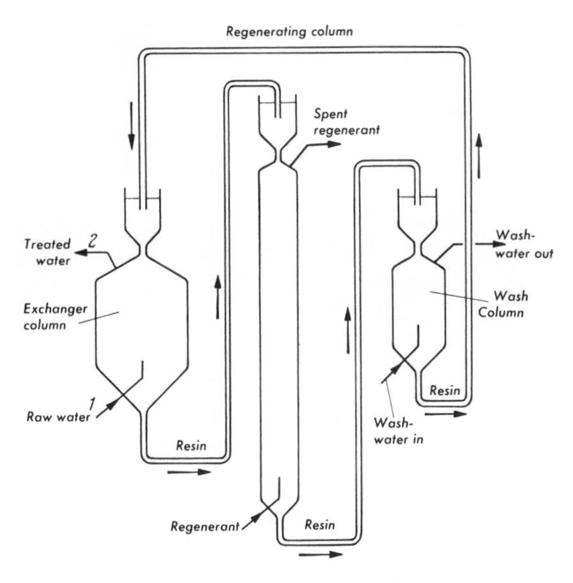

Figure 57. Principle of the Asahi process.

analogously. The special advantages and general applicability of this process, like other continuous ones, may be the future pattern of ion exchange technology [78, 451].

4.1.3.3 The Fluicon process

In the Fluicon or MAN process, a fluidized bed is used so that a true continuous operation can be obtained. As shown in Figure 58, the installation consists of a loading column, a regenerating column, and a wash column which are controlled by a single automatic valve in the brine discharge controlled by the resin level in the regenerating column. The feed streams and chemicals are metered by metering pumps.

Fluicon installations are being constructed for a performance of 10 m^3/h and are offered in standard designs for up to 100 m^3/h. As softening plants, they assure a uniform quality of water with low consumptions of chemical and washwater. In fully automatic operation, a final water quality of up to 0.02° German hardness can be obtained. The fluidized bed stages prevent nonuniform flow and channeling. The space requirement is small because of their compact construction.

4.1.3.4 Other continuous ion exchange processes

In addition to the above three processes, which were described in somewhat greater detail, a number of other continuous ion exchange techniques have been proposed. It has been recognized that the Hig-

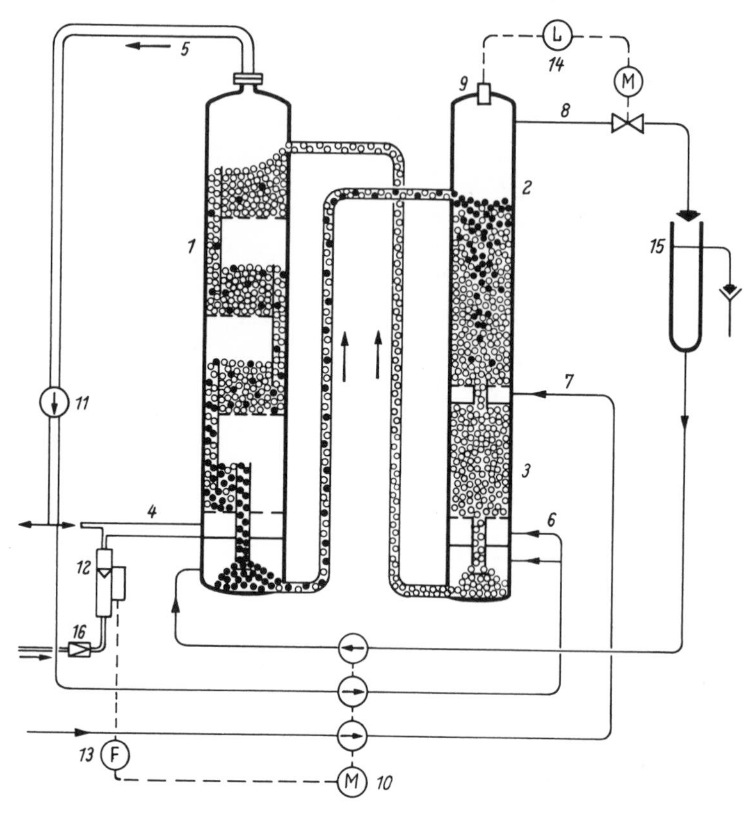

1 loading column
2 regenerating column
3 wash column
4 raw water
5 softened water
6 washwater
7 regenerating water
8 brine
9 level gauge
10 metering pumps
11 circulation pumps
12 flow meter
13 regulator
14 regulator
15 hold tank
16 pressure reducer

Figure 58. The Fluicon process.

gins and Asahi processes have been most successful; nevertheless, alternatives must be developed for both which should be adapted to special requirements, particularly in hydrometallurgy for the treatment of ore slurries and waste effluents.

The Higgins and Asahi processes both operate by the principle of a moving exchange bed. Other processes which essentially operate only with different exchange columns have been developed on the basis of this principle. Beyond this, however, simpler equipment can also fulfill the same purpose of continuous operation; although their efficiency is lower, their flow rate is increased. Slater compared not only the hydrodynamic properties of two continuous exchange units [558] but also offered a detailed summary of the development of continuous exchange processes [557]. As noted earlier, the mass transfer coefficients determine the development of a continuous exchange system [551]. However, after a continuous plant has been constructed, observations for several years are needed to obtain a view of the true cost savings [221, 390]. The continuous ion exchange processes described above serve for the purification of solutions. The next step is the continuous separation of ions on ion exchangers; basic research has been conducted on this subject and research centers have obtained results. The separation of sodium and potassium in 0.1 *N* HCl-solutions is a simple example of the possibility of an industrial solution to such problems [123]. In another system, ionic pairs of K^+-Li^+, K^+-Na^+, and Na^+-Li^+ have been separated [196].

4.2 Water Treatment with Ion Exchangers

For a long time, water treatment was the sole application of ion exchangers in technology; it is still the most important one today. Among the above processes, the columnar or fixed-bed system in a batch operation is used almost exclusively. To achieve continuous treatment, several filters are connected in series and are switched alternately to loading and regeneration. Because good water is beginning to be a scarce raw material, we are being forced to resort to the use of raw water. A consideration of this idea, which would have been inconceivable at an earlier time because of the constitution of raw water, clarifies the reasoning behind the rapid development of ion exchange for water treatment.

In view of the requirements set for water quality, several treatment processes have been developed. These are:

1. water softening
2. partial desalination (softening and decarbonation)
3. full desalination (without and with desiliconization)
4. mixed-bed treatment.

Depending on the intended use of the treated water, a suitable process must be selected and adapted to the existing conditions. An introductory summary of the possible improvements of raw water by the listed processes is shown in Table 21, which shows the composition of the water obtained by the various treatment processes.

Even when water is to be treated with ion exchangers, it must satisfy some requirements; municipal water departments generally deliver a water quality which can be used directly for further treatment in exchanger columns. When large consumers of water make use of their

Table 21

Water treatment of the same raw water by different processes

	Raw water	*Softening*	*Partial desalination*	*Full desalination*	
Total hardness, °dH	15	0.1	0.1	0	0
Carbonate hardness, °dH	9	–	–	–	–
Noncarbonate hardness, °dH	6	–	–	–	–
Bound HCO_3^-, mg/l	196	196	5–10 mg/l CO_2	5–10 mg/l CO_2	0
SO_4^-, mg/l	51	51	51	0	0
Cl^-, mg/l	115	115	115	0	0
SiO_2, mg/l	7	7	7	7	traces
Evaporation residue, mg/l	426	426	273	nearly exclusively SiO_2	practically 0

A, without desiliconization; B, with desiliconization

own wells or draw their supplies from rivers or lakes, this water usually is sufficiently clear and colorless for direct ion exchange treatment only when it has been supplied by deep wells; in other cases, the procedure requires preliminary treatment by coagulation, etc.

The selection of the most suitable process and ion exchanger material for water treatment and purification is based on the water analysis and the intended application. Among identical types, it is necessary to distinguish between weak and strong electrolytes since the efficiency of the regenerants is higher in weakly acid than in weakly basic systems. Consequently, weak electrolytes are used preferentially whenever possible. However, this is not always feasible. Carbonic acid and silicic acid, for example, are too weak to be exchanged by a weak base anion exchanger, so that strong base anion exchanger must be used for the removal of these anions. Other factors then to be considered in a given type are the porosity and particle size. The necessary quantity of exchanger and thus the required size of the equipment are determined on the basis of the capacity, which also determines the quantity of necessary regenerant, the desired hourly output, and the total performance between two regenerations. In plant design, consideration must be given not only to the general operating conditions such as temperature but also to seasonal variations of the raw water composition.

For the power plant chemist who must purify the feed water of steam boilers, it is of primary importance to investigate whether the use of ion exchange is feasible and economical as a treatment process. The same is also true for the treatment of wastewater or water conditioning for a manufacturing process. If the reason for ruling out ion exchange is the malfunctions which might occur, perhaps a suitable pretreatment might eliminate this drawback and then the benefits of ion exchange processes could be realized [302, 546].

Ion exchange is most frequently used to treat water for plants serving to generate steam, but it is by no means limited to these. Manufacturing plants exist where a distributing system with fully desalinated water is installed parallel to a potable water system, so that water fully desalinated by ion exchange is continuously available as a separate water quality. The textile industry is a separate sector where the quality of water is of great importance. Poor water leads to poor dyes, poor hand, and other defects in the products. Water treatment by ion exchange therefore is an important factor for total production, is highly valued as such, and is carried out on the basis of requirements [271, 643]. Other branches of industry where fully desalinated water is used daily cannot be enumerated. However, these include not only the manufacture of finished products, but also intermediates and half-finished goods which frequently cannot be manufactured satisfactorily even with potable water.

An introductory note also needs to be made regarding the treatment of brackish water or seawater for the recovery of potable water. Since the deionization of water with ion exchangers generally is possible only when the total dissolved materials content is not more than 600 ppm, ion exchangers cannot compete with membrane and distillation processes in this field. The costs of regeneration are too high and the deionized water requirements for washing purposes are too important. The final future uses of special weak-electrolyte resins in this water treatment sector remain a task of further research work [377]; the first results have already been achieved in the Desal process for the treatment of brackish water. Otherwise, however, ion exchangers have not become of great significance in sea water desalination, nor do they seem to be seriously considered. This problem therefore will not be discussed further here [313, 435].

4.2.1 *Softening*

Water softening serves for the removal of calcium and magnesium ions known as hardening salts, causing the hardness of water which is expressed in hardness degrees.

When water is passed over an ion exchange bed in a softening process, its Ca- and Mg-ions are exchanged for the Na-ions of the cation exchanger and are thus removed by a simple method. The pure water obtained contains only $NaHCO_3$, Na_2SO_4, NaCl and silicate, which do not produce boiler scale since they are readily soluble in water; they deposit only with complete evaporization of the water.

Customarily, the exchanger units between 0.3 and 4 m diameter and with exchanger beds of 1–3.50 m almost all operate by the descending flow method. After backwashing for the removal of suspended particles, regeneration takes place also in descending direction. The strong acid polystyrene–sulfonic acid resins, which are used almost exclusively, have a hydraulic behavior which makes them responsible for the notable L/D ratio, especially in the larger softening plants. These resins are used in the Na^+-form and after exhaustion are regenerated with salt solution of 5–10 wt.% NaCl/l. Depending on the conditions in different countries, this permits the use of tax-free salt, frequently as a product denatured by Crystal Ponceau 6R. This leads to a relation between the regenerant consumption and the current plant costs. A lower regenerant requirement (economy regeneration) leads to lower operating costs and higher capital investments, while a higher regenerant requirement reverses this ratio. With weight rates of flow of 15–60 $m^3/m^{-2}/h^{-1}$, the capacity as a rule can be utilized to more than 85% compared to corresponding laboratory units. The remaining sodium and calcium chlorides in the exchanger bed after regeneration are removed by elution with 5 to 6 times the exchanger volume. To make the process more economical, it is also possible to use hard water or water of lower hardness, although this again leads to a loss of a part of the regenerated capacity in the upper layers of the bed. A final important point in the operation of softening plants is connected with the iron content of the raw water, which in the case of higher concentrations, can lead to considerable irreversible capacity losses; this can then be remedied only by special treatment.

To afford a maintenance-free transition between loading and regeneration, more recent softening plants have been extensively automated. As an example, Figure 59 shows the diagram of a fully automatic volumetrically controlled Berkefeld* softening plant. The two parallel filters permit an uninterrupted supply of softened water. After a predetermined volume of water has been discharged for regeneration of one of these filters, the water meter switches the unit to backwashing and then successively to regeneration, elution, and finally again to operation. The second parallel filter also begins to operate at the

*All systems cited by the name Berkefeld refer to the tested designs of Berkefeld-Filter, GmbH, 31 Celle.

start of regeneration. It is possible to operate both filters simultaneously and continuously. The operating and regeneration cycles are selected by a time shift so that regeneration of both filters never coincides, *i.e.*, the output of the other filter can still be used while one of the others is being regenerated. Instead of starting regeneration by the volume of water, it is also possible to control the system on the basis of time or a hardness analyzer which switches the system when a residual hardness leak is detected.

Figure 59. Fully automatic volume-controlled Berkefeld softening installation.

4.2.2 *Partial Desalination*

In partial desalination, which is used particularly for raw water with a higher degree of carbonate hardness, a cation exchanger in the H-form is applied so that the water is softened and bicarbonate is simultaneously released in the form of carbon dioxide.

4.2.2.1 The "starvation" process with sulfonated coals

Coal ion exchangers are sulfonated and oxidized so that they contain sulfonic acid, carboxylic acid, and phenolic groups with different acid strengths; and in an ion exchange cycle with acid, they are therefore entirely or partly present in the H-form as a function of the quantity of regenerant. Thus, the SO_3^--groups can be deprived of their activity by partial regeneration ("starved") leading to the name "starvation" process.

These coal ion exchangers, after regeneration with acid, can be used for the removal of carbonate hardness, a process in which calcium is primarily bound and carbon dioxide is released; the latter can be eliminated by subsequent outgassing. This results in a very economical partial desalination, which may be of interest for the production of soft and clear water from very hard raw water–if not alone at least in combination with other ion exchange stages.

4.2.2.2 Partial desalination with carboxylic ion exchange resins

Weak acid cation exchangers of the carboxylic acid type in the H-form serve for partial desalination in which the cations of carbonate hardness are exchanged for H-ions, with the result of a decarbonation and dealkylinization of raw water. The carbonate hardness is converted into carbon dioxide and can be outgassed. Sulfates, chlorides, and nitrates are not influenced in the raw water.

Among the carboxylic cation exchangers built up on different matrices, polyphenol–formaldehyde condensation products are particularly suited for partial desalination. Although they have a lower total capacity, the exchange rates and selectivities for bivalent hardness salts are so favorable that the disadvantages of the capacity are economically more than compensated by the same regenerant requirements.

In practical operations of partial desalination plants, an acid-resistant lining of tanks, pipelines, and switching systems must be provided. Since acid condensation exchangers (which have medium strength) based on phenol–formaldehyde, can also exchange Na-ions, and since the differential degree of swelling between the H- and the Na-form can amount to up to 100%, this volume expansion and its influence on the stability of exchange particles and the mechanical resistance of containers must be taken into account in the case of water of high Na- concentration as a risk leading to extensive material damage. These effects can be counteracted by a lower consumption of regenerant and possibly backwashing to loosen the exchanger bed during loading. In the case of very hard water, no practical difficulties need to be anticipated–even with the "starvation" process–when weak carboxylic acid cation exchangers are used.

Figure 60 shows a Berkefeld partial desalination plant which removes the carbonate hardness from water by means of weak acid exchangers so that the carbonate ions appear in the water in the form of free carbon dioxide. This plant operates in the beverage industry.

A chlorination step and activated carbon treatment for dechlorination and refining follows the exchanger on the left. The carbon filter is provided for backwashing and is equipped with a central switching system. The purification effect of the exchange filter is recorded and indicated by a residual carbonate hardness monitor. Such a partial desalination plant can also be combined with other types of treatment, for example, a CO_2-trickling filter.

Figure 60. Berkefeld partial desalination installation with chlorination and activated carbon treatment for the beverage industry.

4.2.3 Full Desalination

For full desalination, *i.e.*, removal of all ions contained in raw water, the latter must be treated with a cation exchanger in the H-form and an anion exchanger in the OH-form. The simplest possibility for achieving this is a two-column process in which water is first conducted over a strong acid cation exchanger which is regenerated with acid and then over a strong base anion exchanger which is regenerated with sodium hydroxide solution.

Polystyrene–sulfonic acid serves as a strong acid cation exchanger which is used in the H-form. To evaluate the efficiency and economy of the process, the selectivity for Ca^{2+}-, Mg^{2+}- and Na^{+}-ions and their concentration ratios in the raw water, the nature of regeneration with sulfuric acid or hydrochloric acid and the operating rates are decisive factors. Another important factor is the effect known as leakage or slip, which refers to the appearance of Na-ions in the treated water which impairs its quality. Sodium leakage is the result of true ion exchange, ion exchange with secondary reactions, adsorption processes, and rather significantly, ion-exchange chromatographic processes [495]. The Na-ions appearing as a result of leakage are not those loaded on the exchanger with the raw water; rather, these Na-ions originate from residual counterions from the preceding cycle. In cation exchangers regenerated with sulfuric acid, Na-leakage is difficult to control. The situation becomes considerably simpler if approximately 5% hydrochloric acid is used for regeneration since this leads to a practically complete regeneration because the influence of Ca-ions remaining on the resin in subsequent exchange processes can be eliminated and Na-leakage can be mainly suppressed. Complete regeneration with H_2SO_4 is hardly possible because of the precipitation of calcium sulfate. The choice of acid used for regeneration depends only on cost. In Germany, hydrochloric acid is sufficiently inexpensive and is therefore used almost exclusively. In other cases, compromises must be made between costs (both plant and operating) and regenerant consumption and the degree of purity of the fully desalinated water. The removal of anions from the water of the cation exchange filter, including the carbonate and silicate ions, is carried out in a filter with a strong base anion exchanger. The fully desalinated water then contains only fractions of ppm SiO_2 and an amount of NaOH which depends on the leakage of the cation exchange filter. Strong base quaternary trimethylammonium exchangers (Type I) can be considered as anion exchangers that supply a water of highest purity even though their effective capacity is relatively low. The strong base anion exchangers of Type II in the same application also furnish water which is free of CO_2 and SiO_2, but they have a higher effective capacity because of their weaker basicity. The lower stability of exchangers of Type II, however, is more likely to lead to silicate leakage. However, since the removal of carbonate ions represents a high stress on the capacity of anion exchangers in both cases, the use of two-column processes with a strong acid cation exchanger and strong base anion exchanger is more widely accepted for large-scale water treatment plants than might be assumed because of the simplicity of this method.

If the removal of anions from the water of the cation exchanger is carried out on a filter with a weak base anion exchanger, one obtains a water which is free from inorganic acids but contains silicate and is supersaturated with carbon dioxide. The CO_2 can be removed to a large extent in an outgassing tower. Furthermore, small residues of sodium chloride are found. If the quality of the water is acceptable for its intended purpose, weak base anion exchangers with their high effective capacity and high regenerability will be preferred for full desalination. Whether further investigation of their suitability from the standpoint of economy will lead to the choice of an anion exchanger with secondary or with tertiary amine groups depends only on the expected hourly output of the filter. In small plants with low outputs the polyamine resins continue to have their advantages, and anion exchangers with tertiary amine groups will be preferred for large-scale installations in which high weight rates of flow are expected.

The application of this process is primarily determined by economic factors such as plant and chemical costs which can be held within tolerable limits in accordance with the constitution of the raw water by a suitable modification of the overall process and the use of certain inventive connections of the individual filters. Pretreatment of the raw water by other, less costly methods also plays an important role here.

In spite of the diversity of such plants and the custom-designed, extensive manufacturing programs of special firms that adapt their production to the wishes of individual operators, a few typical systems do exist. These also are often determined by whether a desiliconation is to be performed or not during full desalination.

Figure 61 shows the flowsheet of a full desalination plant with cation and anion exchanger filters connected in parallel loops. In this system each individual bed can be operated to the limit of its capacity, although this requires a more precise control with suitable instrumentation and well-trained operating personnel. Simpler servicing and higher operating reliability are offered by a full desalination plant with separately operating filter trains, as shown in Figure 62. In this case, it suffices to measure the conductivity and silicic acid at the end of each train and to switch off and regenerate a respective exhausted train.

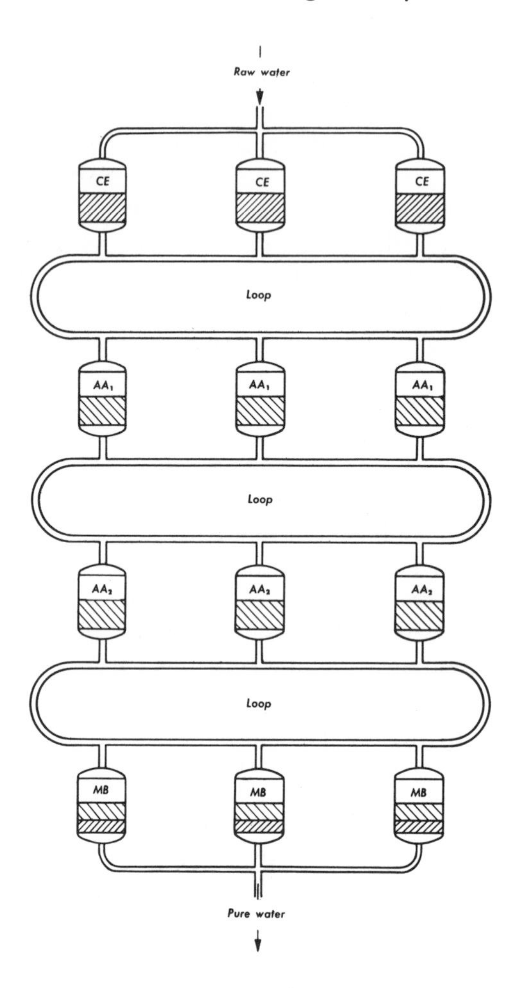

Figure 61. Full desalination in a separated filter system on loops (Bran and Lübbe, Hamburg).

The end of each system is followed by a mixed-bed filter. In the past attempts were made to work entirely with mixed-bed exchangers, but this plan was abandoned because of the complicated design of such filters and for other reasons, so that mixed-bed exchangers for the present are located only in the last stages, at least in larger plants. However, their function may differ. They may simply serve as safety elements which follow the desalination plant proper in order to catch any irregularities and natural "leakage," or they may be programmed into the plant as an indirect means for desiliconation and removal of residual salts.

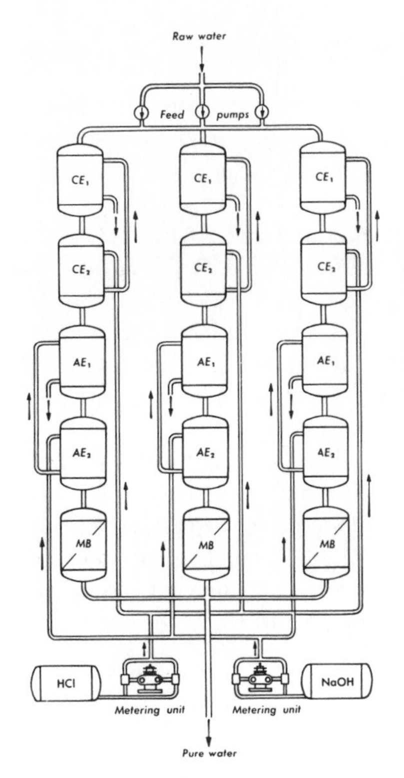

Figure 62. Full desalination in a triple series system. CE_1, strong acid cation exchanger; CE_2, strong acid cation exchanger; AE_1, weak base anion exchanger; AE_2, strong base anion exchanger; MB, mixed-bed filter (Bran and Lübbe, Hamburg).

1 Raw water
2 Feed pumps
3 CE_1
4 CE_2
5 AE_1
6 AE_2
7 Metering unit
8 Pure water

A simple time schedule can be used for the operation of a full desalination plant, such as that shown in Figure 62, for example, in which an operator without special chemical training simply opens or closes the proper valves in chronological order. Inasmuch as the exhaustion of one train is indicated by an increase of conductivity before the mixed-bed filter, it is also possible to operate a full desalination plant on a semi- or fully automatic scale by pneumatic or hydraulic controls [444]. Since silicic acid can appear already behind the exhausted strong base filter without leading to an increase in conductivity, the firm of Bran and Lübbe (Hamburg), for example, has installed a "silicometer" for controlling the desiliconization, which measures the silicic acid at 10-minute intervals with direct indication and recording on a fully automatic basis. The reader is referred to the special literature for other practical control methods of full-desalination installations [616]. Conductivity measurements for the control of full desalination plants are the subject of numerous publications [520, 596].

4.2.4. *Mixed-Bed Treatment*

In the mixed-bed filter, the cation exchanger in the H-form and anion exchanger in the OH-form are present side-by-side in a thorough mixture so that on infinite chain of full deionization units is formed. Consequently, as even better deionization is achieved than with the application of an alternating sequence of cation exchanger and anion exchanger filters. The reason resides in the fact that in the mixed bed, cation exchange and anion exchange are coupled into a single irreversible and therefore complete reaction.

Since mixed-bed treatment with a strong acid and strong base resin results in ultrapure water with a conductivity of 0.05 μS/cm and a silicate content of 0.002 ppm in small plants and of 0.2–0.5 μS/cm and 0.02–0.05 ppm silicate in large plants, it is understandable that mixed-bed filters are used where the highest requirements are set for water purity. It is also possible to connect two mixed-bed filters in series or one mixed-bed behind a full desalination process.

The problem in the management of mixed-bed filters resides in their regeneration. For this purpose, the exchanger particles need to be hydraulically separated to permit a separate treatment with the regenerant solutions. The installation of mixed-bed filters became possible only when the production of ion exchangers with a sufficiently different specific gravity was successful. As shown by Figure 63, which illustrates

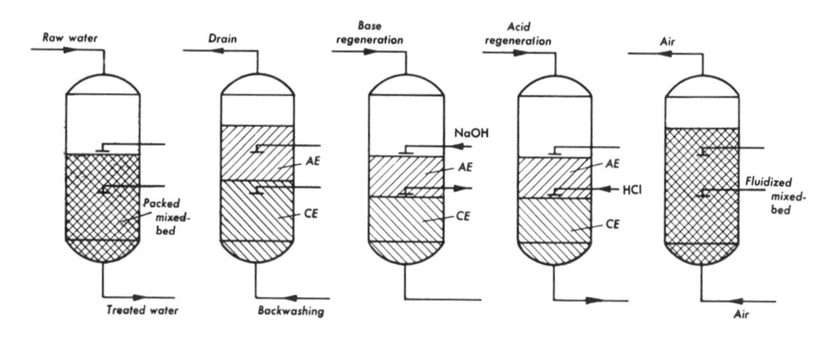

Figure 63. Regenerating cycle of a mixed-bed with separate regeneration.

a regenerating cycle of a mixed-bed filter, the lighter anion exchanger particles accumulate over the cation exchanger. Both layers are then separately regenerated. After regeneration, they are mixed again by compressed air and the mixed-bed is ready for loading. In small filters, regeneration of the cation and the anion exchange layer can also be performed simultaneously as shown in Figure 64.

Mixed-bed filters are characterized by small space requirements, low operating costs due to smaller amounts of regenerant and water, a high discharge rate of ultrapure water, constant readiness for operation, and finally by a low initial investment. It can be easily understood, therefore, that this type of desalination was quickly accepted

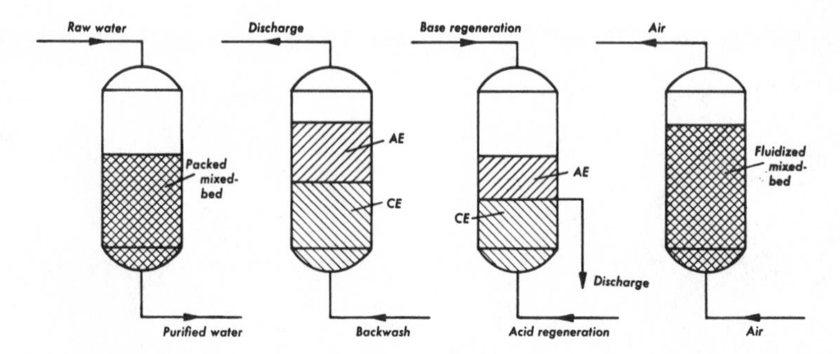

Figure 64. Regeneration cycle of a mixed-bed filter with simultaneous acid-base regeneration.

for smaller plants and laboratories. Dimensioning and packing of a mixed-bed filter is based on the origin and constitution of the raw water, the quality requirements, and thus the intended application as well as the average daily pure water requirements. Specialty firms offer mixed-bed filters which are suited for diverse applications. Such mixed-bed filters, belonging to the mixed-bed section of a Berkefeld full desalination plant, are shown in Figure 65. They supply two charges of 15 m^3/hours of desalinated water to feed a bonderizing line. A part of the control board can be seen on the right in the figure. If coalescence or lumping of the resin occurs by electrostatic attraction of the oppositely charged active exchange group when the mixed bed is started, this can be remedied by an intermediate chemical treatment. A very efficient product for this purpose is Primal ASE-60 of Rohm and Haas, Philadelphia, Pennsylvania, of which 60 g are used in 1350 ml of water per 100 ml of mixed bed with air fluidization. This is followed by normal backwashing and regeneration.

4.2.5 Combined Water Treatment Processes with Ion Exchangers

The means of softening, partial desalination, full desalination, and mixed-bed treatment represent processes in water treatment which in the form of well-defined concepts are methods by which raw water can be transformed into purer water of given quality. When these processes are investigated on the basis of their design, we recognize that they are founded on all individual processes which can be attributed to a certain ion exchanger type in a certain form and whose common characteristic is their industrial performance in filter tanks. A strong acid cation exchanger in the Na-form serves for softening and a weak acid cation exchanger in the H-form for partial desalination, while a strong acid cation exchanger in the H-form, a strong base anion

Figure 65. Berkefeld mixed-bed filter as a part of a full desalination plant.

exchanger in the OH-form, or a weak base anion exchanger in the OH-form is used for full desalination, and a cation exchanger and anion exchanger are simultaneously applied as a special form in mixed beds.

While softening and partial desalination are absolutely single processes, full desalination represents the individual process of a strong acid cation exchanger in the H-form combined with an anion exchange process for which a weak or strong base anion exchanger is selected depending on the desired result, whether with or without desiliconation and decarbonation. The latter can often be performed more suitably by outgassing. Full desalination thus is no longer a single process but is already a combined water treatment technique with ion exchangers.

This combination of possible individual processes based on types and forms of ion exchangers can be further extended, particularly with the addition of decarbonation by outgassing as a function of requirements, and thus can lead to combined water treatment processes with ion exchangers, of which a large number are in use. The selection of a given combination depends on the needs and purposes of the purified water and the quality of raw water; it must be emphasized that different types of raw water will also lead to different qualities of pure water with the same combined water treatment process. The application of special ion exchanger types, such as the macroreticular forms for the removal of colloidal silicate, or in the future, perhaps of ion exchangers, represents a further enrichment of the possible combinations.

The design of combined processes for a given purpose on the basis of a certain type of raw water and with consideration of the most economical utilization of regenerant is the broad field of ion exchange applications engineering for water treatment, in which softening, partial desalination, full desalination, and mixed-bed treatment represent, for practical purposes, only the basic fundamentals. However, almost all other possibilities can be estimated, tested, and finally realized on an industrial scale with nearly equal ease on the basis of a precise knowledge of these fundamental single processes.

4.2.6 *Condensate Treatment*

In modern ultrahigh-pressure boiler plants, the requirements made of the boiler feed water are so high that even the condensate with a solid content of less than 0.5 ppm is no longer sufficient. In comparison, water treated with ion exchangers and used to make up losses of turbine condensate is more purer. The condensate, which may be contaminated with cooling water and corrosion products from the steam generator, the turbine, the preheater, or the pipelines as a result of leaks in the tube bundles, therefore must be purified before it is used again; filtration and deionization with ion exchangers are used for this purpose.

To the extent to which the ion exchanger section is involved, only mixed-bed filters were used originally. But because of the occurrence of an iron oxide sludge, it was necessary to precede this treatment with a cation exchange filter. This filter, however, was contaminated by coarse and colloidal particles to such a degree that it was not possible to dispense with a previous mechanical filtration on gravel filters, gravity filters with activated carbon, or cellulose filter aids. With a preceding gravity filter, a mixed-bed filter then was found sufficient to bind the iron compounds.

The size of a mixed-bed filter depends on the necessary condensate throughput. The operation of the exchanger plant is the same as in full desalination. If difficulties occur in the cation exchanger because of special contamination, alternate washing with water or a water-air mixture may serve as a remedy, followed by a subsequent 24-hour acid treatment. A mixed-bed filter needs to be monitored with a control of conductivity and silicic acid as well as periodic checks of the copper, total iron, and alkali concentrations.

Condensate treatment developments show an increasing trend toward the use of mixed-bed filters with a high flow rate and the use of ion exchange powders [618]. The remaining problems are known and can be solved. The advantages of these processes might also be utilized in such a way that a filter with ion exchange powder will be accompanied by a mixed-bed filter operating at a high rate as a reserve unit which is coupled with two or more condensate loops and will be operated whenever a leak occurs in one of them.

4.2.7 *Ultrapure Water by Ion Exchange*

Water of the highest purity can be produced in a simple way and in any desired quantities by ion exchange processes in their present state of development. A plant with one cation exchanger and one anion exchanger filter is already able to lead to a water quality which corresponds to freshly distilled water according to DAB 6 with 10 μS. However, some applications require ultrapure water in which not even the slightest traces of ionic components resulting from cation leakage and a pH shift, and no nonionic contaminants, suspended colloidal materials, microorganisms, or pyrogens are any longer present.

The problem of neutral and ultrapure water with regard to ionic components can be easily solved by a multicolumn system. If a cation-anion exchanger combination is followed by another cation exchange filter, neutral water can be obtained as soon as the system is started. In a column system consisting of one cation exchanger, one anion exchanger, and a mixed-bed filter, a strictly neutral water quality with a conductivity of <0.1 μS is obtained. With a continuous operation of such plants, care must be taken only that organic contamination and the formation of iron and silicate precipitates on the resins are prevented and that regeneration take place correctly [634].

However, although ion exchange methods have been accepted by most pharmacopoeias for the preparation of ultrapure water, the removal of bacteria and pyrogens is still necessary for some medical applications. Water containing bacteria is not sterile and can therefore not be used in operating rooms and for pharmaceutical purposes. Pyrogens are metabolism products of bacteria which remain in water in the form of colloidal substances even after sterilization and lead to a pyrogen reaction when such water is used in injection solutions. Efforts have been made repeatedly to produce sterile water by ion exchange for medical purposes and pyrogen-free water for pharmaceutical applications; the results obtained still call for reservations.

According to bacteriological studies of ion exchange plants, Brantner [80] arrived at the conclusion that when untreated raw water is processed in small-scale plants with a combination of softening and small-scale sterilization, the product requires additional distillation for pharmaceutical purposes. According to Shimolin and Bochkareva [549], a laboratory installation consisting of a strong acid cation exchanger (Kationit KU-2, H-form) and two anion exchangers (Anionit PE-9 and Dowex 2, OH-form) with a flow rate of 25–30 liters/hour, results in a nontoxic, desalinated water without pyrogenic bacteria and reducing substances, with a dry residue of $<0.001\%$ and a pH of 6.6–7.2. According to the extensive research of Saunders [516, 517], this problem in the production of ultrapure, nonpyrogenic, and sterile water for injection solutions can be controlled only by sterilizing the equipment regularly with a nonionic compound such as formaldehyde, and chlorinating the water followed by at least 15 hours of storage before sending it through the column. When these and other conventional precautions are observed, a nonpyrogenic product is obtained for several weeks with a capacity loss of 10%. By the further application of gamma-radiation in a special sterile cartridge, additional progress can be made in obtaining sterile, nonpyrogenic, ultrapure water. The use of activated charcoal in the systems for the production of ultrapure water is possible only to a limited degree since carbon particles will release water-soluble impurities after some time. These difficulties can be easily overcome with the macroreticular ion exchanger Amberlite XE-238, which was finally selected by Saunders [518] in a system for the production of sterile, nonpyrogenic water in which the final purification stage delivered 400 liters of normally deionized water.

A third problem in the production of ultrapure water is that of the polyamines. These represent degradation products of anion exchangers which reach the water in chemically undetectable traces but which become noticeable by the formation of foam in the deionized water which is more stable than in distilled water. The surface tension of deionized water contaminated with polyamines is lower than that of distilled water. To remove the polyamines, the deionized water is conducted over a column of high-purity activated carbon in which the polyamines are removed and the surface tension of distilled water is obtained.

In addition to the use of ultrapure water in medical and pharmaceutical applications, it is also needed in very large quantities in some branches of industry. Consequently, considerable efforts have been expended to design combinations of purification stages which will satisfy the severe demands. The core of such installations is an ion exchanger combination; a large number of data are available on its function [113]. Special firms with sufficient experience to deliver such installations are active in nearly all countries [429]. The Serva water purification program [544] includes a number of possible combinations

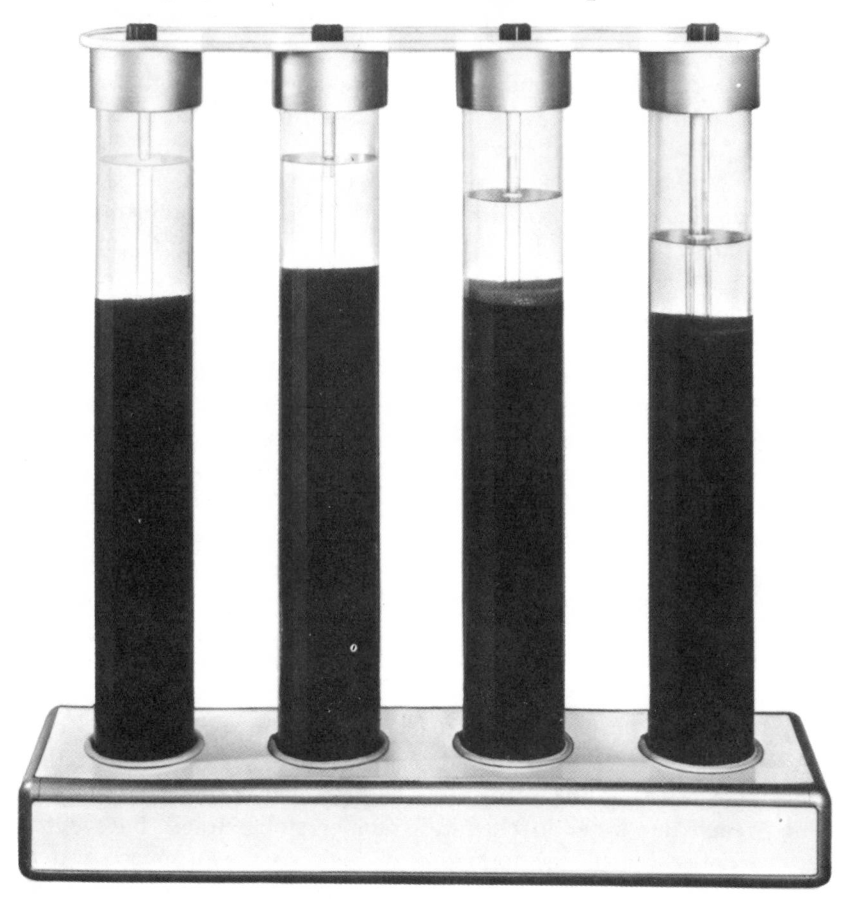

Figure 66. Column arrangement of a Serva KCAM ion filter. Water quality: conductivity $< 0.1\ \mu S$, dry residue < 0.6 mg/l, strictly neutral pH.

for water treatment with ion exchangers for the non-industrial sector; Figure 66 shows the Serva ion filter KCAM. It consists of a carbon filter as well as a cation exchanger, anion exchanger, and mixed-bed column. The customer service offered by the manufacturer facilitates its continuous operation since automatic regenerating systems can be supplied for the regeneration of the cation and anion exchange column as well as the mixed-bed column so that manual regeneration is completely unnecessary (Figure 67). If such an assembly is installed in a hospital, however, a certain degree of decentralization in favor of several smaller units is recommended. If a water treatment plant is planned for a new construction, a separate system for each level of the building is preferable to central heating or hot-water supplies. The distributing lines become shorter so that bacterial recontamination is less likely and the costs therefore become more favorable. In a central system any malfunction will shut down the complete ultrapure water supply, while bypass possibilities exist with the use of several units.

Figure 67. Automatic regeneration with a Serva automatic regenerator.

4.2.8 *New Ion Exchange Methods for Water Treatment*

The development of the industrial application of ion exchangers is just as—and perhaps even more—evident in fields other than water treatment, which is the main application of ion exchangers in general. New developments compared to the classical methods can be considered as newly developed processes as well as new fields of application. Among the newly developed techniques, the countercurrent process, mixed-bed process, continuous processes, and the Powdex process have been described elsewhere; they are joined by the Desal and the Sirotherm processes, while new fields of application are the treatment of potable water on one hand, and of brackish water on the other.

Desal (Kunin) process. On the basis of research on weak base anion exchangers and their regeneration with carbon dioxide, Kunin and Vassiliou developed a combined ion exchange process which has been named the "Desal" process because of its great suitability for desalination [368, 370, 376]. The process takes place in three steps:

Carbonation: conversion of weak base anion exchange resin into the bicarbonate form

Alkalization: conversion of the NaCl content of water into $NaHCO_3$ on this anion exchange resin

Dealkalization: removal of cations on a cation exchanger and outgassing of the CO_2 formed.

It is based on the use of the unusually weak base anion exchanger, Amberlite IRA-68, which can be regenerated with ammonia with 100% efficiency. Three columns are connected in series, the first and last of which contain Amberlite IRA-68, and the middle column is packed with the weak acid cation exchanger IRC-84. Because of the high regenerant efficiency and low costs of NH_3 and H_2SO_4, the use of raw water as washwater, the small degree of ion leakage, and inversion of the process in the sequence of the columns depending on the cycle, this technique can be used for the desalination of brackish water

Figure 68. Industrial Desal installation.

containing up to 3000 ppm of salts. Figure 68 shows an industrial Desal plant [99].

Sirotherm process. In the search for inexpensive possibilities for obtaining potable water with 500 ppm NaCl from brackish water containing up to 2000 ppm NaCl, Weiss and others [627] investigated a method in which the exchange of Na^+- and Cl^--ions takes place on a weak acid cation exchanger and a weak base anion exchanger and in which the salts of these weak exchangers are hydrolyzed under the influence of heat according to the equations:

$$KA \cdot COO^-Na^+ + H_2O \underset{}{\overset{\text{heat}}{\rightleftharpoons}} KA \cdot COO^-H^+ + Na^+OH^-$$
$$AA \cdot N(CH_3)_2H^+Cl^- \rightleftharpoons AA \cdot N(CH_3)_2 + H^+Cl^-$$

It is reported that the process leads to a more dilute and a more concentrated eluate, which only need to be kept separate by alternating the loading of cold and warm salt water without the use of chemicals. The applicability of this technique will not become apparent until some shortcomings have been corrected. However, while it was tested, a number of basic findings were obtained on the behavior of weak electrolyte ion exchangers [73].

Potable water treatment. The treatment of potable water has not yet found an application in general water supplies, but the need for this in the future cannot be ruled out when we consider that certain ions appear to be harmful to health in higher concentrations. Such problems have been investigated in connection with the nitrate ion [514]. The correct application of an ion exchange process for the removal of nitrate ions from potable water also furnished proof that complete removal of these ions is possible with a strong base anion exchanger of Type II [190]. Since this complete removal leads to other modifications of the chemical composition of drinking water, a denitration of a partial stream appears desirable. The removal of fluoride ions, of which higher concentrations are harmful to health, is not practically feasible with anion exchangers because of their low selectivity, but a process with cation exchangers in which fluoride ion removal is possible after loading with aluminum sulfate may become of interest [58].

Denitration may be necessary for the treatment of brewery water, since a nitrate concentration of more than 5 ppm may act as a yeast poison. A Berkefeld system for this purpose is shown in Figure 69, in which a performance of 10 m^3/h is obtained by anion exchange. The regeneration of this plant, which may be considered to be similarly applicable for potable water preparation, is carried out with common salt.

Figure 69. Berkefeld ion exchange installation for nitrate substitution.

4.3 Problem of Organic Fouling of Ion Exchangers in Water Treatment

In the operation of water treatment plants with ion exchangers, a phenomenon occurs which is known as organic fouling. This organic contamination is an irreversible fixation of organic materials to the ion exchange resin. Since surface water is being increasingly used as industrial production expands, the removal of organics has become an important partial problem in full desalination with ion exchangers.

The reasons for organic water pollution may be of a natural nature; the phenomenon then becomes evident in humic and fulvic acid concentrations, the ratio of which is subject to seasonal variations. However, detergents and oily as well as amine-like components may be present as a result of industrial effluents. Finally, organic contaminations may be produced in the plant itself by decross-linking of the cation exchangers that precede the anion exchange filters, particularly by way of copper-catalyzed oxidative influences which then have the same effect.

Organic fouling is noticeable in anion exchangers. Purely mechanical effects may occur or chemical influences may act as resin poisons. The mechanical effects become evident in the plugging of the exchanger bed with insoluble organic components; poisoning must be attributed to chemical reactions with dissolved organic materials. The influences of colloidal components take an intermediate position and are most difficult to correct. In all cases, organic fouling becomes apparent in long washing periods, capacity losses, decreasing pH, and an increase in the conductivity of the deionized water. With strong base anion exchangers, the situation can be understood most easily. If such an exchanger has irreversibly adsorbed organic materials, the active functional groups are blocked and the capacity consequently drops. Since Na-ions can be adsorbed at the same time, the washing times become longer. Weak base anion exchangers, with their low capacities with short exchange periods, are damaged just like strong base exchangers, but the actual influence is more difficult to recognize. In mixed-beds, damage by organic contamination is recognized most rapidly, since mixed-bed filters deliver such a high quality of water that any loss of the water is immediately detected [93]. A reason for this is assumed to be an interaction of the acid used for the regeneration of cation exchangers in the mixed-bed with the humic acids exchanged by the anion exchanger.

Three methods are being investigated to control the phenomenon of organic fouling—pretreatment of polluted water, regeneration of the damaged resins, and development and use of anion exchangers with structures on which the uptake of organic contaminants will be reversible.

The pretreatment of polluted water with other, more inexpensive methods prior to ion exchange treatment naturally should always be used whenever possible. It should never become necessary to replace entire ion exchanger beds just because the raw water was not filtered or, in the case of colloidal components, not prepurified by well-known and proven coagulation methods.

The same is true for the regeneration of plugged ion exchangers in which energetic fluidization and backwashing frequently can correct the mechanical block. If a pronounced conductivity rise is observed, it is possible to remedy the situation by a rapid acid treatment of the anion exchanger followed by warm common salt solution or 10% common salt solution with 1% caustic soda; or in the case of very extensive contamination, treating with 0.25–0.5% hypochlorite solution [269]. With a certain quality of raw water, it may be appropriate to carry out this treatment at regular intervals of one to several months.

Organic fouling can be prevented most effectively by the use of exchangers which reversibly desorb the contaminants. To the extent to which this reversibility can be obtained by the structure, macroporous anion exchangers should be selected in their various forms provided their lower capacity is acceptable [283]. The isoporous ion exchangers with the same favorable pore structure have the additional advantage of a higher capacity. If macroporous anion exchangers are preferred for polluted water, the weak base types offer the additional opportunity for modifying the functional groups so that they attain the efficiency of strong base exchangers; this has been accomplished with the commercial types Amberlite IRA-93, De-Acidite H/IP and Lewatit MP-60 [413]. Another processing step is to layer Amberlite IRA-93 on the strong base anion exchanger in the same filter; this leads to an efficient technique for the treatment of organically polluted water. In a typical case, Amberlite IRA-93 reversibly takes up about 80% of the organic pollutants even after many operating cycles. Naturally, this technique can be applied only in two-column processes consisting of a cation exchanger and an anion exchanger.

It is more difficult to eliminate organic fouling in the smaller mixed-bed filters. In this case, it is important to select only a strong base anion exchanger fraction in such a way that its structure permits a reversible elution of organic contaminants. Otherwise, it must be preceded by a two-column system which sufficiently removes the organic compounds in the anion exchanger stage. Another possibility is the use of prefilters (so-called scavengers) with a strong base anion exchanger in the Cl-form in front of relatively small mixed beds; some commercial types of these scavengers are available for this purpose.

4.4 Wastewater Treatment (Decontamination)

The most diverse chemical and industrial processes result in waste effluents which can no longer be used in the same process and must therefore be discharged in some form. These effluents contain contaminants, and as long as they are present in small concentrations they may be entirely equivalent to raw water and can be properly treated for industrial reuse. It is therefore natural that ion exchange processes are employed for waste effluent treatment and that the process shares many characteristics with conventional water treatment. Nevertheless, waste effluent treatment deserves a section of its own because of two other aspects, *i.e.*, it is performed because its components are too valuable or because they are too toxic or too dangerous in other ways.

While ion exchangers hold a dominant position in water treatment, they find sharp competition in other processes in waste effluent treatment. It may be that some hopes that were held for ion exchangers in waste effluent treatment could not be realized, but the development of continuous processes alone is proof that further advances may still be expected in this area. The large number of metal-containing waste effluents produced in hydrometallurgy are so closely related to the hydrometallurgical processes themselves that they will be discussed separately in Section 4.5. Decontamination of radioactive effluents, however, is a field of application for ion exchangers which has assumed a firm position in radiation chemistry. Other waste effluent treatments by ion exchangers will, then, be briefly pointed out.

Radioactive effluents (decontamination). Most closely related to water treatment are questions of the removal of radioactive ions (decontamination) by ion exchange. This is carried out by techniques which represent the most common ones in nuclear engineering. The purpose is, first of all, to remove radionuclides present in the primary circuit of light- and heavy-water reactors to prevent radiolysis of the water and corrosion of reactor components. The deionized water to be treated in these cases is decontaminated economically only by means of ion exchangers, so that the latter have been used for this purpose since the beginning of reactor technology. Generally, mixed-bed filters which can be operated at a high throughput are used. In the decontamination of salt-containing water, such as the effluents from reactor installations, isotope laboratories, and hospitals, deionized water is the end-result of full desalination since ion exchangers have no selectivity for isotopes; the ion exchangers retain the radioactive ions and all salts dissolved in the raw water.

The question arises here whether the exhausted exchangers should be regenerated or should be used only once and then stored, possibly for hundreds of years, to allow decay of the radioactivity. After consideration of all advantages and disadvantages, regeneration is preferred [415, 417]. On the other hand, in the processes using ion exchangers in reactor engineering, regeneration is not recommended. Usually the exchangers are flushed from the filters and stored at suitable sites or burned. The questions of suitable ion exchanger types for reactor technology and of the efficiency and costs of different ion exchange processes are too complex to be discussed exhaustively here. The highest demands are made of exchanger materials with regard to purity, and precise data concerning their loading form are necessary. It should therefore be mentioned that some firms do market so-called nuclear grades of exchangers.

A special question of decontamination is the recovery of potable water from radioactively contaminated surface waters as might be the case after disasters or nuclear weapons explosions. Laboratory studies have shown that it may be possible to remove all radioactive contaminants in a relatively short time [380]. For practical purposes, installations have been constructed in which full deionization represents the last stage of a combination of precipitation, filtration, and ion exchange, so that decontamination takes place and potable water is recovered from which no further radiation hazards need to be feared [232]. Water containing 1000 to 10,000 times the permissible concentration of radioactivity for drinking purposes can be treated; in this case, only a single use of the ion exchangers is appropriate.

The seven years of decontamination experience gathered at the Karlsruhe Nuclear Research Center in the treatment of radioactive effluents is worthy of note [352]. The ion exchanger installation, which consists of five columns for cation exchangers and anion exchangers as well as a mixed-bed filter with a packing of a total of 3.5 m^3 resins in the H^+- and OH^--form, was compared with chemical precipitation and evaporation processes. The economic limits of application of ion exchange processes became clearly evident. Although organic exchangers are excellently suited for decontaminating low-salt water, they cannot be used economically for radioactive effluents which usually contain 1–2 g/l of salt and only rarely less. If ordinary radioactive effluents are to be conducted through ion exchangers, they require previous flocculation or at least filtration because of their high contamination. The exchangers are regenerated after a throughput of 70 m^3 each, the regenerating solutions are concentrated in a thin-film evaporator, and the residues are encapsulated in cement. Under favorable conditions, up to 99.99% of the radionuclides can be removed from the water with a volume reduction of 1:50. According to these findings, ion exchange processes are economical only for the decontamination of optically clear, low-salt effluents.

Other waste effluent treatment processes. Because of the increasing cost of industrial water, another object in the search for suitable wastewater treatment processes is that the effluents be purified to the point that they can be reused for the same process. It is the general view that this applies to *laundry effluents,* which should be treated so that they can be used again for laundry purposes for economy reasons [339]. Beyond this, legislative regulations, which will require a treatment of laundry wastewater in the future, should not be neglected. Research into the use of ion exchangers has been carried out in connection with these problems [2, 611]. Strong base anion exchangers were used for the adsorption of anionic detergents. Although the resin is highly selective with a high capacity, a polar solvent such as methanol nevertheless must be used with an inorganic acid for regeneration. With weak base anion exchangers in the Cl-form or OH-form, the "merry-go-round" principle, which might lead to a more economical and more practical process since it would require no organic solvents, has been used. In any case, laundry effluents must be pretreated prior to ion exchange treatment.

An ion exchange filter does not need to be the central unit of a wastewater treatment plant. As indicated by the results of cupric ammonia treatment in butadiene synthesis [69], the process may be used as the final purification stage to reduce the copper concentration in its exchange for ammonium ions to such a degree that it amounts only to 2 mg/l at the end of treatment; after further dilution with 3–4 times the amount of river water, the wastewater can be discharged to a central sewage system.

According to Schtanikow [112, 533], virus-contaminated water can also be purified by ion exchangers. Polio virus is inactivated in water after the latter is loaded on Anionit AV-17, OH-form, followed by filtration on Kationit KU-2, H-form.

The literature naturally contains many other possibilities for the use of ion exchangers than have become accepted in practice for reasons of economy [476]. An exchange process cannot always compete with distillation or a precipitation process in a design calculation, if only because in distillation and precipitation, the costs are independent of the concentration of contaminants up to a certain degree, while in the case of ion exchange, they are directly proportional to the degree of contamination.

4.5 Metals and Ion Exchangers

Metals have been investigated most thoroughly with ion exchangers. An almost innumerable number of investigators have been occupied with the behavior of all kinds of metal ions in the most diverse solutions and solvent mixtures on cation and anion exchangers. For laboratory purposes, this research has resulted in a large number of possibilities for separations, concentrations, and purifications. Separations in most cases take place by ion exchange chromatography, which is discussed in detail in Section 5.4. These separations must be considered mainly nonindustrial processes, with the exception of the rare earths. Concentration and purification can be carried out preparatively in the laboratory or from industrial points of view in plants of the metals industry. The metals industry in this case refers to hydrometallurgy, the special characteristics of which will be discussed here from the standpoint of ion exchange.

The application of ion exchange technology in hydrometallurgical practice includes winning, purification, and concentration of metals from aqueous solutions, regardless of their origin. Thus, it deals with the winning of metals from solutions obtained by ore treatment as well as with the recovery of metals from hydrometallurgical effluents. However, even though such a large number of data are available on metal ions and ion exchangers, the commercial process of metals winning and refining by ion exchange has remained limited to just a few.

The reason for this must frequently be sought in the cost aspect. This subject has been repeatedly considered. If a metal is so valuable that its price covers the costs of winning with ion exchangers, the situation is simple. In other cases, however, a metal may be so toxic that it must be removed from waste effluents. If ion exchange then is the most inexpensive compared with other processes, it again has prospects for development and utilization [490]. In the following description, these relationships will not be fully considered; rather, the possible industrial application of ion exchangers in the metals industry will take first place.

Uranium. A number of ion exchange installations are operating today for the recovery of uranium from low-uranium ores and minerals because of the importance of this element in nuclear energy. The original attempts with cation exchangers failed, since iron, aluminum and other heavy metals were sorbed together with the uranyl ion UO_2^{++}. Only when it was discovered that the UO_2^{++}-ion undergoes a stepwise formation of anion complexes with sulfate:

$$UO_2^{++} + nSO_4^{--} \rightleftharpoons UO_2(SO_4)_n^{2-2n} (n = 1,2,3)$$

was it possible to consider a separation with anion exchangers. The development of selective exchangers for the uranium sulfate complex as against chloride, nitrate, bisulfate, and the iron sulfate complex, led to the rapid spread of this process. After dressing of the ores containing 0.1–0.5% uranium and leaching with sulfuric acid, the solution is passed over a sans filter and is then conducted into a series-connected battery of anion exchangers, as shown in the flowsheet of Figure 70. Since the exchanger is selective for uranium, this element finally is exchanged alone by iterative chromatographic effects. The loaded filter is then eluted with concentrated chloride or nitrate solutions, so that the exchanger bed is simultaneously returned to its original loading state. In the further course of the process, uranium is precipitated from the eluate with ammonia, filtered, and dried. The results of these processes are satisfactory [26]. They operate with yields of 98% and result in a concentrate containing more than 80% U_3O_8.

For a long time, this was the only method in the entire world by which uranium was recovered, and it also represented the only industrial application of ion exchangers. Uranium is also recovered by the

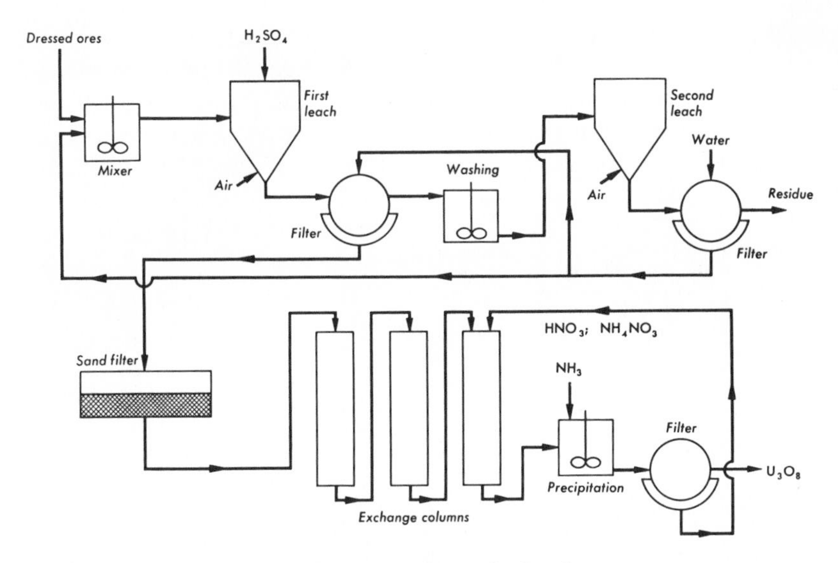

Figure 70. Flowsheet of an ion exchange facility for uranium recovery.

Porter-Arden process [93], which was the first continuous large-scale technique and which is applied in Canada for the industrial manufacture of uranium.

Uranium can also be recovered with liquid ion exchangers. Through the use of di-(2-ethylhexyl)phosphoric acid (D2EHPA) as a liquid cation exchanger, the Dapex process was developed in which uranium is extracted with a 0.1-*M* solution of D2EHPA in kerosene and is stripped with concentrated inorganic acid, soda solution, or MgO suspension and then processed by conventional methods. The reactions on which this process is based can be written as follows in simplified manner:

$$(R{-}O)_2P(=O)OH + UO_2^{++} \rightleftharpoons (R{-}O)_2P(=O){-}O{-}UO_2{-}O{-}P(=O)(O{-}R)_2 + 2H^+$$

$$(R{-}O)_2P(=O){-}O{-}UO_2{-}O{-}P(=O)(O{-}R)_2 + 2\,Na^+ \rightleftharpoons 2\,(R{-}O)_2P(=O){-}O{-}Na + UO_2^{++}$$

Although a hundredfold concentration increase can be obtained, iron takes part in an unfavorable competing reaction which can be prevented only by diluting the uranium stock solution. However, if a liquid anion exchanger of the weak base amine type is used, the Amex process is obtained in which uranium is extracted with 0.1 *M* exchanger solution; the sulfate leach solution is stripped with soda solution, common salt solution, ammonium nitrate solution, hydrochloric acid, sodium hydroxide, or MgO slurry; and processing then continues. In simplified form, the reaction takes place according to the equation:

$$2R-\underset{R}{\overset{R}{|}}N-HCl + [UO_2(SO_4)_2]^{-2} \rightleftharpoons R-\underset{R}{\overset{R}{|}}N-H-UO_2SO_4-H-\underset{R}{\overset{R}{|}}N-R + 2Cl^-$$

$$R-\underset{R}{\overset{R}{|}}N-H-UO_2SO_4-H-\underset{R}{\overset{R}{|}}N-R + 2NaCl \rightleftharpoons 2R-\underset{R}{\overset{R}{|}}N-HCl + [UO_2(SO_4)_2]^{-2} + 2Na^+$$

The liquid anion exchanger is regenerated with aqueous ammonia and recycled.

A combined process of ion exchange resins and liquid ion exchangers exists in the form of the Eluex process [182], with a strong base anion exchanger and a tertiary amine in kerosene. This process is used very economically in several plants in the United States in combination with the RIP process for uranium ore treatment.

Thorium. Like uranium, thorium forms negative anionic complexes and consequently can also be isolated with ion exchangers. If it occurs together with uranium, it is simultaneously adsorbed by a synthetic ion exchange resin and is frontally displaced again in the course of loading. A subsequent anion exchange column then exchanges and concentrates it, and finally it is eluted again. This process is industrially as valuable as the winning of uranium. In this case, competing developments were taking place which finally led to the decision in favor of liquid ion exchangers, so that thorium is recovered with these [134] but in complete analogy with the uranium process.

Gold. Attempts have been made in Germany even during the Second World War to recover gold from waste materials with the first ion exchange resins. Efforts in about 1955 to recover gold by a process similar to that used for uranium had negative results. Detailed research findings on the subject have been published [137]. The recovery of gold from waste effluents, in which it is present as a cyanide complex $[Au(CN)_2]^{-1}$, is possible with ion exchangers because of the high selectivity of anion exchangers for this complex even when only trace concentrations of gold are present beside high concentrations of other salts. However, since this high selectivity also leads to elution problems, it has become the general practice to dry the loaded exchanger and recover the gold by combustion in view of its high value compared to that of the exchanger material.

Platinum. Platinum is concentrated on ion exchangers from waste effluents in which it is present as the $[PtCl_6]^{-2}$ complex. Since it is difficult to elute because of high selectivity, as in the case of gold, the final recovery takes place by drying and burning the exchanger material.

Silver. Since the prices for silver and silver salts are continuously increasing because of their industrial and particularly photographic applications, processes for the recovery of silver wastes always are of high interest. Techniques for the recovery of silver from waste effluents by means of ion exchange have hardly been used. The method of exchange resin burning would not be applicable for silver for reasons of cost, although various practical procedures for elution are known. The treatment of silver residues in the photographic industry which produces silver-containing waste effluents in the manufacture and development of films is gaining increasing attention. The treatment of these wastes is more complicated than that of other silver-containing wastes because they contain not only dissolved silver thiosulfates but also colloidal silver chlorides, bromides and iodides. While colloidal silver in the past was centrifuged and the solute was complexed in the form of $[Ag(CN)_2]^{-1}$ to be fixed on a weak base amine resin, silver thiosulfate complexes, $[Ag_2(S_2O_3)_3]^{-4}$, can also be exchanged on a strong base anion exchanger of Type I such as Amberlite IRA-400 in the OH-form and eluted with 4% NaOH plus 2% NaCN. For the recovery of the colloidal silver fraction, macroreticular strong base anion exchangers represent a decisive advance just as they are useful for the removal of colloidal silicate.

Chromium. Since chromium has many applications in metals finishing and is highly toxic on one hand but relatively valuable on the other, processes for its recovery on ion exchangers have moved into the foreground. The cations occurring as impurities in exhausted chromate solutions are removed on cation exchangers, while the chromic acid of the rinsing baths can be recovered on anion exchangers.

The flowsheet of Figure 71 shows a complete chromating facility with chromating and passivating units. The solutions to be treated are conducted from both parts of the plant into a dilution tank since chromate solutions can be treated with ion exchangers only if the chromic acid concentration is not higher than 100–120 g/l. From there, the solution is loaded on the ion exchanger consisting of a strong acid cation exchanger, where iron, aluminum, manganese, and copper ions are exchanged. The purified dilute chromic acid is conducted into a reservoir and from there is directly returned to passivation or is brought

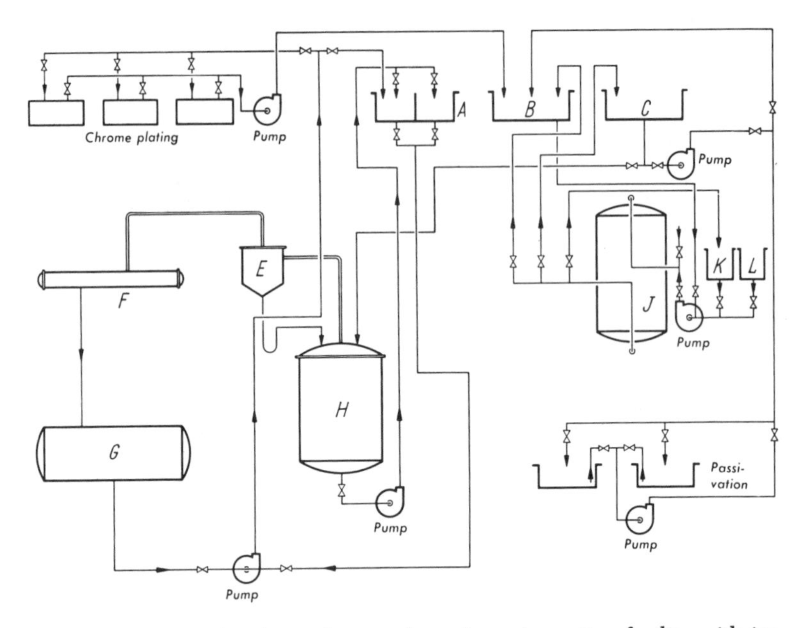

Figure 71. Flowsheet of a complete chromate coating facility with ion exchange unit.

A reservoir	D settler	H evaporator
B dilution tank	E condenser	I exchange filter
C reservoir	F receiver	K, L regenerant acid tanks

to the concentration of the chromate solution in an evaporator. After concentration, this portion is transported to a reservoir and used for chromating as needed. This leads to a pure, cation-free chromic acid, and the need for discarding parts of exhausted chromate solution is eliminated.

In a similar process, the chromic acid of rinsing baths can be recovered on a strong base anion exchanger [23]. Since rinsing baths contain traces of metal ions in addition to chromic acid, a cation exchange filter is used as a first step, according to the flowsheet of Figure 72, where the effluents are freed from metal ions. Subsequently, the chromic acid is concentrated on the anion exchanger. Following elution with 15% sodium hydroxide solution results in chromic acid in the form of Na_2CrO_4, which in turn can be recovered on the cation exchanger after it has been regenerated with sulfuric acid and can be further concentrated in an evaporator. If the chromic acid-containing effluents are initially treated with NaOH, the anions $Cr_2O_7^{--}$ or $Cr_3O_{10}^{--}$ form from the CrO_4^{--} anion, so that the capacity of the anion exchanger for chromium is increased by a factor of 2 or even 3.

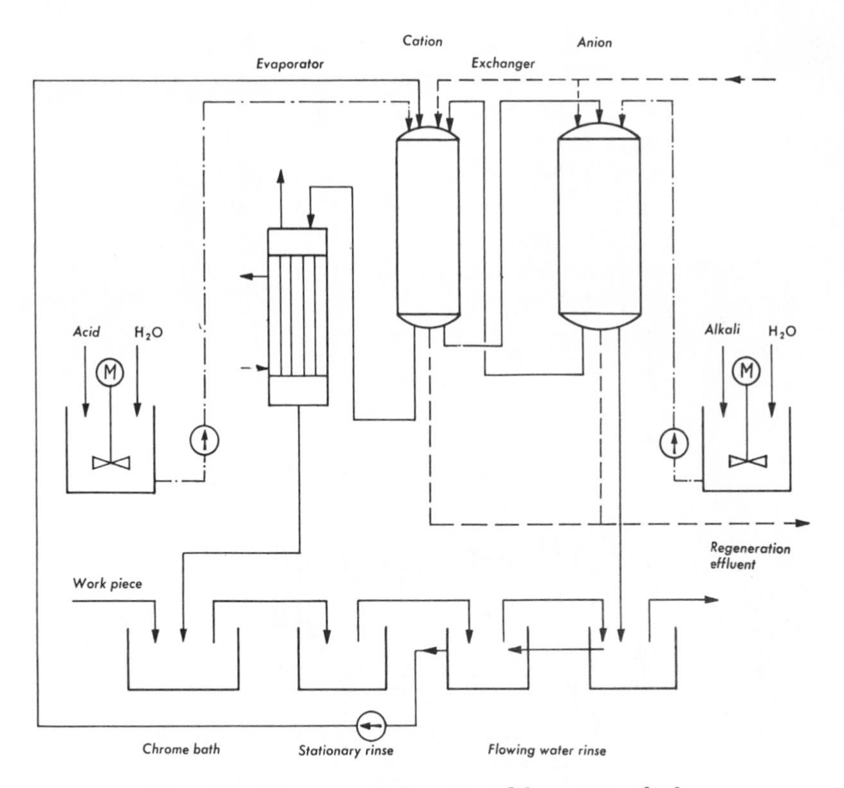

Figure 72. Recovery of chromic acid from rinsing baths.

The significance attributed to ion exchange processes in electrochemistry is demonstrated by another series of procedures for the same problem [363, 364].

Processing of chromium-containing effluents, which must be carried out because of the toxicity of chromium and may possibly also lead to a reuse of the water, can be combined with chromium reduction and cyanide oxidation. According to the first well-known process, the metal salts remaining after precipitation and filtration are removed with a specially prepared cation exchanger. The life of this exchanger depends on the efficiency of the preceding treatment steps. In general, treatment by the batch process is more effective and economical than continuous neutralization.

In contrast, a cyclic ion exchange plant represents an improvement. In this system, the necessary rinsing solutions are carried in a cycle and the metal traces as well as anionic metal complexes contained in them are concentrated in an ion exchange installation. The removal of cyanide is not absolutely necessary since no concentration takes place in the cycle. The regenerants of the ion exchange filters are treated in a stationary detoxification installation which can be used simultaneously for the treatment of exhausted electrolytes. The advantages of this technique reside in a smaller overall plant and a significant improvement of the electrolyte because the total water of the cycle is deionized. Fresh water to make up water losses therefore is suitably added to the cycle upstream of the ion exchange plant [97, 522]. This principle has

Figure 73. Berkefeld full desalination installation for the rinsing baths of electroplating plants. Performance: 5–12 m^3/h.

also been realized in the Berkefeld cycling plant shown in Figure 73. In this system, the used rinse solution is conducted from the right through the five filter tanks, *i.e.*, gravel filter, activated carbon filter, cation exchanger, anion exchanger and, finally, mixed-bed filter. The treated water flows to a compression unit which transports it at 7 atm to the automatic rinsing systems and to the spraying and washing installations. Regeneration takes place by jet aspirators from the polyethylene bottles (shown on the right in the figure) which are directly connected to the system as a shipping unit.

Nickel. Nickel from the rinsing baths of nickel-plating baths is reported to be recovered in the same manner as chromium for similar reasons. A process has been described in which a cation filter and an anion filter are used consecutively so that nickel is recovered from the rinsing baths and the sulfate, chloride, and borate anions are removed. The deionized water can be used for the preparation of fresh nickel-plating solutions or as rinsing water or boiler feed water. The cation exchanger is regenerated with sulfuric acid, and the nickel sulfate obtained in the eluate is recycled to the process.

Copper. The recovery of copper from fiber spinning solutions with ion exchangers is an industrial process which has been described in simplified form as a preparative demonstration experiment in Chapter 3, Section 3.3. The process has been further developed over the years. One of the first continuous countercurrent ion exchange plants was installed with the weak methacrylic acid cation exchanger and has been operating satisfactorily for years. The removal of copper from waste effluents of an alkaline degreasing and copper-plating process is a practical ion exchange method, according to the findings of Furrer [199] and Tallmagde [570] which requires no further explanation.

Zinc. In the manufacture of rayon, zinc sulfate is used as a hardener in coagulation baths together with sulfuric acid so that washing of the fibers results in a waste effluent consisting of sulfuric acid and zinc sulfate containing about 100 ppm Zn. By treatment with a strong acid cation exchanger, the wastewater problem can be solved and zinc can be recovered with a value which usually is equivalent to the costs of the treatment process. The original H_2SO_4–$ZnSO_4$ solution can be used for regeneration of the exchanger [61]. The possible variations of ion exchange technology for the same problem are demonstrated by the use of a fluidized bed for the treatment of zinc waste effluents [325, 326].

Iron. In hydrochloric acid solution, iron forms a complex $FeCl_4^-$ which is exchanged on an ion exchanger with a high selectivity. This process has been investigated for the treatment of pickling effluents [17] and was economically applied in a plant [76]. A Berkefeld ion exchange installation for the iron removal in phosphoric acid pickling baths with a performance of 900 l/h operating in a recycling system is shown in Figure 74. The recycling process is carried out in such a way that the free acids are formed again in the water cycle from all of the salts present in the solution, *i.e.*, in contrast to the earlier examples on a cation exchanger. Preferably, the process involves the removal of iron in addition to calcium, magnesium, aluminum, and trivalent chromium. The materials which cause exhaustion and aging of the bath are continuously removed by recycling of the bath contents since such a filter can recycle and regenerate several baths per hour [422].

Figure 74. Berkefeld ion exchange installation for iron removal of phosphoric acid pickling baths by the cyclic process. Performance: 900 liters/h.

4.6 Manufacture of Sugar and Ion Exchangers

A number of well-developed processes have become known for sugar refining, and yet ion exchangers have also found their place here as a supplement or even replacement of conventional processes. The initial technical difficulties which had to be overcome reside in the alternate handling of sugar solutions of high viscosity and regenerants of low viscosity. Since these and other problems—such as the instability of resins to colorizing components—have been solved in time, the development of sugar refining with ion exchangers today is concentrated on reducing the regenerant costs and the problem of cooling the sugar juices to decrease inversion.

In the *cane sugar industry* in countries overseas, the application of ion exchangers extends to the treatment of cane sugar juices, refining of cane sugar, and the production of sugar syrup. The cationic components of cane sugar juices (sodium, potassium, magnesium, calcium, iron, and silicates) and the anionic components (inorganic anions, amino acids, fats, waxes, and organic acids) are either demineralized on cation exchangers in an Na or NH_4 cycle (analogous to water softening) or are removed on strong base anion exchangers and weak acid cation exchangers in that order. A certain part of the colorants is also separated either by true ion exchange or by general adsorption. The sugar yield is thus increased by 3–6% with a reduction of the ash content of 95%. In cane sugar refining, an ion exchange process is advantageously used after carbon treatment for final refining to remove the last color components from the partially decolorized cane sugar so that the last coloring components can be removed with strong base macroreticular or gel anion exchangers.

The production of sugar syrup ("liquid sugar") rules out crystallization and thus eliminates a very important purification step so that ion exchange processes are highly appropriate here for a reduction of the ash content and elimination of small quantities of organic acids. For this purpose, a mixed-bed consisting of a strong base anion exchanger and a weak acid cation exchanger in a volume ratio of 2:1 is recommended; it results in the desired syrup qualities.

In general, however, the largest field of application for exchangers exists in the *beet sugar industry*. Two processing possibilities can be briefly characterized—demineralization and deionization of sugar syrup. In the former, the calcium and magnesium salts are exchanged for sodium by ion exchangers, thus preventing scale in the evaporators and pipelines. In the latter process, the yield of molasses is reduced by deionization of the sugar juice and the sugar yield is thus increased. Moreover, the use of macroporous anion exchangers permits decolorizing of sugar juices and syrups, and this is done on the largest industrial scale [25, 165]. This is the main field of application of ion exchangers in the beet sugar industry since economy reasons make an increase of the sugar yield of little interest. In most cases a carbon filter follows the ion exchanger for decolorizing purposes, but there are also installations in which an ion exchanger is used alone as the decolorizing stage with suitable precautions. Far more is known today about the nature of the colored components and their exchange properties [129], especially in The Netherlands, than at the time when the first ion exchangers were used to decolorize sugar syrups. On the basis of such systematic research, a number of commercial ion exchangers can be used as decolorizing resins. Amberlite IRA-900 alone or the combination of Amberlite XE-258 with Amberlite IRA-401 S leads to excellent results. The ion exchange resins decolorize sugar syrups as effectively as activated carbon but are simpler to use in the process.

Although sugar syrup decolorizing by ion exchange is a tested and accepted process, the further expansion of this technique is progressing slowly for the most diverse—and sometimes only conservative—reasons. There is greater interest in the treatment of liquid sugar syrups such as those used in the beverage industry, *i.e.*, for decolorizing and demineralization. The Amberlite acrylic-based resins and the weak base macroreticular Amberlite exchangers as well as the Amberlite adsorbents XAD-1 and XAD-2 can be used generally as ion exchange materials for this purpose. The same materials also serve to process xylose solutions for pharmaceutical applications.

4.7 Other Industrial Applications of Ion Exchangers

In addition to the sugar industry, ion exchangers are used in other areas of food technology [209], for example, for the treatment of milk, dairy products and fruit juices as well as for the stabilization and clarification of wine. Ion exchangers have indeed become fairly widespread in the *wine industry* [419]. The advantages reside in the treatment by a stable columnar process, so that uniform wine qualities are obtained, and in the possibility of removing potassium down to any desired value during wine stabilization. The cation exchangers at the same time also remove a part of the nitrogen compounds so that turbidity due to albumin is prevented. For the adjustment of the acid content an anion exchanger can be used simultaneously, since many wines require a reduction of the acid concentration after stabilization. Thus, a series-connection of one cation and one anion exchange filter results for the treatment of wines; together with ancillary equipment for the regeneration and washwater pipelines, these units form a complete ion exchange facility. Changes in the cation content are obtained by the form in which the cation exchanger is used (H^+- or Na^+-form) and by the use of exchangers with different degrees of cross-linking [156]. Even in a plant consisting of a single strong acid cation exchanger to stabilize argol, the final sodium content must be controlled carefully since too high a percentage leads to an unpleasant soapy flavor in champagne bases [141]. If the wine becomes too sour by ion exchange treatment, the champagne obtains a steely and hard flavor which is also a disadvantage.

In *milk and dairy products,* treatment with ion exchangers can facilitate the manufacture of low-sodium milk or the production of condensed milk and other dairy products by increasing the Ca-content. By treating with weak base anion exchanger in the OH-form, according to an American patent, the shelf-life of milk is reported to be improved without a loss of flavor [607]. Extensive investigations have been carried out for the removal of radioactive fallout products, particularly strontium-90, from the standpoint of possible nuclear disasters; an abundance of data and satisfactory results have been published on the subject [89, 301].

In the field of industrial chemistry the treatment of *formaldehyde solutions* for the manufacture of drugs, disinfectants, plastics, textile finishes, etc. is being carried out in numerous plants with ion exchangers for the removal of ionic impurities. For example, the weak base anion exchanger Lewatit MP 60 is well suited for the removal of formic acid. If metal ions are also to be removed, the strong acid cation exchanger Lewatit S 100 is used first, followed by Lewatit MP 60.

Industrially, ion exchangers can also be used as *dryers* [647] and for the *treatment of gases* [188]. If the impurities present in gases are capable of ion exchange and can be converted into an ionic form by hydrolysis, the process in principle is the same as in an aqueous medium provided the system has an appropriate moisture content. The purification of sulfur dioxide has been investigated repeatedly with different exchanger materials [126, 354]. The high resistance at the operating flow rates was found to be a disadvantage of the process. Detailed studies on waste gas treatment for the elimination of inorganic gases and organic compounds have led to a process, after many years of testing, which is now considered an economical and industrial possibility for the decontamination of waste gases [590]. The process which is carried out with a weak base anion exchanger is based on the exchange adsorption from the gas phase; its mechanism has been described.

Other industrial applications of ion exchangers could still be enumerated, but the analogy of the processes will be quickly recognized. Therefore, we will only briefly mention the use of ion exchangers for the purification of glycerol, phenol, acrylic acid and citric acid. In the field of organic chemistry and with the use of macroporous ion exchangers, adsorption processes play an important role which has been investigated in some cases and has been interpreted in a way similar to adsorption on activated carbon [243].

Chapter 5

Ion Exchangers in Analytical Chemistry

The suitability of ion exchangers for analytical purposes has been known for a long time (Siedler 1905); but their application was limited to isolated instances. Only with the studies of Samuelson [508, 509], who made use of organic exchangers, did ion exchangers become more widespread in analytical chemistry. He and his co-workers published a large number of studies dealing with theoretical and practical problems of ion exchange application in analysis. Griessbach made the suggestion to use synthetic ion exchange resins in chromatography. The separations and identifications of uranium fission products and rare earths as a part of the Manhattan Project and later success in the analysis of amino acids are other important chapters in this historical development.

Today, several thousand studies deal with the application of ion exchange in analytical chemistry. In the following an attempt will be made to give a brief description of the fundamental applications of ion exchangers in analytical chemistry.

5.1 Pretreatment of Analytical Solutions

In analytical work certain demands are made of the quality of water. Although water with a conductivity of $0.07 \cdot 10^{-6}$ can be obtained by ion exchange, some precautions must be taken in its treatment by the mixed-bed process. New ion exchanger samples should not be used. Generally, it is advisable to regenerate anion exchange resins several times with sodium hydroxide or sodium chloride for the removal of nitrogen compounds fixed to the exchanger due to the production process. A corresponding technique is used for cation exchangers, which are subjected to a hydrogen–sodium cycle. After complete regeneration, the resin particles must be carefully washed. Polyethylene or Plexiglas should be used for the columns. Care must be taken in packing the bed to prevent channeling along the wall or in the interior. Although the demineralized water obtained by an ion exchange process contains fewer ionic components than distilled water, organic components may remain in it and may lead to erroneous analytical values as a result of complexing with cations. If such complications need to be kept in mind, it is often advisable to make use of triple-distilled water; otherwise, the possibilities described in Chapter 4, Section 4.2.7, for the production of ultrapure water probably will also be sufficient for most analytical purposes. Other substances which are used as analytical reagents can also be purified of ionic components in the same manner as water. This is generally true for the solvents used in titration in nonaqueous media and for the removal of carbonate ions from standard alkaline solutions and metal ions from standard acid solutions. Gelatin is used as a protective colloid for the nephelometric determination of sulfate. Ion exchange can be used for the preparation of sulfate-free gelatin with a method developed by Honda [282]. As a result, the correction of analytical values with blank values is eliminated. The purification of hydrogen peroxide solutions is carried out in analytical procedures in which sulfate and phosphate stabilizers would interfere [110].

The concentration of ions from extremely dilute solutions on ion exchangers is a convenient and valuable aid for the formation of concentrated ion solutions in a small volume, as described earlier. No further discussion of this subject is necessary. The methods are simple and lead to relatively little contamination. In further illustration, a few examples can be cited. According to Inczédy [286], the enrichment of trace concentrations on ion exchangers is particularly suitable for water analysis. Talvitie and Demint [579] used a strong acid cation exchanger in the NH_4^+-form for a radiochemical determination of strontium-90 in water after complexing the Sr^{90} with ammonium ethylenediaminetetraacetate. Fluoride [306] and iodide [100] can be enriched and determined in water with anion exchangers. Other examples are the determination of molybdenum and vanadium with chelate ion exchanger in seawater [497], enrichment of iodine-131 from milk [300], determination of cesium in silicates [456], or special procedures for the determination of metals by X-rays [402] or by "precipitation-ion exchange" [502], demonstrating that the most diverse types of ion exchangers can lead to the same result with different techniques.

The preparation of standard solutions has already been discussed in Chapter 3, on preparative applications of ion exchangers. The reader is referred to that chapter.

For the removal of interfering ions, ion exchangers can be used in a dual way in analytical chemistry. Anions such as phosphate, borate, tartrate and oxalate, which interfere in some analyses, can be separated by fixing the cations which are to be identified qualitatively, particularly those of groups III, IV, and V of the periodic system, on a strong acid cation exchanger and eluting the anions in the form of their acids with 0.01 *N* HCl. After possible reduction of iron with ascorbic acid, the cations can be eluted with 4 *N* HCl and subjected to a conventional separation procedure. In contrast, interfering cations in the analysis of organic acids, in particular, are separated first by binding on a strong acid cation exchanger, followed by analyzing the acids in the washwater or concentrating them by reloading them on an anion exchanger, eluting them with formic acid, and determining them only after driving off the formic acid. A number of analytical procedures which operate according to this principle have been developed and described in the literature. A simple illustration is the separation of acetamide from ammonium acetate and its determination. NH_4^+ is retained from an aqueous solution on Amberlite IR-120 exchanger in the H^+-form in a column of 350 × 15 mm, while acetamide is eluted with water and can be determined in the eluate according to the Kjeldahl method [310].

A final example of the pretreatment of analytical solutions is the separation of electrolytes from nonelectrolytes. In principle, this amounts to deionization and is used when the nonelectrolyte is to be analyzed and the ionic components represent an interference. In the technique the analytical solution is first loaded on a cation exchanger in the H-form, and then on an anion exchanger in the OH-form or on a mixed-bed which can be used to greater advantage. The cations are thus exchanged by the cation exchanger and the forming acid is bound by the anion exchanger. This technique has found application primarily for urinalysis and for desalting of protein solutions.

5.2 Determination of the Total Salt Concentration

For the determination of the total salt content of a solution, the salt solution is passed through a cation exchanger and the forming acid is titrated. This is a highly precise method and can also be used for the characterization of standard solutions. Even if the salt to be analyzed is insoluable, the technique can still be used provided the forming acid is water-soluble. In this case the suspension is shaken with the exchanger, and the acid in the filtrate is titrated.

Strongly acid cation exchangers of the sulfonic acid type are used. During the operation, and particularly during washing of the column, the process should be carried out somewhat more slowly to obtain a quantitative elution of ions which are bound by general adsorption. With regard to the purity of distilled water and of CO_2, the same applies as to customary acid-base titrations. If the starting solution is acid, this original acid content must be determined in a separate experiment.

It is a prerequisite for the analysis that the salt react quantitatively with the exchanger, forming a stable acid that can be analyzed by alkalimetry. Consequently, this method cannot be used at all for hydroxides and only to a limited extent for carbonate, bicarbonates, and sulfites with the use of certain special procedures.

However, the technique is readily applicable to chlorides, bromides, iodides, nitrates, perchlorates, chlorates, sulfates, phosphates, bromates, iodates, periodates, borates, acetates, oxalates, salts of organic acids, and esters in simple solutions as well as in medical and pharmaceutical preparations. Samuelson reports a series of instructions to which reference should be made once more.

As an example, we shall describe the determination of some nitrates and perchlorates or rubidium and cesium according to Samuelson [511]. In an exchange column of 40×6 mm, a strong acid cation exchanger with a particle size smaller than 0.3 mm is transformed into the H-form and is carefully washed to neutrality. Of the analytical solution, 15 ml is loaded on the column and washed with 15 ml water. About 10 minutes is needed for the two steps. After elimination of CO_2, the filtrate is titrated with 0.5 *N* NaOH against Methyl Red. Table 22 shows the accuracy of the method.

Table 22

Analyses of rubidium and cesium salts on ion exchangers

Salt	*Given (meq)*	*Found (meq)*	
$RbNO_3$	0.162	0.162	0.162
$CsNO_3$	0.162	0.163	0.163
$RbClO_4$	0.187	0.187	0.187
$CsClO_4$	0.187	0.187	0.187

Examples in organic chemistry for the principle of total ion exchange as an analytical application of ion exchangers can be given in a similar way. In the performance or development of such procedures consideration must be given to the large size or organic ions and the influence of the exchanger pore size, the frequently slight water solubility of organic compounds, and of the nonaqueous or mixed solvent systems used for ion exchange as well as the tendency of organic molecules to be adsorbed on the exchangers with the result of more difficult elution. Simple organic acids such as formic acid, acetic acid, lactic acid, citric acid, succinic acid, novalgin, etc. thus an be liberated from their sodium or calcium salts on a cation exchanger and determined quantitatively. An example is the pharmaceutical application of the technique to determine barbituates, in which different dosage forms are loaded on Amberlite IRC-50 in dimethylformamide and are titrated in the eluate with 0.1 *N* sodium formate solution against Azoviolet [617].

For the determination of the alkali salts of organic acids, the exchange of anions for hydroxyl groups is carried out correspondingly on anion exchangers. The base content of the acids can thus be determined by a simple acidimetric titration. Procedures based on this principle have been developed particularly for the analysis of pharmaceutical substances; it is possible to determine tertiary amines on a microscale and alkaloids after precipitation with Reinecke salt in acetone.

5.3 Separation of Ions of Opposite Charge

Ion exchange techniques for the separation of ions of opposite charge is a simple cation-anion separation in which either the cations are separated first on cation exchanger or the anions on anion exchanger. Although it hardly needs to be mentioned because of its simplicity, the method has found important applications in the separation of ions of opposite charge which interfere with each other in an analysis.

However, this procedure has become far more important for the separation of metals, particularly when it is possible to transform one or several metals into stable anionic complexes and subsequently to perform a simple cation-anion separation. This can be achieved by converting the metal into an oxidation state in which it forms an oxo- or aquo-complex with the solvent or by the addition of a complexing agent. The latter may be a simple inorganic anion or a complicated organic complex former.

As an example of the first case, we may cite the analysis of crude phosphates requiring the determination of iron, aluminum, magnesium, calcium, and P_2O_5. The phosphates interfere with the analysis of the metals and the metals with that of the phosphate. In the ion exchange procedure, which can be readily used here, the sample is dissolved in acid and passed through a cation exchanger in the H-form. The cations are exchanged and the phosphate is found quantitatively in the filtrate so that it can be analyzed without difficulties. For the metal determination, the metals are quantitatively eluted from the exchanger with hydrochloric acid and are further purified for their individual identification.

An example of the second case is the separation of aluminum and titanium, which can be used for the determination of aluminum in titanium and its alloys. If both metals are present in 0.75 *N* HCl solution and the solution is passed through a strongly acid cation exchanger, titanium is eluted when the column is loaded with additional acid of the same normality and can be determined separately from aluminum determination, the metal is eluted from the column with 3 *N* HCl.

The effects that can be obtained by a further refinement of the method, particularly with regard to pH, can be demonstrated by the following example. Ethylenediaminetetraacetic acid, which is known as a complex former, binds different cations selectively on anionic complexes depending on the pH; these complexes then are not retained in a cation exchange column. In this manner, it becomes possible to separate lanthanum and thorium at pH 2.2, sammarium and iron at pH 1.8, yttrium and scandium at pH 1.35, and magnesium and aluminum at pH 4.0. For this purpose, the cation exchanger in the column is treated with the appropriate buffer solution until the eluted solution has the same pH-value as the inflow. An approximately 0.05 *M* solution (10 ml) in which the cation is to be determined is treated with 30 ml buffer solution and 15 ml 0.05 *M* EDTA solution. The desired pH is adjusted by the addition of acid or ammonia. The solution is then charged on the column and is eluted at a flow rate of 8–10 ml/min. The column is washed three times with 30–40 ml buffer solution. The bound ion is eluted by 90 ml 4 *N* HCl.

The same principle forms the basis of a method for the separation of metal ions in which a certain cation forms a complex of negative charge; this complex is sorbed by an anion exchanger and is thus extracted selectively from the external medium, while the uncomplexed cation passes through the anion exchanger.

This will be illustrated by an example. In contrast to bismuth, uranium(IV) forms sulfate complexes at pH 1.0–1.5. For the separation of these two metals, a strong base anion exchanger in the SO_4-form is used. The uranium removed from this ample solution can be eluted from the column with dilute perchloric acid (1:1) and analyzed by spectrophotometry.

5.4 Ion Exchange Chromatography

5.4.1 *Principles*

Ion exchange chromatography is a method for the separation of ions of equal charge on ion exchangers by the different affinity of ions for the exchanger material. It is an important analytical method and has led to interesting results in inorganic as well as in organic chemistry. From the standpoint of the different forms of chromatography, ion exchange chromatography is a liquid-solid technique in which the ion exchanger represents the solid phase [160].

The separation effect of ions in ion exchange chromatography is based on the selectivity of the exchanger materials, *i.e.*, on the differences of the exchange potential of cations on cation exchangers and of anions on anion exchangers. As the ions are selectively sorbed, certain ions on the other hand displace each other in a consistent manner depending on the position of the equilibrium prevailing between the solid ion exchanger and the ions forming a part of the total system, their concentrations and, to a lesser degree, the temperature.

For the present, all types of ion exchangers have been used in ion exchange chromatography. While it can nearly be said that synthetic ion exchange resins have given practically equal service in ion exchange chromatography and industrial applications, other types have been developed almost exclusively for ion exchange chromatography. The various paragraphs describing ion exchanger types have already made reference to the special properties which make them particularly useful for chromatography. Special experience is frequently necessary for the correct selection of an ion exchanger for chromatograhic separation; this has been fully discussed by Kohlschütter for silica gel [327].

The routes by which the separation of ions bound to an ion exchanger takes place may differ. Accordingly, different techniques must be distinguished on the basis of the procedures used for ion exchange chromatography—*frontal chromatography, displacement chromatography*, and *elution chromatography*, on one hand, and *ion exchange chromatography with organic solvents*, on the other.

Strictly speaking, ion exchange chromatography with complexing agents and with organic solvents are not separate procedures since they are mainly carried out according to elution chromatography. For didactic reasons, however, it seemed appropriate to subdivide the subject further since the special characteristics of these important and very frequently used techniques can thus be made more easily understood.

The simplest chromatographic procedure is *frontal chromatography*. It consists of the continuous injection of the solution which contains the components that are to be separated at the head of the column. The individual components are chromatographically separated in order of increasing selectivity while migrating through the column. The appearance of individual ions in the eluate is followed by any suitable detection method.

The separation of sodium and potassium by frontal chromatography on a cation exchanger in the H-form is shown in Figure 75. While the solution of sodium and potassium is continuously charged on the exchanger, the H^+-ions are displaced from the exchanger according to

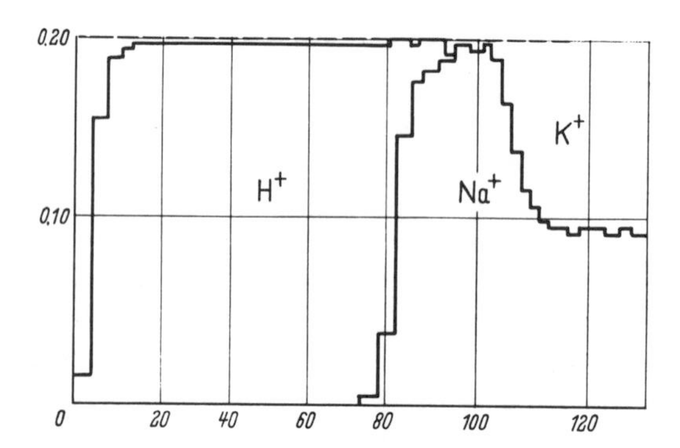

Figure 75. Separation of sodium and potassium by frontal chromatography (according to D. Reichenberg, unpublished).

increasing selectivity from H^+ over Na^+ to K^+, exchanged first for Na^+-ions and K^+-ions, and moved forward in a sharp front. Of sodium and potassium, Na^+ is sorbed least by the exchanger, and it therefore migrates most rapidly from the start; in an enriched zone, it immediately follows H^+. Na^+ then emerges in pure form from the column until the upper sections of the column have established an equilibrium with K^+-ions. Since solution with Na^+- and K^+-ions is continuously fed into the column, so many K^+-ions have already been extracted from the upper layers of the separation column that further exchange is no longer possible, and the solution finally passes through the column without a change in composition.

As shown by this example, frontal chromatography furnishes only one pure fraction—the component of the starting solution which is least retained (Na^+). The second component (K^+) can also be detected by the appearance of its front, but it is always contaminated with the first component. The degree of separation of the first component can be increased by lengthening the column, but in the long view this method is not suited for quantitative and preparative separations. However, frontal chromatography can be used for qualitative analyses.

In *displacement chromatography,* the ions to be separated are charged on the column and exchanged for the counterion of the ion exchanger near the upper end of the column. To perform a chromatographic separation, this step is followed by washing with an electrolyte solution which contains an ion that is preferentially sorbed by the ion exchanger and consequently displaces all other ions.

As an example, it shall be assumed that the ion exchanger was in the H^+-form and that sodium and potassium were to be separated,

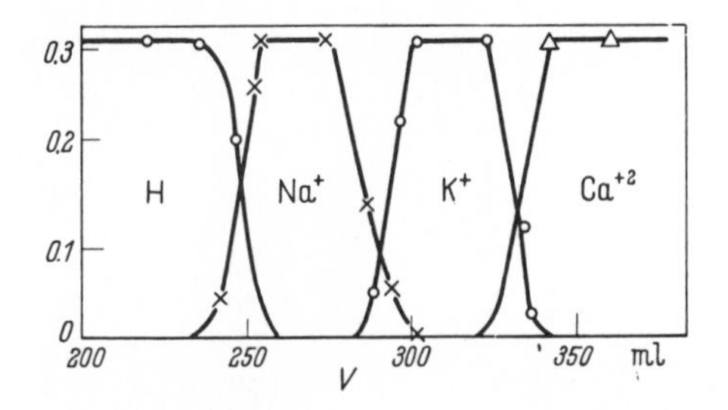

Figure 76. Separation of sodium and potassium by displacement chromatography [132].

with the selection of a calcium chloride solution as the displacement medium, as shown in Figure 76. The selectivity of the ions involved is $H^+ < Na^+ < K^+ < Ca^{++}$, i.e., the first counterion is bound most weakly and the one serving for the displacement is bound most strongly to the exchanger. The chromatogram develops as a result of the displacement of all other counterions from the uppermost layers of the column by the calcium ions and their advance in front of the latter in the course of development of the chromatogram with the formation of a sharp front. Depending on the sequence of their selectivity, the H^+, Na^+, and K^+ ions are bound more or less strongly by the exchanger and displace each other; in this manner they fall into order, forming the zones with self-sharpening fronts. The ions which are to be separated, Na^+ and K^+, appear in the eluate in the order of their selectivity. A certain degree of zone-overlapping cannot be avoided.

This method furnishes pure and mixed fractions, with the latter containing the two ions which are adjacent in their selectivity sequence. If desired, the mixed fraction can be purified further by repeating the entire separation. The method is therefore mainly suited for preparative separations.

In *elution chromatography,* the ions to be separated are charged on the column as in displacement chromatography and are exchanged for the counterion of the ion exchanger at the top of the column. For the performance of chromatographic separation, the column is then washed with an eluant containing the same ion which initially formed the counterion of the entire exchanger. Although the eluant with its counterion overtakes the bound ions, which have a higher affinity for the exchanger, a separation of the ions nevertheless occurs in the course of further elution, and their sites are occupied by the counterion of the eluant and thus by the counterion of the original loading state of the exchanger. The ions which are to be separated form zones in the course of their migration through the column as a function of the degree of binding on the exchanger. The zones broaden increasingly during their migration and emerge from the column together with the ion of the eluant.

As an example, the separation of sodium and potassium on a cation exchanger in the H^+-form with the use of HCl as eluant can be considered. Figure 77 shows the result of a sodium and potassium separa-

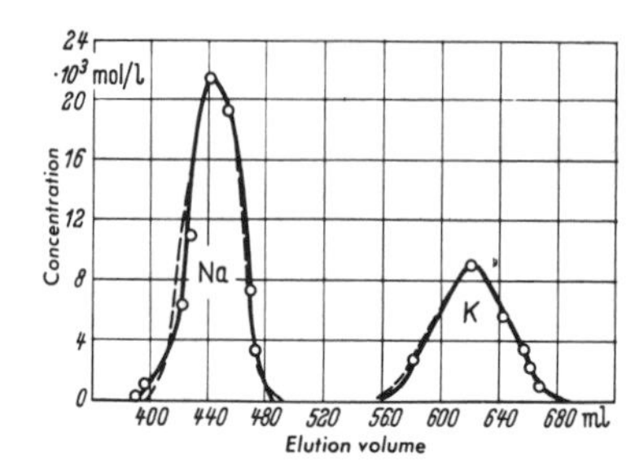

Figure 77. Separation of sodium and potassium by elution chromatography [54].

tion. The two ions appear as sharp symmetrical bands in the eluate together with the eluant HCl. In this separation the first 370 ml eluant are discarded, after which 160 ml are collected in fractions which contain sodium and finally 190 ml containing potassium. At the end of such a separation, pure eluant again emerges from the column. The ion exchanger itself does not change its loading state during the entire separation and it can therefore be used immediately again without a special regeneration. The separation takes place in the selectivity sequence of Na $<$ K. The conditions to prevent overlapping of the bands must be determined empirically or theoretically.

Thus, under suitable conditions elution chromatography furnishes pure fractions of the components which are to be separated and which appear together with the eluant. The degree to which this circumstance proves to be an interference naturally depends on each individual case. The main field of application of the technique is for analytical separations.

A variant of elution chromatography is *gradient-elution chromatography*, consisting of the use of a continuously changing eluant which is prepared in special equipment outside of the column.

The gradient-elution technique in particular suppresses the broadening of the chromatographic zones, known as tailing, and thus improves separations. Alm, Williams, and Tiselius [10] gave a detailed description of the general principles of this method. Numerous other authors later made use of it in ion exchange chromatography and applied it increasingly, especially when the acid concentration of the eluant needed to be increased for the elution of more strongly bound components. In practice, one proceeds so that the eluants are consecutively charged into a mixing vessel provided with a magnetic stirrer. If the mixing vessel first contains a solution of one concentration and a solution of higher concentration is continuously added, the concentration will first increase sharply and finally will slowly approach that of the added solution. The composition of the eluant can be madified by different methods by a suitable adjustment of the quantities in which the solutions flow into and out of the mixing vessel [162].

Equipment for gradient elution may have different designs but all of these operate by the same principle. A reservoir with contents of constant composition is connected with a mixing vessel, the contents of which have a different concentration. The volume of the mixing vessel is dimensioned such that a certain development of the gradient is obtained. The concentration of the solution in the two vessels represents the limit values of the gradient. When the solution from the reservoir enters the solution of the mixing vessel, an elution gradient is produced since the continuously changing eluant from the mixing vessels enters the separation column until all the solution of the mixing vessels has been consumed and the eluant has the composition of the solution in the reservoir.

The variables available for the development of a given gradient are the compositions of the two solutions in the reservoir and in the mixing vessel, the volume ratios of the two vessels, and the flow rate ratio of the reservoir to the mixing vessel and of the mixing vessel to the separating column.

These variables can be determined mathematically more or less accurately and have led to the design of a number of different systems which are described in the literature. Figure 78 is a schematic repre-

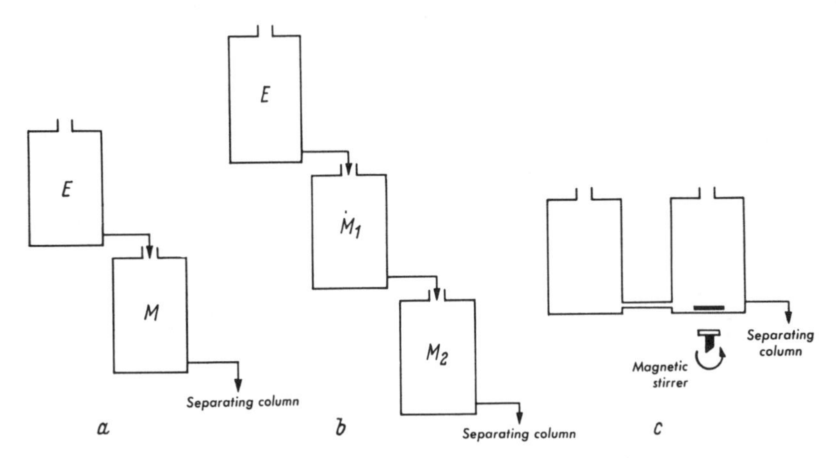

Figure 78. Arrangement of equipment for gradient elution (schematic). (a) Makeup vessel and mixing vessel of constant volume; (b) makeup vessel with two mixing vessels M_1 and M_2 of contant volume; (c) from two communicating vessels.

sentation of only three possibilities. Gradient-elution instruments have been extensively mechanized and automated [498, 530]. Gradient development devices are available on the market to set up a complete chromatographic system (for example, Serva-Elumant).

Without further discussion of the many possibilities of a theoretical treatment of gradient elution, the advantages which can be observed empirically can be understood readily in a study of the distribution coefficients and separation factors, which are explained in Section 5.4.2.

If the relationship between the distribution coefficient K_D, which represents the basis of the total chromatographic process, and the elution gradient is considered, very important information is already apparent. First of all, this relationship demonstrates the possibility of influencing asymmetrical bands, *i.e.*, those which are customarily called bands with a diffuse or sharp trailing or leading front. In the case of bands with a diffuse trailing front, we speak of tailing, which easily leads to overlapping and therefore to a smeared separation of adjacent bands. The appearance of a chromatographic peak depends on the isotherm on which it is based and thus on the distribution coefficient K_D of the chromatographic sample. If K_D is concentration-dependent and increases with increasing concentration, the zones of higher concentration migrate more rapidly than those of lower concentration. The result is a band with a sharp leading front and tailing. If K_D decreases with increasing concentration, the zones of high concentration migrate more slowly than those of low concentration. This leads to a band with a smeared leading front and a sharp trailing front.

In elution chromatography (and this is true for adsorption as well as ion exchange chromatography), the distribution coefficients are functions of the eluant concentration. For the special case of ion exchange chromatography in which the distribution coefficient is dependent on the acid concentration used, this also results in a pH-dependence. Thus, for example, if K_D decreases in its concentration-dependence with increasing acid concentration, then an increase in acid concentration will result in a decrease of tailing in accordance with the above considerations.

When we consider these conditions on the basis of the diagram of Figure 79, it can be easily understood how tailing is suppressed as a result of a gradient developed along the column. The substance which forms the smeared trailing front is continuously displaced toward the center of the zone. Consequently, the zone becomes increasingly symmetrical and in the ideal case, a Gaussian curve is recorded for its elution from the column. Generally, in practice it is necessary to use fairly steep gradients to prevent tailing.

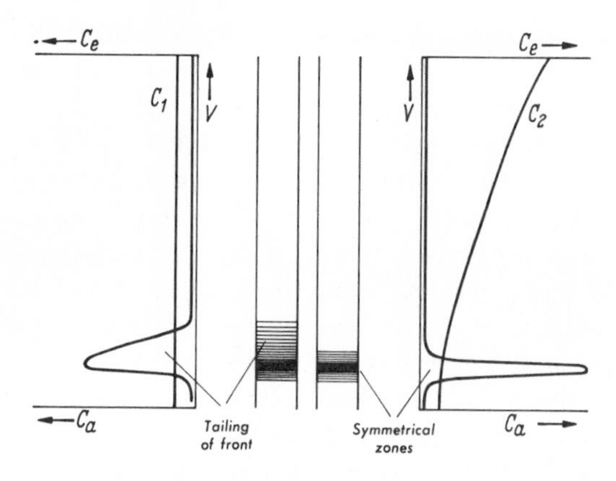

Figure 79. Illustration of chromatographic zones and eluant curve. V, volume of eluant; C_1, curve of constant eluant; C_2, curve of gradient; C_a, concentration of adsorbed substance; C_e, concentration of eluant.

The second application of gradient elution is in the case when two samples have very different K_D-values so that they will be widely separated during elution. By the application of gradient elution the subsequent component is moved more rapidly through the separating column, and the total time for development of the chromatogram is reduced. Naturally, one may not forget that this also involves the risk of a poorer separation by overlapping of successive components.

This already indicates the influence of gradient elution on the separation factor α. Two components with the separation factor $\alpha = 1$ cannot be separated by gradient elution even if their distribution coefficients vary in the same direction. On the other hand, if the elution gradient influences the distribution coefficients in the contrary direction, α will differ increasingly from 1, thus improving separation when the components move through the separating column. If only one of the two K_D-values of two components is charged, the separation by gradient elution will be sufficient of α thus becomes sufficiently different from unity.

In accordance with the nature of ion exchange chromatography, the components of a mixture are separated if their selective behavior with respect to the exchanger is sufficiently different. It has been found, however, that numerous ions whose chromatographic separation is of interest exhibit the same selectivity. A considerable amount of time is necessary to arrive at sufficient separations in that case. Compared to the quantities of sample available for the separation, long columns and large quantities of eluant need to be used. In such cases, the technique of *ion exchange chromatography with complex formers* has been very successful; the ions of the mixture form complexes of different strength which then exhibit the required selectivity differences.

This method will be illustrated with the example of the separation of sodium and potassium. Schwarzenbach reported that in connection with sodium and potassium, uramyldiacetic acid forms stable complexes with lithium and sodium but not with the other alkali ions. The compexing behavior of these two ions in uramyldiacetic acid as a function of pH shows a sufficient complex formation at a pH of >9. However, as a result of this behavior, the operating conditions in the column are predetermined, since the exchangers cannot be used in the acid form because of the pH-dependence of complex formation. Consequently, they need to be transformed into a suitable salt form where the affinity for the exchanger must be lower than the affinity of the alkali ions which are to be bound on the column. For example, the NH_4^+-ions cannot be used since the affinity of NH_4^+ for the exchanger is higher than that of Li^+ and Na^+; on the other hand, dimethylamine and tetramethylammonium hydroxide satisfy the set conditions. Figure 80 shows the result of an Na/K separation on a fulfonic acid exchanger

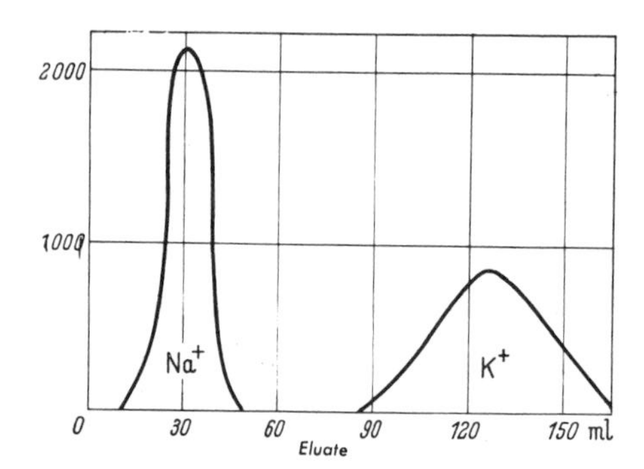

Figure 80. Separation of sodium and potassium with uramildiacetate as complex former [109].

(Amberlite IR-120) with dimethylammonium–uramyldiacetate. It is worthy of note that small quantities of eluant are needed compared to other methods, so that the separation is highly accelerated.

As such, there is no limit to suitable complex formers, and their choice in an individual case depends on the state of research in the field of complex chemistry. For the present, the most commonly used organic complex formers have been citric acid, lactic acid, ethylenediaminetetraacetic acid, nitrilotriacetic acid, α-hydroxybutyric acid, thenoyltrifluoroacetone, etc., while inorganic complexing agents are the halogen and pseudohalogen complexes as well as the complexes of nitric acid, phosphoric acid, sulfuric acid, and perchloric acid.

In practice, the procedure involves charging the exchanger with the ions in the complexing solution and then washing it with the complexing solution or subjecting the ions which are bound to the exchanger to chromatography with the complexing solution. However, even pure solvents can act as complex formers, as in the case of hydrochloric acid, for example. Depending on the conditions, the concentration of complex former and the pH-value of the solution, a displacement or elution chromatography process finally results.

Ion exchange chromatography with organic solvents is a technique which most closely resembles partition chromatography. It is known from equilibrium experiments concerning the distribution of mixed aqueous solvents on ion exchangers that in the case of a mixed solvent, the resin phase is primarily aqueous and the surrounding solution is primarily organic. This creates the conditions for partition chromatography between two phases.

As in the case of the previously described techniques in ion exchange chromatography, we will again discuss the separation of alkali metals as an example. Figure 81 shows the results of a separation of lithium,

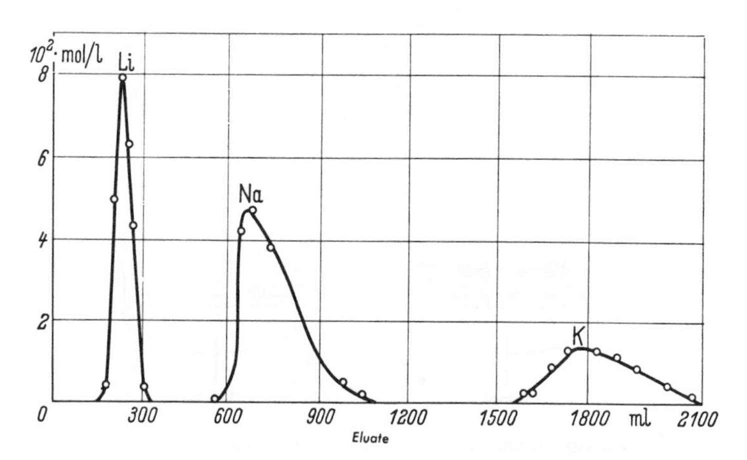

Figure 81. Separation of lithium, sodium and potassium by ion exchange chromatography with organic solvents [136].

sodium, and potassium. The ion exchanger utilized was pretreated with a mixture of acetone and water (4:1) and the chlorides of the above alkalies were then loaded on the exchanger. To develop the chromatogram, the column was washed with 0.7 *N* HCl in the same acetone–water system. In accordance with the increasing tendency toward ion pair production of $K^+ < Na^+ < Li^+$ and low dissociation of lithium salts in acetone, lithium appears immediately without having participated in the exchange process. The other two elements appear in the sequence of sodium and potassium according to increasing selectivity but with a considerably larger quantity of necessary eluant, since ion exchange is inhibited by a distribution between water and acetone. This actually negative result—since a chromatographic separation should be carried out in the shortest possible period of time—demonstrates the influence of the organic solvent on the separation.

Ion exchange chromatography is used to advantage when a superimposition of partition on ion exchange leads to a better separation of those substances (*e.g.*, organic acids) whose molecular structure with a superimposition of hydrophilic and hydrophobic properties allows the partition coefficient between an aqueous and an organic phase to play a role. Even when two metals form the same complex, a better separation can be obtained with ion exchange chromatography by partition with the use of organic solvents; this is also true for substances (acids and bases) with only a limited water solubility [65].

In addition to the above techniques, other effects of ion exchange have also been utilized for separations by ion exchange chromatography. A brief review of these will follow below.

Even the phenomenon of *general adsorption on ion exchangers* can be used for chromatographic separations. Aromatic compounds in particular have a high adsorption affinity. Moreover, transition phenomena exist between a weak ion exchange and a strong general adsorption, with the latter effect being capable of leading to a reverse separation since molecular adsorption and desorption are very slow processes compared to ion exchange.

The *ion exclusion method,* also known as leading electrolyte method, serves for the separation of ionized substances from nonionized or only weakly ionized substances when both are present in aqueous solution. Customary ion exchange resins are used, both cation and anion exchangers, and highly ionized exchangers generally produce better results than those of low ionization. It is only important for the components that are to be separated to be inert with respect to the exchanger material. Moreover, three other demands must be made of the exchangers: (1) The fixed ion concentration in the interior of the resin particles must be high; (2) the water content of the exchanger must be sufficiently high so as to result in a sufficient capacity; and (3) equilibrium must be rapidly attained, *i.e.*, diffusion must be as high as possible.

For an understanding of the processes occurring during ion exclusion, it must be kept in mind that a column packed with ion exchange resin contains three phases—the solid network of resin particles, the liquid phase within the particles (resin liquid), and the liquid which surrounds the particles (dead space). The behavior of ionized and nonionized molecules in these three phases differs. While organic nonionized components tend to distribute in equal concentration over the resin liquid and the dead space, the concentration of ionic substances remains higher in the dead space and lower in the resin liquid as a result of the Donnan membrane effect. Thus, in a mixture containing ionized and nonionized components and eluted from the top of an ion exchange column by means of water, the ionized component will reach the base of the column first, since essentially it needs to displace only the dead space of the column. The nonionized component, on the other hand, must displace the dead space as well as the resin liquid, so that it will appear in the eluate only when the ionized component has already emerged from the column. A true chromatographic separation which finds its theoretical explanation in the plate theory has thus been achieved.

Ion exclusion has proved to be a method for the chromatographic deionization or at least removal of the major fraction of ionic components from organic products with the simple technique of ion exchange. Typical separations include those of acids and salts, of glycerin, alcohol, and amino acids, of highly and weakly dissociated substances (hydrochloric acid and boric acid), of mono-, di-, and triethanolamine or mono-, di-, and trichloroacetic acid, etc. The ion exchange resin should always be used in the same ionic form as that of the ionized component of the substances to be separated. The operating conditions which influence separation are flow rate, usable quantity, ionic strength, concentration of nonionized component, temperature, and size of nonpolar molecules. General data can be given only to the extent that the flow rate can be very high compared to that of other techniques of ion exchange chromatography. The volume of usable quantity must be smaller than the volume of resin liquid; thus, the degree of separation increases with a decreasing ion concentration while the concentration of nonionized component is less influential and may be higher. The influence of temperature is not great; however, large molecules generally cannot be separated because of their size and relative immobility. Operating conditions for a given problem must be developed for each case.

The *ion retardation method* is used for the separation of water-soluble substances with amphoteric ion exchangers in a column process.

The exchangers employed for the purpose adsorb cations and anions. However, since the functional groups are very closely packed, they partially neutralize each other so that the adsorbed ions are only weakly retained and can be displaced again by water.

Salting-out chromatography leads to the chromatographic separation of water-soluble nonelectrolytes on ion exchange resins as the stationary phase and aqueous salt solutions as the mobile phase. The mechanism of this process is an alternative solvation of the functional groups of the exchanger end of the components which are to be separated.

Chromatography on specific ion exchangers takes place when the exchanger is rendered more or less specific by a suitable modification of the active groups of the exchanger material. The principle of the method was developed by Samuelson and was tested with the separation of carbonyl compounds on bisulfite ion exchangers. A redox chromatography on redox exchangers can also be tested on the basis of similar considerations.

Separations based on the sieve effect are made possible by the pore structure of the ion exchange materials. With a total sieve effect, the entry or penetration of one of two components of different size is completely impossible, while with a partial sieve effect both components diffuse in or through the molecular sieve at different velocities. This results in possibilities of applying ion exchange resins as ion sieves as well as molecular sieves.

5.4.2 Theory

As in every scientific discipline, it is also important in ion exchange chromatography to estimate the available possibilities on the basis of theory and thus to derive the feasibility of separations. Since elution chromatography is the most commonly used technique among those described, only its theory will be more fully discussed here. In principle, two theories are used—the plate theory and the theory of the elution constant; these will be briefly summarized below.

In accordance with the definition of chromatography in general, the *plate theory* is based on the distribution coefficient K_D for a substance B which is in equilibrium distribution between an exchanger and a liquid; the coefficient is defined by:

$$K_D = \frac{\text{quantity of } B \text{ per unit of weight of exchanger}}{\text{quantity of } B \text{ per unit of volume of the liquid}} = \frac{\overline{C}}{C}$$

A graphic plot of K_D leads to three types of isotherms, one linear and two nonlinear, since the distribution coefficient may or may not be independent of concentration. Figure 82 shows these three possibilities. The same figure depicts the resulting respective chromatographic bands compared to the isotherms; thus, a symmetrical band corresponds to isotherm A, a band with a sharp leading front corresponds to the convex isotherm B, and a band with a sharp trailing corresponds to concave isotherm C.

According to investigations on the behavior of ions between an exchanger and its surrounding liquid medium, the distribution coefficient on ion exchangers is linear for a low external concentration and low capacity. In ion-exchange elution chromatography, an ion which is to be separated represents less than 3% of the resin capacity. In other words, under the conditions involved, the isotherm range in which linearity has been approximately preserved is always present.

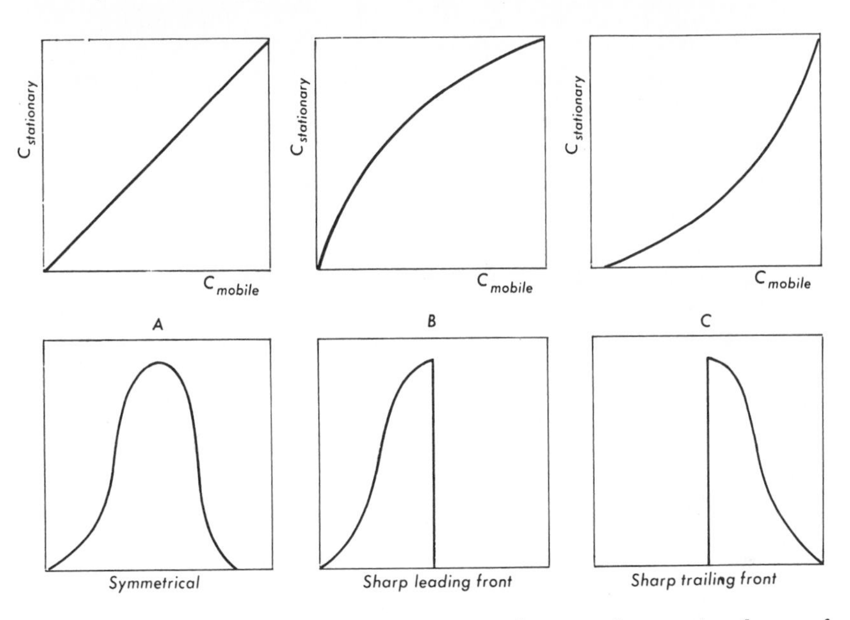

Figure 82. The three possible exchange isotherms and respective shapes of the elution bands.

On elution of an ion from a column, the volume of eluant required depends on the rate with which the ion band migrates down the column. On the other hand, this volume depends on the distribution of the ion between the exchanger and eluant, or again on the distribution coefficient. Therefore, the band is characterized by retention volume $\overline{V}_i$, which represents a characteristic value for a given exchanger under fixed conditions. The retention volume is the most important parameter in ion exchange chromatography. It can be used in a simple and known manner for a qualitative identification of a component of a mixture. Its significance can be recognized from Figure 83. It follows from this figure that the retention volume $\overline{V}_i$ of a component i is given by the quantity of eluant necessary to elute the band maximum reduced by the volume, the interstitial volume, V of the exchanger column. The interstitial volume V is the volume of liquid, outside of the exchanger

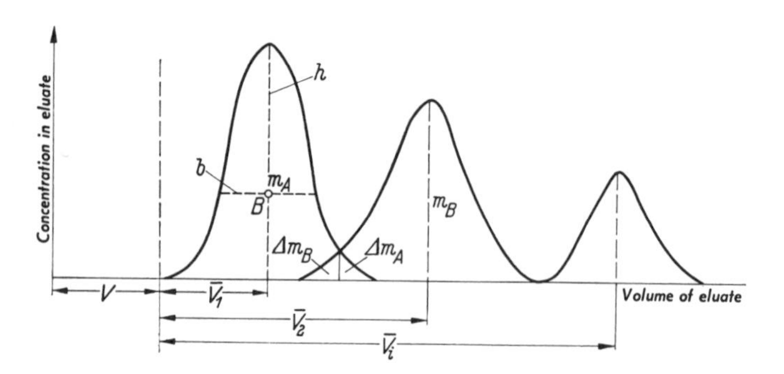

Figure 83. Schematic chromatogram.

particles in the column, which is already present prior to the start of elution in the column and must therefore be subtracted.

In the chromatographic separation of two ions, each with a given retention volume $\overline{V}_1$ and $\overline{V}_2$, the position of the two band maxima with respect to each other is a measure of the separation, as can be seen in Figure 35. Therefore, the higher the ratio of $\overline{V}_2:\overline{V}_1$, the better is the separation effect. This ratio $\overline{V}_2:\overline{V}_1$ is the separation factor α. It is described by the following simple relation to the distribution coefficient:

$$\alpha = \frac{\overline{V}_2}{\overline{V}_1} = \frac{K_{D_2}}{K_{D_1}}$$

For the separation of two components, the value of α must differ from unity. The necessary value of α for a quantitative separation depends on the band width in order to rule out overlapping of two neighboring bands.

The main problem of a separation by ion exchange chromatography therefore is to find the conditions under which α has a useful value or, if this is not immediately the case under simple conditions, to find means of influencing α in a direction favorable to separation.

Since α is the ration of K_{D_2} and K_{D_1}, the latter are determined under the expected elution conditions on the selected resin. The experimental procedure consists of covering a known quantity of exchanger in a small bottle with a known volume of eluant, adding a small quantity of the ion (less than 0.1 meq/g resin), and equilibrating by shaking, suitably at least at two concentrations. As a rule, shaking for 24 hours is sufficient. The α value is then calculated from the analytical values of K_D.

With the exception of the first studies of ion exchange chromatography, all investigations for the development of such separations are based on a determination of distribution coefficients. It is therefore worthwhile to review the procedural possibilities once more. Naturally, the distribution coefficients for cations were first determined on cation exchangers, both in water and in inorganic acid media. The extent of such studies was limited either to specific separations of interest or was expanded systematically to as many elements as possible with a view of developing all-inclusive separation principles.

In the case of inorganic acid media, the distribution coefficient is determined best by plotting its logarithm as a function of molarity, so that diagrams are obtained such as those shown in the examples of Figure 84 which were arbitrarly selected from the literature. These

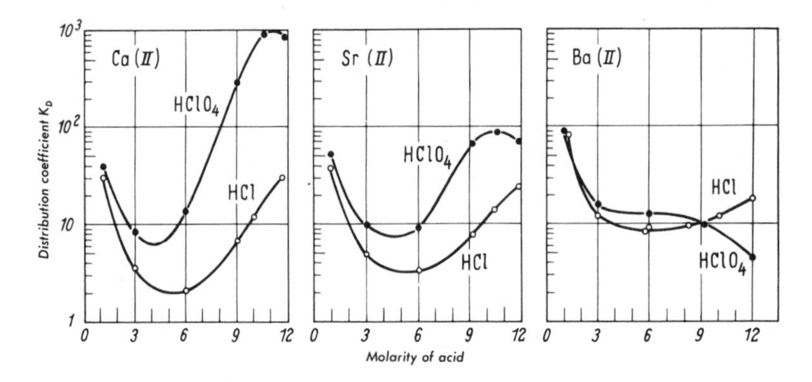

Figure 84. Exchange properties of Ca(II), Sr(II) and Ba(II) from HCl and $HClO_4$ solutions on a strong acid cation exchanger [449].

curves can be used advantageously for a discussion of the exchange properties and furthermore, can be arranged in analogy to the periodic system, as was done by Nelson, Murase and Kraus [449] according to Figure 85. This procedure permits one to compare the results obtained for a given system at one glance and to determine the possibilities of separation. Such extensive data have been compiled to serve as a basis for elution chromatography of cations on cation exchangers in the most diverse media [404, 411, 448, 449, 569, 572]. Naturally, the distribution conditions were also investigated systematically for anions on anion exchangers and the conditions for separation were read from the results. Although the studies are not as numerous on this subject as on cations, the correct method has been demonstrated in a number of publications [114, 195, 281]. The procedure used by Kraus and Moore for the separation of cations on anion exchangers might appear contradictory at first glance [346]. However, later this method became of great importance when a large number of systematic studies were carried out on the exchange properties of cations on anion exchangers.

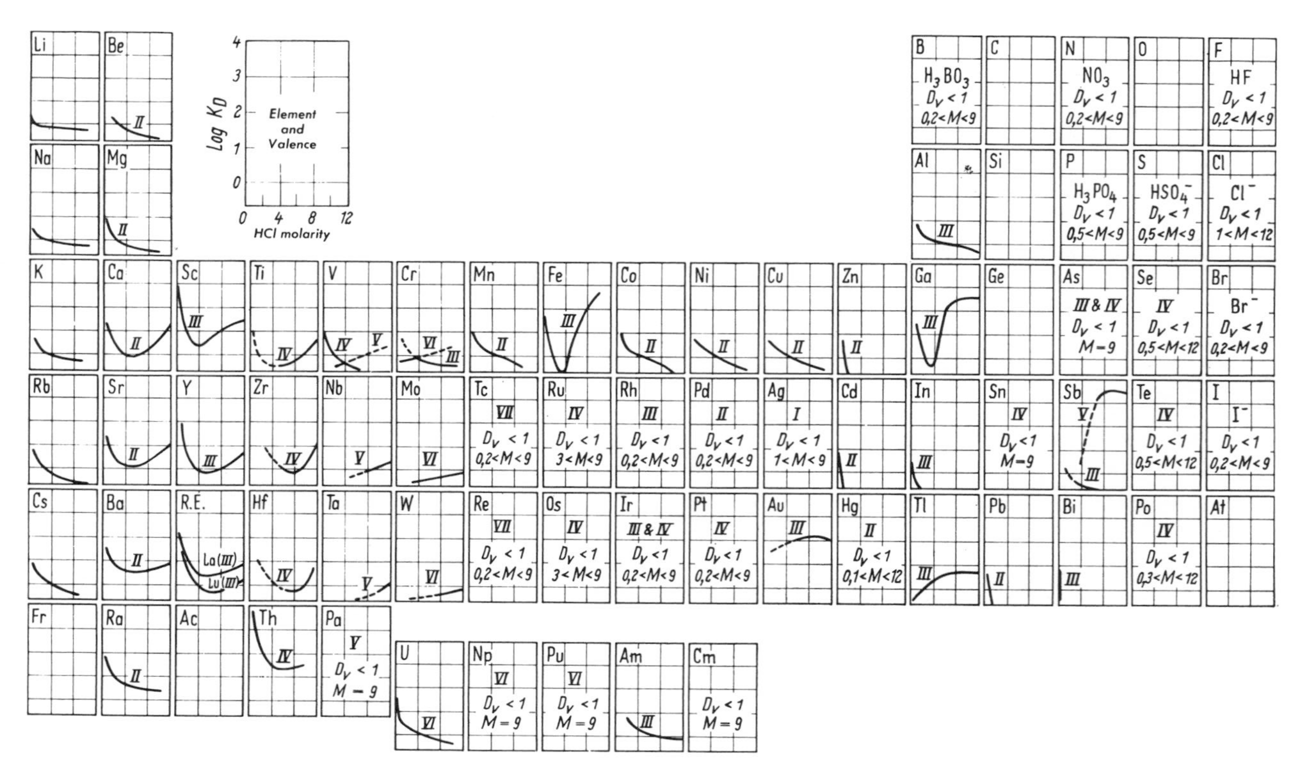

Figure 85. Exchange properties of cations on cation exchangers in concentrated HCl solutions. Chart based on periodic system.

Naturally, such a technique is based on transforming the cation into a complex anion so that it becomes a part of chromatography with complexing agents. In this connection, the detailed study of Kraus and Nelson [348], from which we derived Figure 86, is particularly informative. This illustration represents the same type of survey as that given above for cations on cation exchangers, and it can be used like the latter to work out separations.

In addition to hydrochloric acid, other media have also been investigated with equal thoroughness; their results, whether they refer to all or only to some of the elements, have to be evaluated for their informative value with regard to distribution coefficients so that far-reaching documentation will be obtained to solve separation problems [176, 284, 290, 379, 571, 591].

When the observation that distribution coefficients are favorably influenced by the addition of water-miscible organic solvent was applied to ion exchange chromatography of cations on anion exchangers, an additional possibility was developed to expand first the HCl system [194]. Methyl, ethyl, or isopropyl alcohol takes up metals at lower HCl concentrations, and the comparable distribution coefficients are higher. A systematic extension of this variant furnished additional data for the separation of cations on anion exchangers [11, 171, 193, 629] and led to a large number of very satisfactory separations in analytical as well as preparative chemistry, based on ion exchange chromatography in both cases.

If the distribution conditions of the system in solution which is to be chromatographed are known, the separation is applied to a column. The factors determining the appearance of the elution bands are best determined by plate theory in which the chromatographic column is compared to a fractional distillation column. According to the derivation of Glueckauf [225], the following relation:

$$n = 8\,\frac{(V + \overline{V}_i)^2}{b^2}$$

exists between the width of an eluted peak and the number of theoretical plates of the separation column.

In the above, V is the mentioned interstitial volume of the column, V_i is the retention volume of the investigated peak i, and b is the peak width at point 0.368 of the peak height. This formula therefore permits a calculation of the number of theoretical plates of a separation column from the chromatograms obtained.

If we now divide the length L of the column by n, we obtain a value H, which is usually known as the HETP (height equivalent to a theoretical plate):

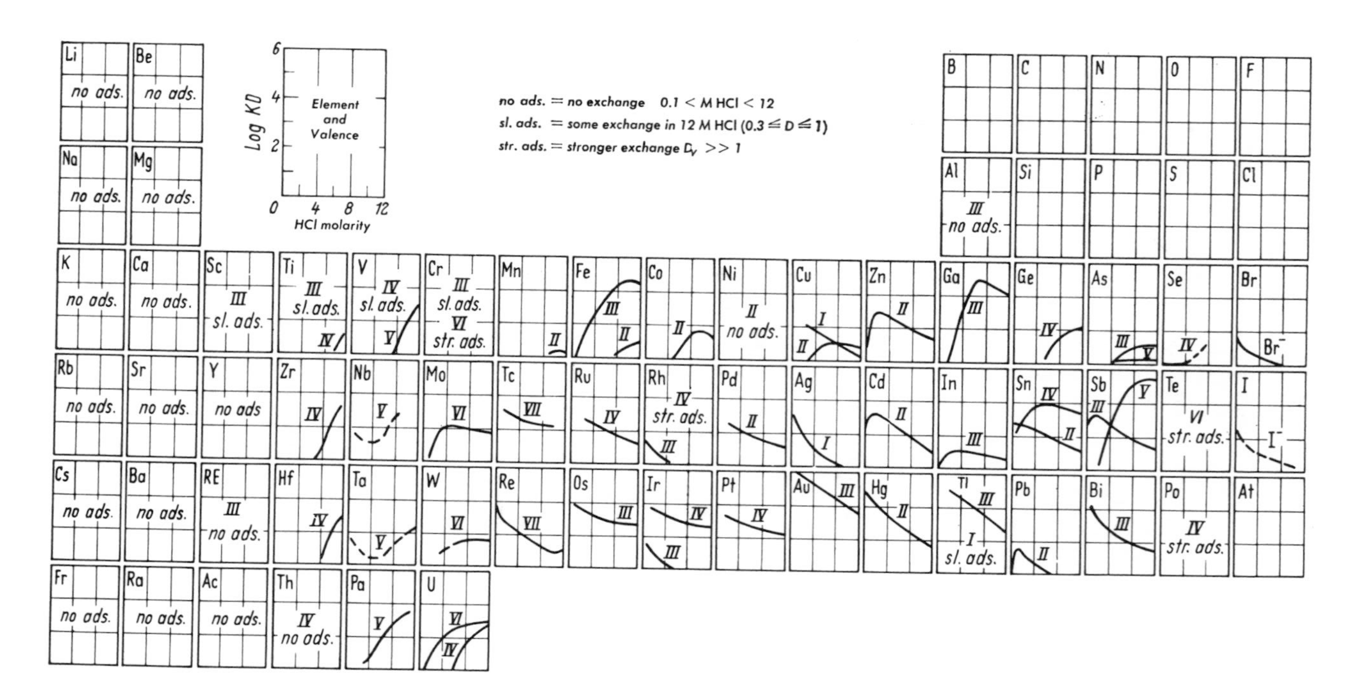

Figure 86. Exchange properties of cations on anion exchangers in concentrated HCl solutions. Chart based on periodic system.

$$H = \frac{L}{n}$$

This relation together with the above equation for n directly indicates that a smaller plate height leads to a better separation, since each peak becomes more narrow and gives less rise to overlapping. Inversely, therefore, the greater the number of theoretical plates provided for a separation column, the more efficient it must be.

However, H is a function of the column variables. Using a material balance, Glueckauf also found that the following equation applies to H:

$$H = 4r + \left[\frac{K_{D_b}}{(K_{D_b} + f)^2} \cdot \frac{0.142r^2F}{\bar{D}}\right] + \left[\frac{K_b{}^2}{(K_{D_b} + f)^2} \cdot \frac{0.266r^2F}{D(1 + 70rF)}\right]$$

where r is the particle radius, K_{D_b} = distribution coefficient in the exchanger bed (approximately 0.3 k), f = fraction of solution volume in the bed (about 0.4), F = linear flow rate in ml per unit of cross section and unit of time, $\bar{D}$ = diffusion coefficient in the exchanger phase ($\approx 10^{-6}$ cm^2/sec), and D = diffusion coefficient in the liquid phase ($\approx 10^{-5}$ cm^2/sec).

This equation also indicates the influence of temperature on a separation by ion exchange chromatography. Since the value H of a separation stage should be as small as possible, the diffusion coefficients in the denominator of the second and third term will influence the separation. The diffusion coefficients increase with increasing temperature. The larger their values, the smaller will be the value H, the larger will be the number of theoretical plates in the column, and the more efficient will be the separation.

Using the above theoretical treatment of the process of ion exchange chromatography according to plate theory, the number of theoretical plates required for a separation can be obtained from a known distribution coefficient.

The *theory of the elution constant* of Kraus and Moore [347], in contrast, makes use of an elution constant E, given by the relation:

$$\mathrm{E} = \frac{l \cdot A}{V}$$

for a quantitative description of the elution process. In the above, l is the migration distance of a band in cm after charging of V ml eluting agent. A is the column cross section in cm^2.

Moreover, the following relation applies to the elution constant:

$$E = \frac{l}{(i + D)}$$

where i is the fraction of solvent in the bed and D is the distribution coefficient in the exchanger bed. Thus, i is equal to K_{D_b} and D to f in the Glueckauf equation. An approximate value of 0.3 K_D was given for D; the following relation is valid:

$$D = \frac{\text{quantity per ml exchanger bed}}{\text{quantity per ml solution}}$$

If we now introduce the elution constant of the first traces E_e and the elution constant of the last traces E_l according to Jentzsch [296a], with the assumption that l is the column length in cm and that the volumes of eluant are measured in ml as volume V_e and V_l which are used to elute the first and last traces of an ion from the column, we obtain the following two relations:

$$E_e = \frac{l \cdot A}{V_e}$$

and

$$E_l = \frac{l \cdot A}{V_l}$$

The separation effects for mixtures can be precalculated from these relations. The separation of two ions, 1 and 2, is quantitative if the following condition is satisfied:

$$E_l^1 - E_e^2 > 0$$

where E_l^1 is the elution constant of the last traces of element 1, and E_e^2 is that of the first traces of element 2.

Theoretical concepts for other chromatographic procedures were also developed parallel to the theory of elution chromatography. A full contribution to the theory of frontal analysis has been published by Persoz [469]. Rachinskii [484] considered the same problems as well as the theory of displacement analysis. Moreover, Dybczynski [169] has noted the importance of the influence of temperature on ion exchange chromatography, and Samuelson [570] has surveyed the variables which influence the separation factors in ion exchange chromatography. In favorable cases the influence of complexing and of the pH, and the concentration of the eluant can be predicted with simple equations; in contrast, little study has been devoted to interaction forces in the resin phase.

5.4.3 *Applications*

The applications of ion exchange chromatography are very numerous in inorganic as well as organic chemistry. Not only analytical but also preparative and industrial problems are solved by this method. Although the industrial applications are no less important than the analytical, only the latter will be discussed. It is readily apparent, however, that no fundamental difference exists among the various applications, even though technical and perhaps technological problems arise in the case of preparative and industrial processes.

In inorganic chemistry the separation of rare earths is one of the most outstanding demonstrations of the efficiency of ion exchange chromatography. Since the development of the separation of lanthanides by ion exchange chromatography, an old dream of chemists has been realized—the quantitative separation of rare earths in one step and in any desired quantity.

The studies of Lindner in Germany and of Boyd, Tomkins, Cohn, Coryell, Marinsky, and Glendenin in the United States, which started at about the same time, were continued to a complete solution of the problem of separating rare earths by the use of ion exchange resins and of citric and tartaric acid as complex formers. To illustrate the work of Boyd *et al.*, Figure 87 shows the equipment used by this group.

As in the above example, the separating column is the core of the entire system in every chromatographic assembly. The column design is adapted to a particular separation problem. Generally, no special

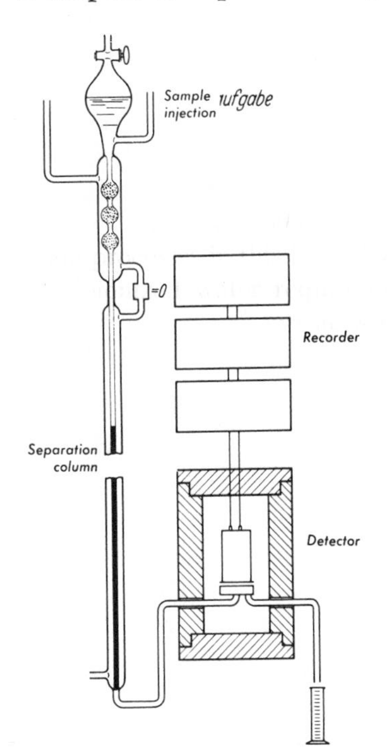

Figure 87. Equipment for the separation of rare earths according to Ketelle and Boyd.

system is used for the sample injection although accessories can be added to the chromatographic equipment to permit the process to proceed automatically and to observe the separation obtained continuously on discharge of the individual components from the column. Thus, if possible a continuous identification should be made for which any suitable chemical or physical properties are utilized. Since these systems permit a continuous identification of the components, they are known as detectors. The measured values received by the detector can be recorded through a recording unit, which leads to the diagram known as the chromatogram. The detector and recording unit are also desirable in ion exchange chromatography and probably will continue to be refined, but in principle, they are not necessary. However, a fractionator or fraction collector is practically indispensable to separate the eluate automatically into single fractions as it emerges from the column. Subsequent studies are then performed with these fractions.

Boyd *et al.* used this equipment in a packing of 97-cm length and 0.26-cm^2 cross section with Dowex 50 of 270–325 mesh particle size at 100°C to separate the rare yttrium oxides at pH 3.28 and the cerium oxides at pH 3.33 by elution with 4.75% citric acid. The high temperature permitted a flow rate of 2 $ml/cm^2/min$, so that the time required for the complete separation could be extensively reduced. The results for the light lanthanides are shown in Figure 88 and those for the heavy lanthanides in Figure 89. Fractionation of the rare earths was so complete that the method could be considered for their quantitative analysis. Thus, it was found that an erbium oxide sample of spectroscopic purity contained 10 ppm thulium after neutron irradiation and according to chromatographic analysis.

A notable improvement of the method for the separation of rare earths by ion exchange chromatography was achieved with the use of gradient elution according to Nervik. This method led to the solution of the problem that the conditions under which the heavier rare earths are rapidly separated result in extremely long elution periods for the separation of the light rare earths, and that conditions under which the light earths are separated with good efficiency elute the heavy ones too rapidly and therefore in insufficient purity. Figure 90 shows the equipment utilized.

In the mixing vessels preceding the separation column, the pH-gradient is produced by metering the solution of higher pH from the upper vessel continuously into the solution of lower pH in the lower vessel. In this case, a 1-*M* lactic solution of pH 3.19 was in the bottom vessel and a 1-*M* lactic acid solution of pH 7 in the top vessel. The metering process was adjusted so that a pH change of 1.167 pH-units was obtained per hour. The magnetic stirrer provides for mixing the

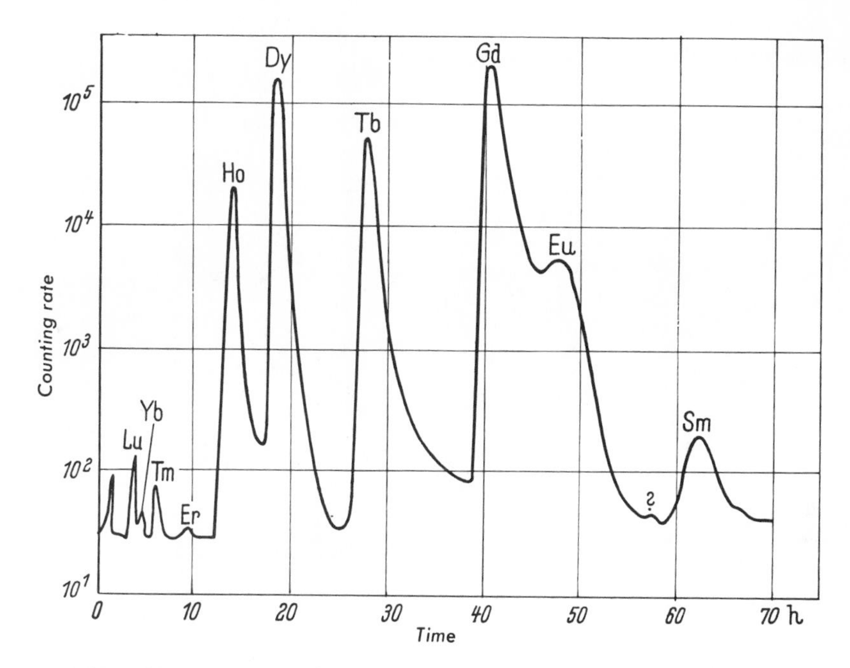

Figure 88. Separation of rare yttrium oxides by ion exchange chromatography on Dowex 50 according to Ketelle and Boyd.

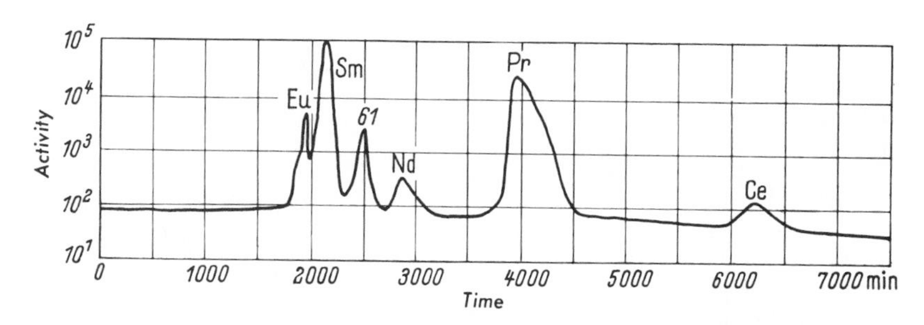

Figure 89. Separation of light rare earths by ion exchange chromatography on Dowex 50 according to Ketelle and Boyd.

eluant before its entry into the separation column. The remaining accessories for heating the column to the operating temperature of 90°C are apparent from the figure.

Under well-controlled operating conditions ,an excellent separation of rare earths can be obtained, as demonstrated by Figure 91. Under special conditions, the technique can also be used for the preparative separation of larger quantities.

After the separation of rare earths on ion exchangers had demonstrated the usefulness of ion exchange chromatography, the technique

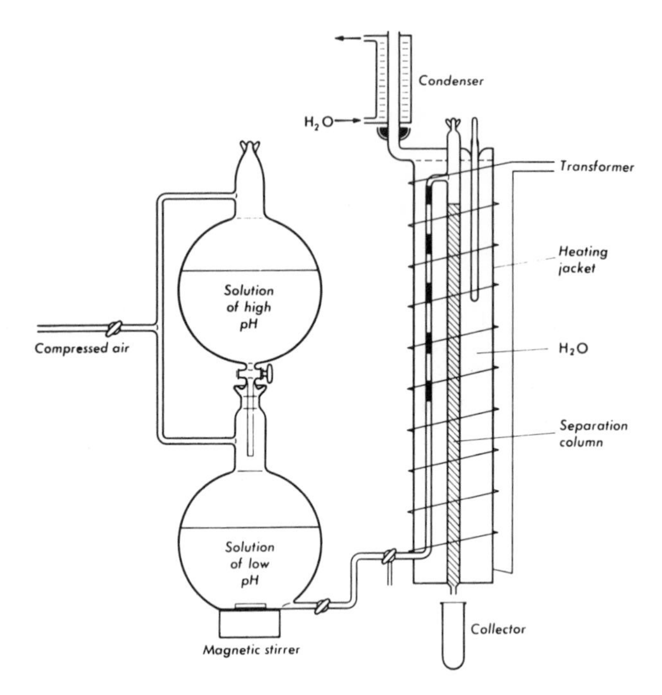

Figure 90. Equipment for gradient elution of the rare earths according to Nervik.

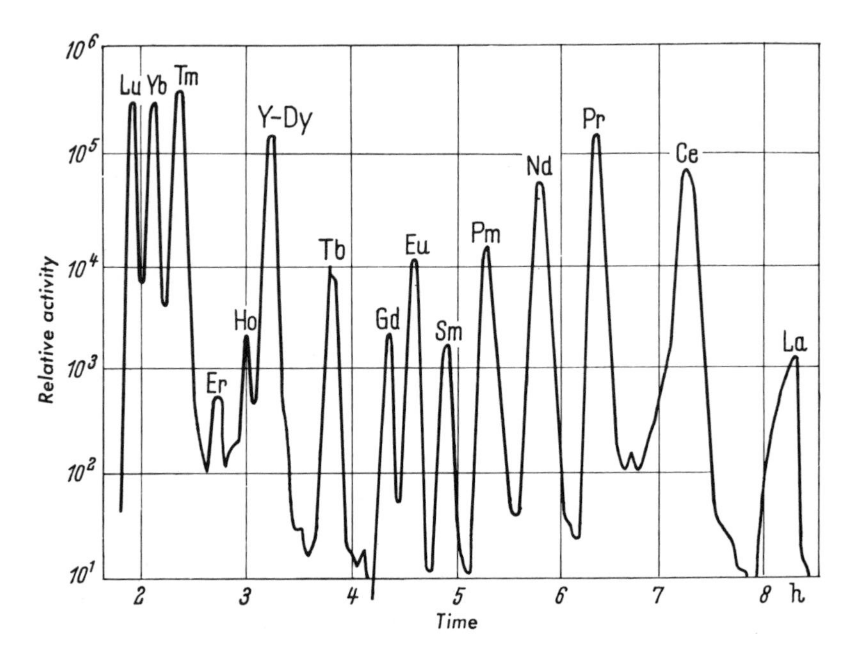

Figure 91. Separation of rare earths by gradient elution according to Nervik.

was also used in other cases. Examples from inorganic chemistry are listed in Table 23. Successful results were also obtained in the separation of actinides. The discovery of heavy transplutonium elements would have been difficult without ion exchange chromatography. In that case, use was made of the analogy between the known separation factors of the lanthanides and the artificial elements, and the position of the elution bands of the unknown elements could be predicted up to atomic numbers of 103.

In organic chemistry, ion exchange chromatography holds a special position among the various methods for the separation and identification of amino acids. Complex mixtures of amino acids can be separated in a relatively simple manner by this technique.

Moore and Stein [441] have published important research in this field since 1950, and in particular have dealt with the selection of exchangers and buffer solutions. One method in which the gradient-elution technique was used will be described in greater detail. For the purpose, a special exchanger was employed—Dowex 50 with 4% instead of the usual 8% divinylbenzene as cross-linking agent.

The resolving power of such an exchanger is excellent, as indicated by Figure 92 for a mixture of 50 components. The separation column consisted of a glass tube of 165 × 0.9-cm length with an inserted glass frit and a heating jacket with water circulation from a thermostat. The exchanger bed itself amounted to 150 cm with the exchanger in the Na-form. The preparation of the exchanger and manipulation of the column require some precautions to permit one to obtain optimum results; these have been described in detail by the authors. For the separation the sample is brought to pH 2.0–2.5 and is injected on the column with a pipette. Additional buffer of pH 2.2 is then briefly injected until the sample has entered the exchanger bed, and finally elution is started with buffer of pH 3.1. The column is located over a fractionator; water of 30 ± 1°C flows through the jacket. The eluate is collected in 2-ml portions with a maximum quantity of 8 ml/h. After serine has appeared, the temperature of the column is raised to 50°C. After twice the volume required for the appearance of the maximum of aspartic acid has emerged from the column, gradient elution is started by a continuous increase of the pH and of the ionic strength of the eluant. For this purpose, a buffer of pH 5.1 is continuously added to the starting buffer in a mixing vessel. The pH-gradient which forms flows from the mixing vessell to the column. For the separation of tryptophan from arginine, the temperature is raised to 75°C after the elution of histidine. The buffer composition is shown in Table 24. The normalities refer to sodium. The buffer solutions are stored in a refrigerator at 4°C.

This well-known Ninhydrin reaction, modified by Moore and Stein

Table 23

Selection of examples of ion exchange chromatography in inorganic chemistry

Components	*Exchanger*	*Eluant*	*Refs.*
Na, K, Rb, Cs	Dowex 50; H	0.5 *N*, 0.3 *N* HCl	[125]
Li, Na, K, Rb, Cs, Fr	Dowex 50; H	HCl + methanol	[450]
Cu, Ag	Amberlite IR-100; H	*N* $NaNO_3$	[534]
Cu, Ag, Au, Zn, Fe, Co, Ni	Amberlite IRA-400	0.2 *N* Hcl, 2 *N* NaCN, acetone + 5% HCl, 2 *N* KSCN	[107]
Be, Mg, Ca, Sr, Ba, Ra	Dowex 50; H	ammonium lactate: 0.55 *M* ph 5, 1.5 *M* pH 7	[434]
Be, Ca, Sr, Ra, Na, Rb, Cs	Dowex 50	2.6, 5.5, 8.7, 12.2 *N* HCl	[146]
Sr-90, Y-90	Dowex 50	1.4 *N* and 3 *N* HNO_3	[486]
Ca, Mg in dolomite	Dowex 50	*M* acetate buffer pH 6	[139]
Ba, Sr, and other elements	AG-50W	3 *N* HCl and alcohol, 3 *N* HNO_3	[570]
Zn, Cd, Fe	Dowex 50	0.5 *M* HCl, acetone	[192]
Bronzes (Mn, Al)	Dowex 1	HCl	[140]
Lanthanides, actinides	Cation exchanger	HEDTA	[168]
U, Th	Amberlite IR-120	*N* HCl, 3 *N* H_2SO_4	[307]
U in Fe-alloys	Dowex 1	6 *N* HCl methylglycol	[334]
U in sea water or monazite	Dowex 1	6 *N* HCl methylglycol	[258]
Np, Pu, U, Zr, Nb, Mo	Dowex 2	HCl, HNO_3	[636]
Am, Cm	Dowex 50	0.1 *M* ammonium titrate	[222]
Silicate analysis	Dowex 1-X8	10.5 *N*, 6 *N*, 0.5 *N* HCl	[649]
Sn, Sb, Pb	Amberlite IRA-400	3% malonic acid, 9 *N* H_2SO_4	[138]
Zr, Ti, Th	Dowex 50	1% citric acid	[96]
Zr, Hf, impurities	Dowex 1	Methanol, 12 *M* HNO_3	[337]
Zr, Hf	Dowex 50	2 *N* H_2SO_4	[399]

Zr, Np, Nb	Dowex 1	HCl, HF	[279]
Zr-95, Nb-95	Sn Phosphate	H_2SO_4	[513]
Polymeric phosphates	Dowex 1-X10	buffered KCl	[55, 231]
Sn, As, Sb	Dowex 2	0.5 *N*, 1.2 *N*, 3.5 *N* KOH	[315]
V, Ti, Fe	Dowex 1	12.1 *N*, 9.1 *N*, 1 *N* HCl	[351]
Nb, Ta	Dowex 1	Oxalic acid, HCl	[564]
Pa, Ta	Amberlite IR-4B	6.5 *N* HF	[648]
Sulfate, sulfite, thiosulfate, sulfide, polythionates	Diaion SA 100	$NaNO_3$-Acetone	[285]
Se, Te	Amberlite IR-120	0.3 *N* HCl	[21]
Cr(II), Cr(III)	Dowex 50	1.4 *N* HCl	[246]
Mo, W, U	Dowex 1	HCl-HF	[350]
Mo, Cr, V in steel	Amberlite IR-120	HCl	[251]
Re, Mo, W	DEAE-Cellulose	NH_4SCN	[294]
F^-, Br^-, Cl^-, J^-	Dowex	*N* $NaNO_3$	[29]
J^-, JO_3^-, $J_2O_7^-$	Amberlite IRA-400	NaOH, *N* KNO_3	[227]
JO_3^-, BrO_3^-, ClO_3^-	Dowex 1	$NaNO_3$	[555]
Mn, Fe, Co, Ni, Cu, Zn	Dowex 1	12.6, 6, 4, 2.5, 0.5, 0.005 *N* HCl	[347]
Tc, Mo, Co, Ag	Amberlite IRA-400, IR-120	potassium oxalate, KOH, NH_4SCN	[252]
Fe, Al, Cr	Dowex 1	3 *N* HCl	[440]
Fe, Co, Ni	Dowex 1-X8	3 *N* HCl; 4 *N*, 0.1 *N* HCl	[328]
Co, selective separation	Dowex 50	HCl-acetone	[333]
Pr, Pt, Rh, Ir	Dowex 2	NH_4OH-NH_4Cl	[421]
Os, Fe, Ni, Cu	Dowex 50	HCl	[401]
Isotopes			[148, 147]

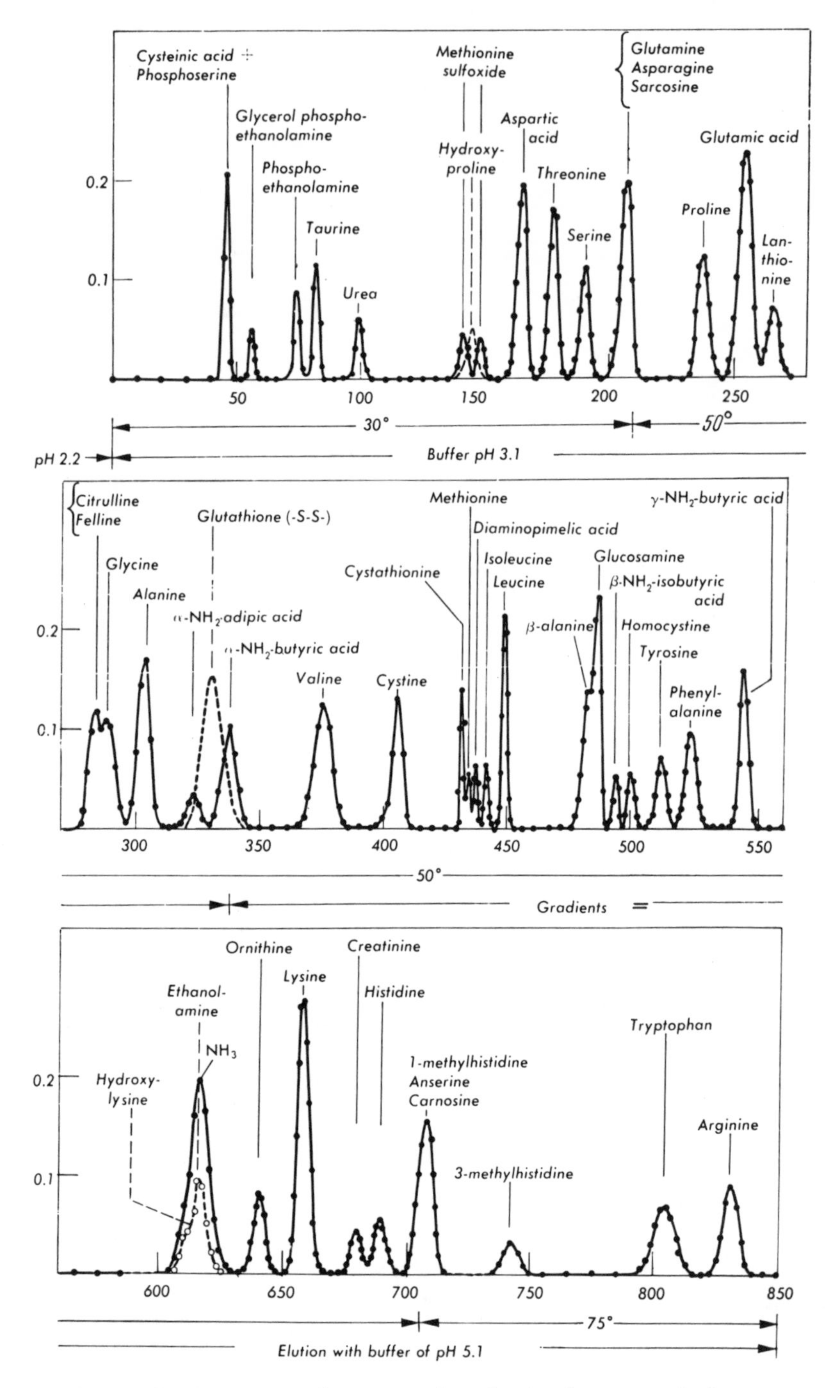

Figure 92. Separation of amino acids and related compounds from a synthetic mixture of 50 components [441].

Table 24

Buffers for the gradient elution of amino acids according to Moore and Stein [441]

Buffer	*Citric acid* · H_2O	*Glacial acetic acid* (*g*)	97% NaOH (*g*)	*Na-acetate* · 3 H_2O (*g*)	*Conc.* HCl (*ml*)	*End volume* (*l*)
pH 2.2 ± 0.03; 2 *N* Na-citrate	105		42		80	5
pH 3.1 ± 0.03; 2 *N* Na-citrate	714		282		393	34
pH 5.1 ± 0.02; 2 *N* Na-citrate-acetate	3570	730	1600	4630		34

[441] for a photometric analysis, serves for the determination of amino acids in the eluate fractions. The reagent consists of 2% Ninhydrin and 0.3% hydrindantin in 3:1 methyl Cellosolve–4 *N* sodium acetate buffer of pH 5.5. Since the buffer is strong enough, it is not necessary to adjust the pH of the eluant with Ninhydrin prior to the reaction.

Finally Moore, Spackman, and Stein developed an improved technique for the chromatographic determination of amino acids on finely powdered and carefully classified Amberlite IR-120 cross-linked with 8% divinylbenzene. The smaller particles permit a higher flow rate of the improved sodium citrate buffers used as eluants. A complete amino acid analysis of a protein hydrolysate can be performed in 48 hours with the fraction collector and in 24 hours with recording. With the use of the latter, which simplifies the detection, the Ninhydrin staining takes place continuously. The mixture of eluate and Ninhydrin flows through a Teflon capillary coil in a boiling water bath. The developing color is measured continuously at 570 and 440 mμ and recorded. Customary methods are used for the evaluation of the chromatograms.

Figure 93 shows an example in the form of a well-proved instrument, the amino acid analyzer BC 200* which was developed as a very compact unit for extremely complicated separation procedures as well as for routine clinical analysis. Its mechanism of action is based on the studies of Spackman, Stein, and Moore. The analyzer is equipped with a fully automatic sample injector for 20 analyses and is generally very easily used for separations of protein hydrolysates, physiological fluids, tissue specimens, and culture media. For a quantitative determination of free or bound amino acids, the instrument is used in hospitals for diagnostic purposes, diet control, and heterozygate and homozygote

*Manufacturer: BIO-CAL Instrument GmbH, Am Kirchenhölzl 6, 8032 Munich-Gräfeling.

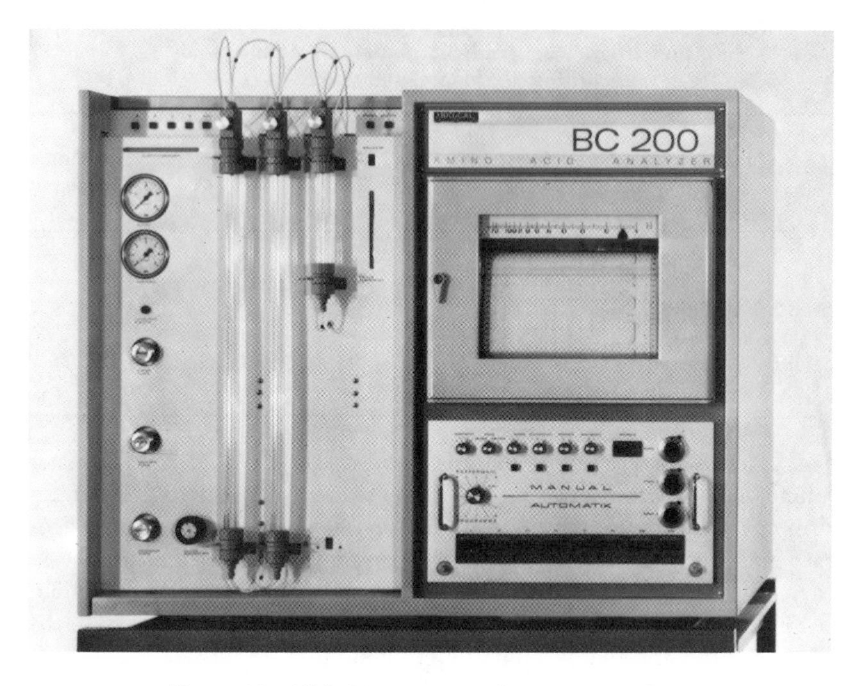

Figure 93. BIO-CAL amino acid analyzer BC 200.

tests. The hydraulic component with the ion exchange columns and the electrical component on the left and right half of the instrument are completely separated the latter consisting of the lower electronic controls and the upper recorder. The instrument can be operated by a one- or two-column technique. Its sensitivity permits a quantitative analysis of 0.05–1.0 μmole of an amino acid in a standard cuvette. The microcuvette with its long optical path permits a determination of up to 0.001 μmole. Thus, less than 1 mg protein is sufficient for an amino acid hydrolysate analysis. With regard to accuracy, a reproducibility of $< \pm 3\%$ with a variation in retention time of $< \pm 2\%$ can be attained with some experience. A number of standard methods have been developed for practical application in analyses of hydrolysates and physiological fluids as well as of plasma and urine specimens. Rapid routine clinical programs are available, for example, for phenylketonuria, tyrosinosis, histidinemia, hypervalinemia, liver function disorders, and leucinosis. Manipulation of such an instrument is simple according to instructions of the manufacturer. The necessary chemicals required for its use are available as a complete reagent kit.*

Table 25 lists some examples of ion exchange chromatography in organic chemistry which can be used as a starting point for literature research.

5.5 Special Applications of Ion Exchangers in Analytical Chemistry

In *microanalysis* it is possible to use ion exchangers as the reaction media for microdetection reactions leading to a technique for spot reactions on ion exchangers [197]. The spot reaction is based on the intense coloring of a few particles of a light ion exchanger; the reaction results from the sorption of ions of characteristics colors from the analytical solution. A precipitation on the resin surface can also serve for such a test. The higher sensitivity and improved selectivity of detection reactions, higher stability of the coloring obtained, and uniformity of the ion exchange resin are emphasized as advantages of this technique.

The instruments used are customary spot-test plates, capillary pipettes, small glass stirrers and spoons as well as microgas generators and microdistillation equipment. Interfering ions are eliminated with the use of microcolumns of 2–3 × 30 mm designed so that the spot test can be carried out immediately on emergence of the eluate from the column, as indicated by Figure 94.

A simple example is the following method for the detection of titanium according to Fujimoto [197]. A few particles of a strongly acidic

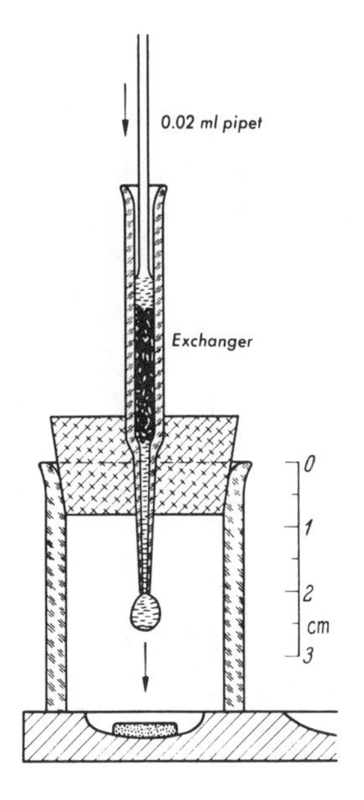

Figure 94. Equipment for a spot test on ion exchangers after previous separation of interfering ions [197].

Table 25

Selection of examples of ion exchange chromatography in organic chemistry

Components	*Exchangers*	*Eluant*	*References*
Glycols	Dowex 1	0.925 M Na_3BO_3	[547]
Phenol, cresols, xylenols	Dowex 50	sodium citrate	[344]
Chlorophenols	Dowex 2	gradient of acetic acid and methanol	[554]
Aldehydes and ketones	anion exchanger in bisulfite form	H_2O	[201]
Fatty acids	Amberlite IRC-tO	acetone–methylethylketone–water	[542]
Organic acids	Dowex 1	Mg-acetate solution	[384]
Orthanilic sulfanilic acid	Amberlite IRA-400	HCl	[566]
Substituted ammonium ions	Zr phosphate	0.1 N HCl	[489]
Amino acids	ligand exchange through Zn		[226]
Peptide	Dowex 1	buffer	[532]
Proteins	DAE- and CM-Cellulose	various buffers	[471, 472, 561]
Collagens	Cm-Cellulose	buffer	[475]
Sugar	Dowex 1	Na_3BO_3	[308]
Sugar phosphates	Dowex 1	borate	[309]
Aldobionic acids/aldonic acids	Dowex 1	Na-acetate	[512]
Nucleic acids		NaCl	[49, 50]
Polynucleotides	Dowex 2-X2		[619]
Nucleotides	Dowex 1	formic acid	[124]
Cumarins	Dowex 50	methanol	[464]
Pyridoxine, ascorbic acid	Amberlite IR-4B	H_2O	[297]

Cobalamines	Amberlite XE-97	H_2O	[424, 425]
Amidase	DEAE-Cellulose	pH-gradient	[561]
Bromelaine	Carboxyl-KA	K-phosphate	[445]
Thyroid hormones	Dowex 1	acetate buffer	[205]
Pituitary hormones	Oxycellulose; Amberlite XE-97	0.1 *N* HCl	[393, 394]
Insulin	KA, H^+	0.2 *M* Na_2CO_3	[396]
Viruses	ECTEOLA-Cellulose	phosphate buffer	[128]
Streptomycin	Amberlite IRC-50		[505]
Insecticides	Dowex 1	HCl-gradient	[477]
Flavones	Amberlite IRC-50	20% isopropyl alcohol–80% water	[202]
Plastic analysis			[531]
Pantothenic acid synthetase	DEAE-Sephadex	0.1 *M* tris-HCl	[353]
Alkane sulfonates	BioRad AG-50	NH_4HCO_3–alcohols	[447]
Polyamides	BioRad AG-50	6% NH_4OH	[460]
Mineral oil products	Amberlyst 29/15	CO_3^-–methanol	[622]

cation exchanger in the H-form are mixed on a spot-test slide with 1 drop of test solution and 1 drop 0.2 N HCl, $HClO_4$ or H_2SO_4; 1 drop of 1% H_2O_2 is added. In the presence of Ti(IV) in the test solution the yellow peroxy-complex forms, which is sorbed by the exchanger with the formation of an intense yellow color. The limit of detection of this test is 0.25 mg with a limiting dilution of $1:1.6 \cdot 10^6$. Cr(III) and Cr(VI) interfere, but after oxidation into Cr(VI), these can be separated in a microcolumn on a strong base exchanger.

Ion exchangers are recommended as *indicators* for analysis in which the results are obtained by colorimetry but in which the test solution at the same time needs to be adjusted to a certain pH. Every dissolved indicator color would interfere with the colorimetric analysis. This can be avoided with the use of ion exchange particles as indicators. Since the indicators have an ionic character, they can be sorbed by exchangers by treating the particles with 0.1% solutions of the indicators in 95% ethanol. For example, to perform a titration only a few particles of the indicator resin are necessary; they can be easily removed from the solution as a separate phase and after washing with water, can be used again. Thymol Blue, Bromcresol Green, and phenolphthalein forms of the Ameberlite IRA-400 and Nalcite SAR exchangers were found to be suitable; the Methyl Orange, Methyl Red, and Congo Red forms produce less distinct end points [385, 433].

Ion exchangers as digestants. Inherently insoluble substances can be brought into solution with the aid of ion exchangers. For example, if barium sulfate and a strong base cation exchanger in approximately stoichiometric ratio are covered with distilled water and stirred, the solution of barium sulfate can be clearly recognized after some time. Silver chloride can also be brought into solution by the following reaction:

$$AgCl + KA{-}H \longrightarrow HCl + KA{-}Ag$$

After exchange, the quantity of cations sorbed by the exchanger can be determined by titration of the hydrochloric acid. In such reactions, the exchanger naturally must be available in the form of a very finely divided suspension so that equilibration between the resin and solution can be obtained in a reasonable time.

Chapter 6

Ion Exchangers in Pharmacy and Medicine

Ion exchangers have also found use in pharmacy and medicine. The methods and applications have become so diverse by now that it appears appropriate to devote a separate chapter to this subject. While ion exchangers in the preceding examples of preparative, analytical, or industrial chemistry served more or less as intermediary agents, they find use as directly active media in medical fields. Thus, they have entered the inventory of drugs as therapeutics for producing specific effects.

Since numerous biochemical systems are also ion exchangers, the knowledge found in synthetic exchange resins can rationally be applied to these systems and can facilitate the interpretation of physiological processes. In some cases it has also been possible to bind biochemically active substances, such as enzymes or redox systems, to ion exchangers and to prepare model substances in this manner.

6.1 Pharmacy

The use of ion exchangers in pharmacy is primarily of a preparative or analytical nature. In addition, water condition has become important [104].

The industrial purification of water will be the subject of a later chapter. Here we shall only deal with its special aspects in pharmacy.

The purity of water plays an important role in the production of pharmaceuticals. The German Drug Book [Pharmacopoeia] describes the preparation and use of distilled water. In the opinion of experts, the latter absolutely should be equated with demineralized water as obtained by ion exchange. Electrolyte-free water can be easily prepared by ion exchangers, and in the view of some investigators it satisfies the requirements of the pharmacopoeia. The remaining objections refer to the presence of bacteria as well as pyrogenic and colloidal impurities which are not removed from water by ion exchange. In particular, users are warned not to employ demineralized water, which chemists consider to be of higher purity than distilled water, for the preparation of injection solutions and ophthalmological drops. These discussions thus lead to the need for a further treatment of demineralized water after ion exchange with the use of asbestos filters for the removal of harmful bacteria and pyrogens. Equipment for these purposes is commercially available [163].

The *preparative application* of ion exchangers in pharmaceutical chemistry corresponds to the examples described earlier in Chapter 3, Sections 3.2 and 3.3. For example, a simple ion exchange is involved if the salts of pharmaceutically valuable organic acids or bases are converted into other salts, as illustrated below:

$$CE^- - K^+ + \text{penicillin}^-Na^+ \rightarrow CE^- - Na^+ + \text{penicillin}^-K^+$$
$$AE^+ - NO_3^- + \text{vitamin } B_1{}^+Br^- \rightarrow AE^+ - Br^- + \text{vitamin } B_1{}^+NO_3^-$$

With regard to the purification of pharmaceutical solutions and substrates, we must emphasize the continuous need for the extraction of electrolytes. This deionization or demineralization is most suitably performed in a mixed-bed to avoid a transitory shift of the pH into the acid or alkaline range, which may destroy acid- or alkali-sensitive substances. The desalting process of glycerin on ion exchangers is already of industrial importance. Amino acid solutions, proteins (plasma protein solutions), and enzyme and virus preparations are always advantageously treated with this method.

The concentration of pharmaceutically active substances and their chromatographic purification probably are the most important preparative applications of ion exchangers in pharmacy. The concentration of *quinine* was described in Chapter 3, Section 3.3, as a model experiment. Other alkaloids can be obtained by similar methods. Thus, *nicotine* can be recovered from the exhaust air of tobacco drying houses by conducting the air through wash towers and filtering the washwater through a cation exchanger in the H-form. The concentrated nicotine is then eluted with ammoniacal alcohol. The isolation of *morphine* from poppy seed shells has been investigated repeatedly [423]. According to Kovka *et al.* [343], morphine is adsorbed on Wofatit F from 90% methanol of pH 4.4 and is desorbed nearly quantitatively from alkaline, aqueous–alcohol solution, provided the desorbant has a higher moisture regain capacity than the solvent from which it was adsorbed. For *cinchona alkaloids*, the distribution equilibria of the sulfates were investigated as a function of cross-linking and particle size on a strong acid cation exchanger for the purpose of their enrichment and separation [305]. Research was conducted with *atropine* for its isolation from plant materials on strong acid cation exchangers [436] and for the development of a suitable dosage form on bentonite ion exchanger

[589]. The isolation of *solanine* was the subject of studies by Patt and Winkler [465]. Since the pharmaceutical industry is interested in the commercial exploitation of ion exchange processes for the isolation of alkaloids and other natural medicinals, many procedures have been described in the patent literature and it may be assumed that a number of others have not been published at all. Research on the subject, then, needs to be carried out in each individual case.

Among the antibiotics, the isolation of *streptomycin* with Amberlite IRC-50, a weak acid cation exchanger, is a historic achievement. This exchanger was developed for the isolation and purification of streptomycin. Its excellent properties permit the direct adsorption of the antibiotic from crude solutions and its elution as a highly concentrated and purified solution [101–103, 298, 612]. This ion exchanger thus has played a significant role in making this important antibiotic available in pure form and at low prices. *Neomycin* can also be concentrated and isolated by this technique. According to another patent, it is claimed that it can also be isolated on a strong acid polystyrene–sulfonic acid resin [605]. For *tubercidin,* ion exchange chromatography with a weak acid cation exchanger represents an improved isolation technique [550], particularly at pH 5.0.

Among other groups of biologically active and therefore pharmaceutically interesting materials, a number of examples of enzymes and hormones can be listed, which were isolated and purified by ion exchange. A detailed study has been made on the ion exchange properties of the vitamin B complex [319, 320]. In a patented technique, a strong base anion exchanger serves for the isolation and purification of tocopherol [609]. Studies on cholate and glycholate binding have furnished not only specific data but also additional information of interest in pharmaceutical chemistry concerning large organic molecule binding on ion exchangers [62].

Ion exchange chromatography of corticotropin is an impressive example of the efficiency of such procedures. Figure 95 shows the chromatogram indicating how pure α-corticotropin can be obtained. While bands II_1 and II_2 have a high corticotropin activity, bands I and III are inactive.

Ion exchangers have found frequent use in the *analysis of medicinal chemicals* for the removal of interfering ions and concentration of test substances [323, 324]. Moreover, methods for the determination of the total electrolyte content of solutions by quantitative H^+- or OH^--exchange have found a broad field of application. Thus, sodium, magnesium, or zinc sulfate can be identified in injection solutions by loading the solution on a cation exchanger in the H^+-form and titrating the eluate with alkali. The same type of analysis is involved in numerous

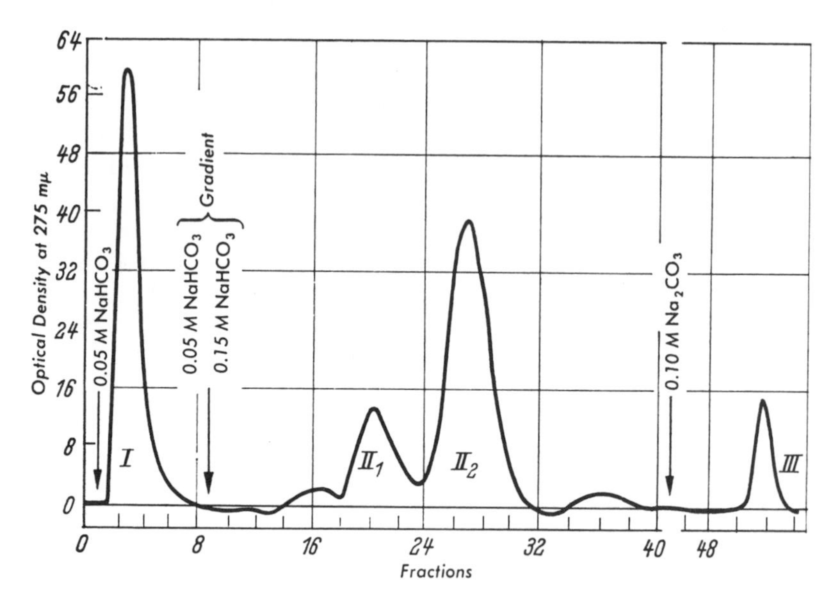

Figure 95. Ion exchange chromatogram of ACTH on Amberlite IRC-50 [395].

methods reported for the determination of alkaloids in drugs and medicinal products [553]. Such techniques have also been used for analgesic, local-anesthetic, sympathomimetic, and spasmolytic salts of weak bases. Furthermore, a large number of ion exchange chromatography procedures exist which furnish quite satisfactory results.

As a first example of their pharmaceutical application, the use of ion exchangers as a tablet-disintegrating agent to induce the disintegration of tablets in gastric juice or in water should be noted. Because of their high degree of swelling, ion exchangers are very well suited for this purpose as are starch, pectin, agar-agar, sodium carbonate, or magnesium peroxide. For example, Amberlite IRP-88 or the K-form of Amberlite IRP-64 as a fine powder, is recommended for this purpose [1]. Ion exchangers are especially useful for the disintegration of hydrophobic or waxy tablets.

Ion exchangers are also used to bind a drug as a counterion and prolong its action, thus releasing the active principle not instantaneously but gradually. The exchanger serves as a depot agent [576]. Since other depot forms of drugs also exist, the future will show the range of application of ion exchangers in this area. This form of a drug is recommended when the active principle is excessively irritating or unacceptable for the patient. For example, if *p*-aminosalicylic acid is bound to a weakly base anion exchanger, it can be used orally without irritating action. Fe^{+2}-ions as counterions in a strong acid cation exchanger of the sulfonic acid type also have proved to be less toxic than ferrosulfate, since the iron ions are only gradually released in the gastrointestinal tract. Moreover, it was observed that the action of sulfathiazole was enhanced when it was bound to the H-form of bentonite.

To protect drugs from the action of gastric juice, they are coated with exchanger materials; substances which are sensitive to acids, such as antibiotics, are presented in this form with particular advantages. For example, penicillin, when coated with ion exchanger material, passes through the intestinal system wothout losses. In general, ion exchangers can be used for the stabilization of tablets and capsules with the fixation of a component of the pharmaceutical preparation to the exhanger, which thus will prevent a harmful interaction with the other components [552]. This increases the shelf life of drugs.

In connection with pharmacy, reference should also be made to the use of ion exchangers in cosmetics [620] where they can be employed for the sorption of skin-irritating components or for the production of antiperspirants.

Table 26 shows a detailed survey of ion exchanger types based on synthetic resins manufactured and purified by special procedures for pharmaceutical applications. The Amberlite ion exchangers of pharmaceutical grade have specified minimum concentrations of iron and heavy metals in the trace range. They are available in powder form with particle sizes of 100–500 mesh and <235 mesh. The Public Health office must approve their use as complexing agents in drug preparations. A reference file of these resins is kept at the United States Food and Drug Administration as well as in governmental health departments of other countries. The table also lists other information for the application of weak and strong acid and weak and strong base ion exchangers.

6.2 Medicine

In medicine, ion exchangers have found applications as preparative media and for clinical-chemical analyses. Beyond this, attempts have been made to use them as therapeutic agents. Since it is very difficult to obtain a clear picture of the actual extent of medical applications, some of the practical directions apparent in the literature will be discussed below.

In the preparation of preserved blood, citrate and dextrose solutions are added to the blood to prevent coagulation. However, these substances have little influence on the stability. But if blood is passed over

Table 26. Amberlite ion exchange
CE = cation exchanger,

	Designation	Matrix	Character	Active Group	Ionic form
KA	IRP-69	Styrene DVB	strong acid	$-SO_3^-$	Na^+
	IRP-69M	Styrene DVB	strong acid	$-SO_3^-$	Na^+
	IRP-64	methacrylic acid DVB	weak acid	$-COO^-$	H^+
	IRP-64M	methacrylic acid DVB	weak acid	$-COO^-$	H^+
	IRP-88	methacrylic acid DVB	weak acid	$-COO^-$	K^+
AA	IRP-58	phenol–polyamine	weak base	$<^{NH}_{NH_2}$	FB
	IRP-58M	phenol–polyamine	weak base	$<^{NH}_{NH_2}$	FB
	IRP-67	Styrene DVB	strong base Type I	$-N-(CH_3)_3^+$	Cl^-
	IRP-67M	Styrene DVB	strong base Type I	$-N-(CH_3)_3^+$	Cl^-

the Na-form of a cation exchanger for the removal of calcium and magnesium ions, coagulation can be prevented. In spite of this advantage, however, this does not solve the problem of the transmission of hepatitis virus with the application of preserved blood. An additional ion exchange technique—treatment of blood plasma with a mixed-bed exchanger—made it possible to extract the hepatitis virus, fibrinogen, and thermally labile proteins, and to obtain a pasteurized plasma protein solution (PPS) containing the total plasma albumin, one-half of the plasma globulins, only a small fraction of the antibodies and enzymes of the initial plasma, and no prothrombin or factors V and VII; this has been demonstrated by studies of the Swiss Red Cross in particular. Clinical trials showed that PPS is an excellent blood substitute with the advantage that it can be stored for up to two years [454, 598].

In clinical-chemical analysis, a test of gastric function without intubation with ion exchangers permits a distinction between anacidity and hypoacidity, normal acidity and hyperacidity [358, 537–540]. For this purpose, the patient receives ion exchange preparations carrying a component which can be easily detected as the counterion in blood or

resins of pharmaceutical quality
AE = anion exchanger

Particle size (mesh)	*Original resin*	*Moisture content (max) (%)*	*pH range*	*Max. operating temperature (°C)*	*Application*
100–500	IR-120	10	0–14	120	vehicles for cationic agents
<325	IR-120	10	0–14	120	which are bases or salts
100–500	IRC-50	10	5–14	120	vehicles for cationic agents which are bases
<325	IRC-50	10	5–14	120	
100–500	IRC-50	10	5–14	120	
100–500	IR-4B	10	0–8	40	vehicles for anionic agents which are acid
<325	IR-4B	10	0–8	40	
100–500	IRA-400	10	0–12	60	vehicles for anionic agents which are acids or salts
<325	IRA-400	10	0–12	60	

urine. This component, which is released in the gastric juice by H^+-ions and absorbed by the body fluids, permits an approximate determination of the gastric acidity. A number of products containing quinine or a cationic dye (Methylene Blue, Azur A, etc.) as the counterion are commercially available. For example, quinine can be determined quantitatively in the urine by fluorescence spectroscopy after 2 hours. This technique makes the frequently unpleasant intubation of the patient unnecessary even though the reliability of the quantitative values is doubted by some investigators.

Ion exchange chromatography procedures have been used for the separation of globulins and plasma proteins, for the analysis of amino acids, and for the determination of porphyrins. Further details and references to the special literature can be found in Chapter 5 on the analytical applications of ion exchangers.

In therapeutic uses of ion exchangers, we should note the treatment of edema with these products [108, 641]. Edemas are the consequence of sodium retention in extracellular tissue fluid, among other things. When a cation exchanger is applied, sodium is removed from the body,

Table 27

Summary of the application of ion exchangers in medicine with the example of Amberlite exchangers (CE = cation exchanger, AE = anion exchanger)

Stage of application	*Field of application*	*Exchanger type*
Introduced applications	ulcers, hyperacidity	weak base AE; Amberlite IRP-58
	adsorption of toxins, infantile diarrhea	weak base AE; Amberlite IRP-58 plus kaolin
	elimination of sodium, hypertension, and cardiac edema	weak acid CA; Amberlite IRP-64 and Amberlite IRP-88
	elimination of potassium, kidney problems	weak acid CA; Amberlite IRP-64
	tuberculosis; PAS-carrier	weak base AE; Amberlite IRP-58 as carrier for *p*-aminosalicylic acid
	diagnostic indicator of gastric acid	weak acid CE: Amberlite IRP-64 salt of Azure Blue
	dietary control, allergy, cough suppression	strong acid EC; Amberlite IR-120 as salt with alkaloids, antihistimine, amphetamine
	tablet disintegrator	K-form of weak acid CE; Amberlite IRP-88
	vaginitis	weak acid CE: Amberlite IRP-64
	vitamin B-12 stabilization	weak acid CE: Amberlite IRP-64
	pruritis	strong base AE: Amberlite XE-235
	ointment additive	mixture of all exchangers; Amberlite XE-87
Clinical investigations	against obesity	weak base exchangers with little cross-linking, which swell in the stomach
	as laxative	weak acid exchangers with little cross-linking, which swell in the intestine
	as sustained-release hypertensive agents	nitrate and nitrate forms of AE
Blood therapy	decalcifying of blood	Amberlite 200
	normalization of blood and transfusion of preserved blood	mixture of all exchanger types

Table 27 (continued)

Stage of application	*Field of application*	*Exchanger type*
	extracorporeal elimination of K+ and NH_4+ in hepatic coma	Amberlite 200
	removal of toxins: for example, elimination of aspirin and barbiturates in extracorporeal transfusion	Amberlite IRA-900
	artificial kindneys	
Planned applications	elimination of common salt	Amberlite IR-120
	all-purpose antitoxin	types will low degree of cross-linking
	aspiring stabilization and buffering	aspirin salt of Amberlite IRP-58
	redox control	Amberlite XE-239
	tablet drying	dry Amberlite types as drying agents
	surgical dressings, gauze	ion exchange fibers for pH control and as drug substitutes

thus leading to a decrease of the osmotic pressure of extracellular fluid and a loss of water and the edema in an attempt by the body to maintain the osmotic equilibrium between the intravascular and extracellular spaces.

Cation as well as anion exchangers are applied for the treatment of hyperacidity [646], heartburn, acute and chronic gastritis, and gastroenteritis as well as acute and chronic peptic ulcer. This produces a neutralization effect, but without suppressing the weak acidity of gastric juice needed for normal digestive activity. A reported advantage in this connection is that a new acid stimulation does not occur during treatment with ion exchangers, as is the case, for example, with the use of sodium bicarbonate.

A number of studies have also been made on the application of ion exchangers in dermatology. For example, the cobalt and copper form of Amberlite XE-64 in ointments, face powders and skin cleansers is reported to be an effective agent against pruritis and dermatitis. Ion exchangers can also be used as deodorants, in which they exhibit a highly adsorbing action for ammonia, urea and organic acids.

When research in the field of application of ion exchangers as therapeutics had reached it speak, a number of commercial products had also become available for medical use (for example, Carbo-Resin, Elutit AS and Elutit HK, Enatrol, Katonium, Masoten, Natrinil, and Resodex as cation exchangers, primarily of the carboxylic acid type, and Basex, Muresin, RAlBis, Resinat, and Styrion as anion exchangers, all of the polyamine resin type). In the meantime, however, these products have been replaced by more effective agents. Cuemid, a product of Sharp and Dohme, continues to be marketed for the treatment of pruritus due to biliary congestion [175].

The firm of Rohm and Haas, Philadelphia, Pennsylvania, which gave the first decisive stimulus to the use of ion exchangers in medicine with the development of their Amberlite types, has published a survey of their exchanger production program that was generally made available to the author. It resulted in Table 27, which offers a summarizing indication of applications already in use, of experimentation carried out in the clinic and laboratory, and of new fields of application of ion exchangers in medicine which may become possible in the future.

Chapter 7

Theory of Ion Exchange

In the course of the treatment of ion exchangers in this monograph, reference has been made several times to theoretical considerations and formulas presented which form a basis for the calculation of a particular problem. However, ion exchange represents a special field of work and deals also with problems in the natural sciences, and it should be recalled that it deals with high molecular weight acids and bases in the gel state. Consequently, it is interesting and important for several reasons to subject ion exchangers and ion exchange processes to a theoretical consideration, first to promote an understanding of these processes, and second to obtain a clear recognition of the resulting consequences and possibly to apply these to the interpretation of other natural processes.

An attempt will therefore be made within the space available to describe the principles of the theory of ion exchange with regard to thermodynamics and transport phenomena.

7.1 Thermodynamics

It is the purpose of the thermodynamics of ion exchange to give a quantitative picture of ion exchange equilibria and derive formulas for the important thermodynamic functions. In this respect, the ion exchanger is considered a concentrated electrolyte or a gel, but not a solid. The limited mobility of fixed ions is justified by the assumption of a Gibbs-Donnan equilibrium. Such equilibria were first studied by Donnan [157–159]. He made use of the simple model shown in Figure 96 in which a vertical line indicates that the salt NaR is located on one side of the membrane which is impermeable to the anion R^-, while NaCl is on the other side. NaCl will then diffuse from (2) to (1), so that an equilibrium state is finally reached as shown by Figure 97. This picture can also be extended to the case of an electrolyte with no common ion, in which only NaR is present on one side of the membrane and only KCl on the other (Figure 98). KCl will then diffuse through the membrane in one direction, but NaCl will also permeate the membrane in the opposite direction. An equilibrium state results, which is to be evaluated qualitatively so that the anion which cannot diffuse through the membrane apparently attracts the cation of a second, quite different, electrolyte whose anion, hower, is apparently displaced to the same degree.

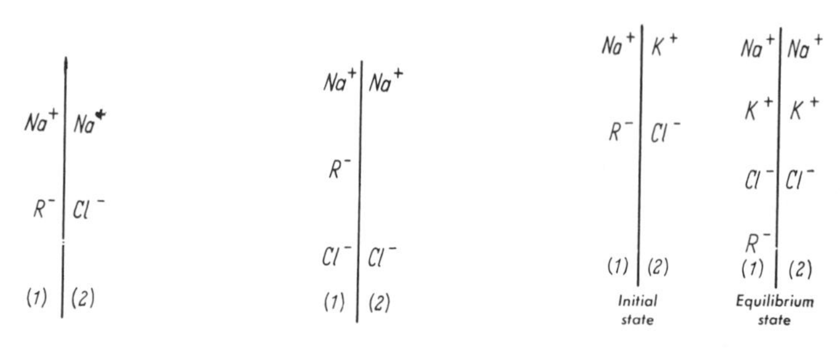

Figure 96. **Figure 97.** **Figure 98.**

This model can also be applied to ion exchange if the interface between the exchange resin and the aqueous phase is considered as a membrane and the resin or the ion fixed on it is considered the indiffusible component on one side. The counterion of the resin matrix is freely mobile. Therefore, if the exchanger is surrounded by an electrolyte solution containing a common or not a common ion, equilibria will form according to the Donnan concept and can be treated according to the principles of membrane theory.

For the case in which the membrane is permeable only to the solvent, the well-known theory of the osmotic equilibrium and derivation of the osmotic pressure results from the membrane theory. In ion exchange, the capillary pressure or swelling pressure plays the role of osmotic pressure which alone affords an equilibrium between the two phases with respect to the entire system.

In the case of ion exchangers, Gregor [235], in particular, noted the significance of the Donnan equilibria, but his original, rather qualitative theories did not agree with the concepts of thermodynamics.

In the following, we shall use the exact theory of ion exchange equilibria developed by Dickel [150] as our basis; this theory takes all influences into account and avoids the oversights and simplifications of others (Argersinger, Davidson, and Bonner; Ekedahl, Högfeldt, and Sillén; Gaines and Thomas; etc.) in some of their studies. Deletions can always be made later in the equations obtained should one find that these terms can be ignored. However, the reader is offered the possibility of estimating the magnitude of the accepted error by a comparison of the literature data obtained through simplified formulas with the exact equations.

The formal thermodynamic treatment of ion exchange equilibria of Dickel is derived from the generally valid Gibbs condition of the equal chemical potentials of ions *J*:

$$\mu'_J = \mu''_J \tag{1}$$

and of solvent *L*:

$$\mu'_L = \mu''_L = \mu'''_L \tag{2}$$

where ′ identifies the exchanger phase, ″ the aqueous, and ‴ the gas phase.

This is based on the assumed relations for the chemical potential in the gas phase:

$$\mu'''_L = \mu^{\circ'''}_L(T, P_0) + RT\ln p_L, \tag{3}$$

where p_L is the vapor pressure of the solvent and P_0 the external pressure applied on the system, as well as for the liquid solvent phase:

$$\mu''_L = \mu^{\circ''}_L(T, P_0) + RT \ln a''_L \tag{4}$$

and an ion *J* dissolved in it:

$$\mu''_J = \mu^{\circ''}_J(T, P_0) + RT \ln a''_J + z_J F_e \Psi''_J \tag{5}$$

In the above equation, a_J is the activity of the respective ion, z_J is the number of its elementary charges, Ψ is the electrical potential in the solution, and F_e is the electrochemical equivalent.

In analogy to the above, and with consideration of the capillary pressure π, we obtain the following for the exchanger phase:

$$\mu'_L = \mu^{\circ'}_L(T, P_0) + \pi v^*_L + RT \ln a'_L \tag{6}$$

$$\mu'_L = \mu^{\circ'}_J(T, P_0) + \pi v^*_J + RT \ln a'_J + z_J F_e \Psi'_J \tag{7}$$

where v^*_L and v^*_J are the partial volumes under one-half of the capillary pressure and P_0 is the external pressure.

Since a system of infinite dilution is selected as the reference system in the solution and in the exchanger phase, and capillary pressure should vanish in the case of the exchanger, the following is valid:

$$a^{\circ'}_L \to x^{\circ'}_L \to 1;\ a^{\circ''}_L \to x^{\circ''}_L \to 1;\ a^{\circ'}_J \to x^{\circ'}_J \to 0;\ a^{\circ''}_J \to x^{\circ''}_J \to 0 \tag{8}$$

where x are the molar fractions defined as follows:

$$x'_L = \frac{n'_L}{\Sigma\, n'_J + n'_L};\quad x''_L = \frac{n''_L}{\Sigma\, n''_J + n''_L};$$

$$x'_J = \frac{n'_J}{\Sigma\, n'_J + n'_L};\quad x''_J = \frac{n''_J}{\Sigma\, n''_J + n''_L} \tag{9}$$

The sum of the number of moles n_J of all ions—the fixed and co-ions in the exchanger as well as the co-ions and ions of the solvent—is found in the denominator. In the further derivations, this sum is abbreviated by n′ and n″, respectively.

If we now refer to the initially indicated identity of chemical potentials and consider the above derivations, the following is obtained for the chemical potentials:

$$\mu^{\circ\prime}{}_J = \mu^{\circ\prime\prime}{}_J;\ \mu^{\circ\prime}{}_L = \mu^{\circ\prime\prime}{}_L \qquad (10)$$

for the equilibrium of the pure solvent with its vapor:

$$\mu^{\circ\prime\prime\prime}{}_L(T, P_0) + RT \ln p_0 = \mu^{\circ\prime}{}_L(T, P_0) + RT \ln 1 \qquad (11)$$

and for the equilibrium of any solution with its vapor:

$$\mu^{\circ\prime\prime\prime}{}_L(T, P_0) + RT \ln p_L = \mu^{\circ\prime\prime\prime}{}_L(T, P_0) + RT \ln p_0 + RT \ln a''_L \qquad (12)$$

or, because of the definition of moisture F:

$$a''_L = p_L/p_0 = F'' \qquad (13)$$

From these relations, the equations for the free energy, the equilibrium and the enthalpy of ion exchangers can be obtained.

The *free energy* of the exchanger-solvent system containing any number of ions J is determined by the equation:*

$$G = G' + G'' = \Sigma\, \lambda'_J\, \mu'_J + \lambda'_L\, \mu'_L + \Sigma\, \lambda''_J\, \mu''_J + \lambda''_L\, \mu''_L \qquad (14)$$

from which we finally obtain

$$G = \sum_{\Sigma J, R, C, L} (n'_i \mu^{\circ\prime}{}_i + n'_i RT \ln a'_i + n'_i \pi v^*{}_i) + \sum_{\Sigma J, C, L} (n''_i \mu^{\circ\prime\prime}{}_i + n''_i RT \ln a''_i) \qquad (15)$$

for the free enthalpy.

The term ΣJ indicates that the summation must extend over all counterions J_1, J_2 . . . , present in the exchanger as well as over the fixed ion R, the co-ion C and the solvent.

---

*The λ-values, which are still unknown, must be determined with the Lagrange multiplier method by means of the auxiliary conditions:

$$\Sigma\, n'_J z'_J = 0;\quad \Sigma\, n''_J z''_J = 0$$

which represent the conditions for electron neutrality and from the equation applying for the swelling volume v_Q which must be present:

$$n'_L v_L = v_Q$$

These conditions undoubtedly are satisfied if we let:

$$\lambda'_J = n'_J;\ \lambda''_J = n''_J;\ \lambda'_L = n'_L;\ \lambda''_L = n''_L.$$

The *equilibrium* is characterized by the fact that the free enthalpy has a minimum:

$$(\delta G)_T = 0 \tag{16}$$

If the conditions for constant quantity

$$n'_i + n''_i = \text{const.} \quad (i = J_1, J_2 \ldots R, C, L) \tag{17}$$

are satisfied here, so that we obtain the following for the stoichiometric values:

$$\nu_i \equiv \nu'_i = -\nu''_i \tag{18}$$

it finally follows that

$$\Sigma\, \nu_i \mu_i = \Sigma\, \nu_i \left(RT \ln \frac{a'_i}{a''_i} + \pi v^*_i \right) = 0 \tag{19}$$

In the case of the *enthalpy* of ion exchange, a distinction is made between the differential heat tone of the ion exchange h^A, on which the temperature variation of the equilibrium depends and which is given by the expression:

$$\frac{\partial}{\partial T} \sum_{\Sigma J, R, C, L} \nu_i \left(\ln \frac{f'_i}{f''_i} + \frac{\pi v^*}{RT} \right) = \frac{h^A}{RT^2} \tag{20}$$

and the integral enthalpy H^A, as the heat which must be supplied to the system, for example, to transform an exchanger containing only ions of species 2 into one containing only ions of species 1, and inversely for the solution. This is given by the expression

$$H^A = -RT^2\, \Sigma\, \frac{\partial}{\partial T} \int_1^2 \left(\ln \frac{f'_i}{f''_i} + \pi v^*_i \right) d\nu_i \tag{21}$$

where f_i is the activity coefficient $f_i = a_i / x_i$.

Having obtained the fundamental equations of a thermodynamic treatment of ion exchange equilibria in this manner, these can be used as a basis for deriving the relations for the special systems of swelling, ion exchange, and ion exchange with general adsorption.

For *swelling*, we must first consider the case where a dry exchanger is combined with pure solvent, in which the free enthalpy

$$\Delta G_Q = G_e - G_a = \sum_{\Sigma J, R, L} n'_i \mu'_{ie} - \sum_{\Sigma J, R} n'_i \mu'_{ia} + n''_L \mu^{\circ\prime\prime}_L \tag{22}$$

(a = before and e = after swelling) or, with substitution of the chemical potentials

$$\Delta G_Q = \sum_{\Sigma J,R} n'_i\left(RT \ln \frac{a'_{i,e}}{a'_{i,a}} + [\pi v^*_i]_e - [\pi v^*_i]_a\right) + n'_L(RT \ln a'_{L,e} + [\pi v^*_L]_e) \quad (23)$$

is obtained. By transforming the sum on the right with the use of the Gibbs-Duhem equation:

$$\sum_{\Sigma J,R,L} n'_i\delta(RT \ln a'_i + \pi v^*_i) = 0 \quad (24)$$

we arrive at the formula:

$$\Delta G_Q = n'_{L,e}(RT \ln a'_{L,e} + [\pi v^*_L]_e) - \int_0^e n'_L\delta\, RT(\ln a'_L + \pi v^*_L) \quad (25)$$

If the exchanger forms an equilibrium with the solvent, the following applies:

$$RT \ln a'_L + \pi v^*_L = RT \ln F' \quad (26)$$

according to our earlier derivations, since no reaction with ions takes place here. If the above Equation (26) is substituted into the preceding equation, the latter will contain only measurable quantities, *i.e.*, the water content of the exchanger and the moisture in the respective state.

As a second case, the free swelling enthalpy

$$\Delta G = n'_{L,e} RT \ln F'_e - n'_{L,a} RT \ln F'_a - RT \int_{F'_a}^{F'_e} n'_L\delta \ln F' \quad (27)$$

follows for swelling from the state of moisture F'_a to the state of moisture F'_e.

The studies of Dickel show further that when the Gibbs-Duhem equation is used as a basis, the values of the *activity coefficients* of the ions contained in the exchanger can also be obtained. In this instance, use is made of the relation

$$a_i = x_i f_i \quad (28)$$

and consideration is given to

$$\Sigma\, n_i\delta \ln x_i = \Sigma\, n dx_i = \Sigma\, nd\left(\frac{n_i}{n}\right) = \Sigma \frac{ndn_i - n_i dn}{n} = 0 \quad (29)$$

In the derivation it was taken into account that the sum of molar quantities n is generally variable in swelling as well as ion exchange.

Thus, for an exchanger containing only a single counterion J, we obtain specifically:

$$\delta\left(RT \ln f'_J{}^{1/z} f'_R{}^{1/z} R + \pi\left[\frac{v^*{}_J}{z_J} + \frac{v^*{}_R}{z_R}\right]\right) = -\frac{n'_L}{n'_{\ddot{A}}}\,\delta(RT \ln F'/x'_L) \tag{30}$$

Because of the electron neutrality, the expressions

$$\Sigma\, n'_J z_J = n'_R z_R = n'_{\ddot{A}} \tag{31}$$

were used with n_E as the number of mole equivalents in the exchanger. Moreover, in place of z_i, we introduce the stoichiometric values according to the equation:

$$\nu'_i = 1/z_i \tag{32}$$

and to abbreviate, we write:

$$n_L/n_{\ddot{A}} = \nu_L \tag{33}$$

For the mean activity coefficient $f\pm$ we have:

$$[f^{\nu}\pm]'_{JR} \equiv f'^{\nu_J}{}_J \cdot f'^{\nu_R}{}_R \tag{34}$$

and, because of the combination of the volumes of oppositely charged ions into one total volume,

$$[\nu\pm v^*\pm]_{JR} \equiv [v^*/z]_J + [v^*/z]_R \tag{35}$$

With this relation, the integration of Equation (30) finally leads to:

$$RT \ln [f^{\nu}\pm]'_{JR,L,e} + [\pi\nu\pm v^*\pm]_{JR,e}$$
$$= RT \ln [f^{\nu}\pm]'_{JR,L,a} + [\pi\nu\pm v^*\pm]_{JR,a} - RT \int_a^e \nu'_L \delta \ln (F/x_L)' \tag{36}$$

We thus have obtained a formula for the dependence of the activity coefficient on the water content (a represents its arbitrary initial state) which is valid not only for the monoionic form of an exchanger but for any mixed form. It is characteristic for this equation that the activity coefficients in the exchanger phase are always related with the expression for the work of cross-linking pressure.

For *ion exchange* between two ions, 1 and 2, according to the equation:

$$\nu'_1 J'_1 + \nu'_{L1} L' + \nu''_2 J''_2 + \nu''_{L2} L''$$
$$\rightleftarrows \nu'_2 J'_2 + \nu'_{L2} L' + \nu''_1 J''_1 + \nu''_{L1} L'' \tag{37}$$

and with completion of the reaction from right to left, we obtain the integral free enthalpy:

$$\Delta G = \sum_{J_1, R, L} [n\mu]'_i - \sum_{J_2, R, L} (n\mu]_i - \sum_{J_1, C, L} [n\mu]''_i + \sum_{J_2, C, L} [n\mu]''_i \quad (38)$$

or with consideration of Equations (4), (5), (13), (26), (28), and (31) to (34):

$$\Delta G = n_{\bar{A}} \cdot RT \left(\ln \frac{[f^{\nu}\pm]'_{1R,0}[f^{\nu}\pm]''_{2C,0}}{[f^{\nu}\pm]'_{2R,0}[f^{\nu}\pm]''_{1C,0}} + \ln \frac{x'_{1,}{}^{\nu_{10}}x''_{2,}{}^{\nu_{10}}}{x'_{2,}{}^{\nu_{20}}x''_{1,}{}^{\nu_{10}}} \right.$$
$$\left. + \ln \frac{[F^{\nu L}]'_{1,0}[F^{\nu L}]''_{2,0}}{[F^{\nu L}]'_{2,0}[F^{\nu L}]''_{1,0}} + \frac{[\pi\nu\pm v^*\pm]_{1R,0}}{RT} - \frac{[\pi\nu\pm v^*\pm]_{2R,0}}{RT} \right) \quad (39)$$

For the discussion of Equation (39), it must be kept in mind that the last logarithmic expression vanishes when $F'_1 = F''_1$ and $F'_2 = F''_2$ and if no water transport takes place during the exchange process, so that the integral free enthalpy is given essentially by the first member with the easily determinable ratio of activity coefficients. For the equilibrium of the ion exchange reaction, Equation (19) furnishes:

$$RT \ln \frac{[a^{\nu}\pm]'_{1R}[a^{\nu}\pm]''_{2C}}{[a^{\nu}\pm]'_{2R}[a^{\nu}\pm]''_{1C}} \left[\frac{F'}{F''}\right]^{v_L} + [\pi\nu\pm v^*\pm]_{1R} - [\pi\nu\pm v^*\pm]_{2R} = 0 \quad (40)$$

in which a is the mean activity, defined analogously to the mean activity coefficient of Equation (34), and $\nu\pm$ and $v\pm$ have the significance of Equations (33) and (35). The activities of this equation refer to the mixed state; the corresponding activity coefficients can be obtained by combining the Gibbs-Duhem equation (24) with the differential of Equation (40), by eliminating one each of the two desired quantities f'_{1R} and f'_{2R}, respectively. First, we obtain for f'_{1R}:

$$\delta \left(\ln [f^{\nu}\pm]'_{1R} + \frac{[\pi\nu v^*]_{1R}}{RT} \right)$$
$$= -\frac{n'_2 z_2}{n_{\bar{A}}} \delta \ln \frac{x'_1{}^{\nu_1}[a^{\nu}\pm]''_{2C}}{x'_2{}^{\nu_2}[a^{\nu}\pm]''_{1C}} - \nu'_L \, \delta \ln [F/x_L]' \quad (41)$$

and a similar equation applies to f'_{2R}. The integration of this equation is now performed so that one begins with a pure exchanger loaded with ion 1 in equilibrium with a solution of moisture $F''_{1.0}$. We further consider ion 1 to be exchanged for ion 2 so that ion 2 is added to the external solution, which is in equilibrium with the exchanger, until the desired state of mixing characterized by 1 R,2 has been attained, where the moisture over the solution is to amount to $F'_{1,2}$. We then obtain:

$$\ln [f^{\nu\pm}]'_{1R,2} + \frac{[\pi v^{\pm} v^{*\pm}]_{1R,2}}{RT} = \ln [f^{\nu\pm}]'_{1R,0} + \frac{[\pi v^{\pm} v^{*\pm}]_{1R,0}}{RT} - \frac{n'_2 z_2}{n_{\bar{A}}} \ln k_{1,2} + \int_0^{n'_2} \ln k_{1,2}\, \delta \frac{n'_2 z_2}{n_{\bar{A}}} - \int_0^{1,2} \nu'_L\, \delta \ln [F/x_L]' \quad (42)$$

and analogously for f'_2:

$$\ln [f^{\nu\pm}]'_{2R,1} + \frac{[\pi v^{\pm} v^{*\pm}]_{2R,1}}{RT} = \ln [f^{\nu\pm}]'_{2R,0} + \frac{[\pi v^{\pm} v^{*\pm}]_{2R,0}}{RT} + \frac{n'_1 z'_1}{n_{\bar{A}}} \ln k_{1,2} - \int_0^{n'_1} \ln k_{1,2}\, \delta \frac{n'_1 z_1}{n_{\bar{A}}} - \int_0^{2,1} \nu'_L\, \delta \ln [F/x_L]' \quad (43)$$

where the following was used in abbreviation:

$$k_{1,2} = \frac{x'^{\nu_1}_1 [a^{\nu\pm}]''_{2c}}{x'^{\nu_2}_2 [a^{\nu\pm}]''_{1c}} \quad (44)$$

A determination of the *activity coefficients* $f_{iR,0}$ of the monoionic exchanger forms is not necessary, since only their ratio appears in Equation (39). If we substitute Equations (42) and (43) into (40), we obtain:

$$\ln \frac{[f^{\nu\pm}]'_{1R,0}}{[f^{\nu\pm}]'_{2R,0}} + \frac{[\pi v v^*]^{\pm}_{1R,0}}{RT} - \frac{[\pi v v^*]^{\pm}_{2R,0}}{RT} = - \int_0^1 \ln k_{1,2}\, \delta \frac{n'_2 z_2}{n_{\bar{A}}} + \int_{1,0}^{2,0} \nu'_L\, \delta \ln [F/x_L]' \quad (45)$$

and when this expression is introduced into Equation (39), we obtain the following for the free enthalpy:

$$\Delta G = n_{\bar{A}} RT \left(\ln \left[\frac{x'_{1,0}}{x''_{1,0}}\right]^{\nu_1} \cdot \left[\frac{x''_{2,0}}{x'_{2,0}}\right]^{\nu_2} \cdot \left[\frac{x'_{L1,0}}{x''_{L1,0}}\right]^{\nu_L} \cdot \left[\frac{x''_{L2,0}}{x'_{L2,0}}\right]^{\nu_L} \right) - \int_0^1 \ln k_{12}\, \delta \frac{n'_2 z_2}{n_{\bar{A}}} - \int_{1,0}^{2,0} \ln [f'_L] \delta \nu_L + \ln \left[\frac{f''_{2C,0}}{f''_{1C,0}}\right]^{\nu_L} + \ln \left[\frac{f''_{L2C,0}}{f''_{L1C,0}}\right]^{\nu_L} \quad (46)$$

The last four terms represent the excess quantities. The integral expressions generally are determined graphically by plotting the measured values of the integrands $\ln k_{12}$ and $\ln f'_L$ versus the integration variables. The last two expressions can practically be disregarded. For the differential heat tone we obtain from Equation (20):

$$h^A = -RT \frac{\partial}{\partial T}\left(-\ln k_{12} + \ln \frac{f''_2{}^{\nu\pm}}{f''_1{}^{\nu\pm}} + \ln \frac{f^{\nu'}{}_L{}^L}{f^{\nu''}{}_L{}^L}\right) \tag{47}$$

For the evaluation, the parenthetical expression is plotted as a function of composition at two similar temperatures and the difference quotient is formed. The integral heat tone can be obtained from this by graphic integration provided direct measurements are not preferred.

With consideration of *electrolyte adsorption,* which usually accompanies ion exchange so that the concentration of counterions appears to be elevated, it is necessary to multiply the mole numbers in Equation (15) in the exchanger phase by the factor $(1+\eta_1)$ and $(1+\eta_2)$, respectively, and the corresponding amount must be subtracted in the solution phase. The relative increases of the molar quantities are η_1 and η_2. In place of Equation (39), we now obtain:

$$RT\ln \frac{[a^{\nu\pm(1+\eta)}]'_{1R,c}}{[a^{\nu\pm(1+\eta)}]_{2R,c}} \cdot \frac{[a^{\nu\pm(1-\eta)}]''_{2c}}{[a^{\nu\pm(1-\eta)}]''_{1c}}$$
$$+ [\pi\nu^{\pm}(1-\eta)v^{*\pm}]_{1R,c} - [\pi\nu^{\pm}(1-\eta)v^{*\pm}]_{2R,c} \tag{48}$$

with introduction of the mean activities by the abbreviations:

$$[a'^{\nu\pm(1+\eta)}]_{iR,C} \equiv a'_i{}^{\nu_i(1+\eta_i)} \cdot a'^{\nu_R}_R \cdot a'^{\nu_c\eta_i}_C;$$
$$[a''^{\nu\pm(1-\eta)}]_{iC} \equiv a''_i{}^{\nu_i(1-\eta_i)} \cdot a''^{\nu_c(1-\eta_i)}_C \tag{49}$$

and the mean volumes by:

$$[\nu^{\pm}(1+\eta)v^{*\pm}]_{iR,C} = \nu_i(1+\eta_i)v^*{}_i + \nu_R v^*{}_R + \eta_i\nu_i v^*{}_C \tag{50}$$

The values of η_1 and η_2 still are some functions of the ionic composition as well as of concentration. For the remaining formulas they are considered to be constant, so that we obtain the following for the activity coefficients:

$$\ln [f^{\nu\pm(1+\eta)}]'_{1R,C2} + \frac{[\pi\nu(1+\eta)v^*]^{\pm}_{1R,C2}}{RT}$$
$$= \ln [f^{\nu\pm(1+\eta)}]'_{1R,C0} + \frac{[\pi\nu(1+\eta)v^*]^{\pm}_{1R,C0}}{RT} - \frac{n'_2z_2}{n_{\ddot{A}}} \cdot \ln k_{1,2,C}$$
$$+ \int_0^{n'_2} \ln k_{1,2,C}\, \delta \frac{n'_2z_2}{n_{\ddot{A}}} - \int_{1R,C,0}^{1R,C,2} \nu'_L\, \delta \ln [F/x_L]' \tag{51}$$

and

$$\ln [f^{\nu\pm(1+\eta)}]'_{2R,C,1} + \frac{[\pi\nu^{\pm}(1+\eta)v^*]_{2R,C,0}}{RT}$$
$$= \ln [f^{\nu\pm(1\pm\eta)}]'_{2R,C,0} + \frac{[\pi\nu^{\pm}(1+\eta)v^*]_{2R,C,0}}{RT} + \frac{n'_1z_1}{n_{\ddot{A}}} \ln k_{1,2,C}$$
$$- \int_0^{n'_2} \ln k_{1,2,C}\, \delta \frac{n'_1z_1}{n_{\ddot{A}}} - \int_{2R,C,0}^{2R,C,2} \nu'_L\, \delta \ln [F/x_L]' \tag{52}$$

Relation (49) applies to the activity coefficients here if f is substituted for the quantities a. The significance of $k_{1,2,c}$ follows from the formula:

$$k_{1,2,c} = \frac{[x^{\nu(1+\eta)}]'_1}{[x^{\nu(1+\eta)}]'_2} \cdot \frac{[a^{\nu\pm(1-\eta)}]''_{2c}}{[a^{\nu\pm(1-\eta)}]''_{1c}} \tag{53}$$

The thermodynamic treatment of the ion exchange equilibrium is purely formal in nature. It must be kept in mind that it does not permit a calculation of exchange equilibria as such; rather, the equilibrium constants must be determined point by point according to Equations (44) and (53) and the moisture according to (13) by means of measurements. The quantity which can be calculated is the activity coefficient. If we also consider the results obtained with regard to the temperature dependence of the equilibria [Equation (20)], the information offered by thermodynamics has been substantially exhausted.

However, the activity coefficients are not only a clear measure of the forces of molecular interaction but also a necessity in general transport theory.

If discrepancies are found between the present theory and the results, this is not a sign of failure of the theory but only that not all factors were taken into account, as mentioned above. This is demonstrated by the studies of Glueckauf [223, 225] on the anomalous behavior of exchange resins during the uptake of electrolytes. The nonuniformity in counterion distribution within the exchanger polymer is justified by the real picture with the local molal counterion concentration M as a consequence of the fixed charges of the exchanger. This concept includes not only the heterogeneities of cross-linking but also those of reciprocal penetration of the chains as well as the change of the chemical potential and thus of concentration, as these occur at distances from the fixed charges of the polymer chains in any one of the larger "pockets" of occluded solution.

The concept developed by Glueckauf can be understood best on the basis of the electron micrograph shown in Figure 99. Accordingly, the (solid) ion exchanger does not represent a homogeneous medium but one in which such zones are imbedded, as can be readily recognized in the micrograph; these zones evidently represent a separate phase from the thermodynamic standpoint. Since they do not contain counterions from the practical standpoint, these "pockets" evidently are filled with external solution, so that a Donnan equilibrium again forms with the surrounding exchanger medium. We thus can let:

$$(\nu_c\bar{m})^{\nu_c}(M + \nu_g\bar{m})^{\nu_g}\bar{\gamma}^{\nu} = k(\nu_c m)^{\nu_c}(\nu_g m)^{\nu_g}\gamma^{\nu}$$

or

$$\bar{m}_c^{\nu_c}\left(\frac{M}{\nu_g}+\bar{m}\right)^{\nu_g}\bar{\gamma}^{\nu}=k(m)^{\nu_c}(m)^{\nu_g}\gamma^{\nu}$$

where ν_c and ν_g are the number of co-ions and counterions forming a molecule of the electrolyte in the aqueous solution, $\nu = \nu_c + \nu_g$, γ is the mean molal ion activity coefficient of the electrolyte, and k is a

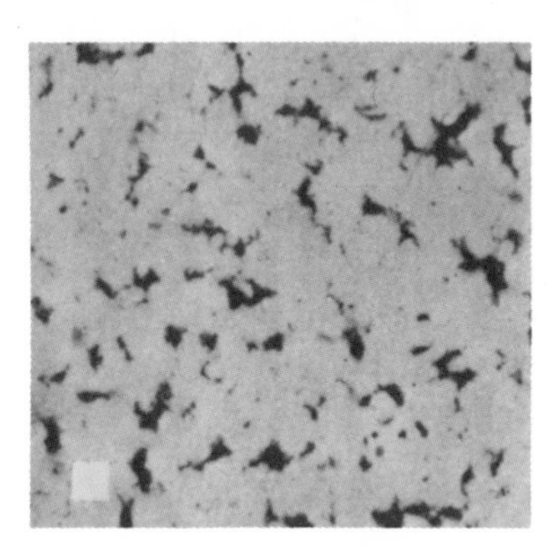

Figure 99. Electron micrograph of thin ion exchange membranes according to Glueckauf.

constant for the influence of the swelling pressure (exp. $\pi V/RT$). The swelling pressure is treated as a constant (near unity) and it is assumed that γ of the electrolyte in the exchanger phase is effectively proportional to γ in the aqueous phase. The emphasis is placed on the word "effective." For example, if the internal solution is 0.01 molar, the value of γ is not different at $M = 5$ (on the average).

Finally, by a number of derivations, the following is obtained for the observed mean electrolyte uptake Q (per g wet resin) and the observable mean uptake concentration of electrolyte m corresponding to this uptake (in mmole/g swelling water):

$$Q \approx m \approx (m)^{(2-z)}$$

where m = molal concentration of electrolyte and z = inhomogeneity constant.

The consequences which result from these inhomogeneities for the treatment of the ion exchange equilibrium are not highly significant, since the mean activity coefficient γ (according to Glueckauf) makes the most important contribution to the distinction of different ionic species. That inhomogeneities exist in ion exchange materials was also demonstrated by Dickel [151] on the basis of water adsorption isotherms on genuinely porous and nonporous exchangers. The two authors cited arrive at practically the same conclusions. According to Dickel, the capillary pressure occurring during swelling of ion exchangers generates a tension in the exchanger network, resulting in a compressive stress with a small water content which becomes a tensile stress with increasing water saturation. Shortly before the end of water uptake, the water enters the uncross-linked "pockets." However, in these pockets the water is practically only under the pressure of the external solution; this was confirmed experimentally and is also thermodynamically plausible. Again, we obtain the picture of a binary phase system formed by the "pockets" occluded in the network proper.

7.2 Transport Phenomena (Kinetics)

While thermodynamics deals only with equilibrium states, thermodynamics of nonequilibrium states has been developed in recent decades for the treatment of transport phenomena. The transport of individual components is assumed to be proportional to the deviation from the equilibrium state. In the case to be treated here, these "driving forces" are the gradients of the chemical potential and of the swelling pressure. Beyond this, however, the electrical potential must also be considered in principle; in the cases in which this is not applied as such (such as at an electrode), this also occurs in the form of the diffusion potential.

The exact derivation of this transport equation was obtained only most recently [220, 240, 242, 580]:

$$J_k = -L_{ks} \operatorname{grad} P - \sum_{j=1}^{n=2} L_{kj} \operatorname{grad} \mu''_j + \frac{f_k}{z_k} I \quad (k = 1, 2 \ldots n-2) \tag{1}$$

where pressure was used here in the first term for thermal diffusion. The quantities μ''_j represent transformed potentials, L_{ks} and L_{kj} are the transport coefficients, f_k is the transfer coefficient and z_k is the charge of the ions.

This general transport equation has not been used for osmotic processes before. Usually, use was made of the Nernst-Planck equation or the formula derived by Schlögl [523] with the application of a decisive simplification:

$$J_k = -L_{ks} \operatorname{grad} P - \operatorname{grad} \mu_k - \frac{D_k z_k C_k F}{RT} \operatorname{grad} \bar{\varphi} \tag{2}$$

In the limiting case $I = O$, $dP = O$, *i.e.*, in the case of pure diffusion, this equation furnishes the Nernst-Planck equation [264]. This contradicts the principles of exact thermodynamics, since although the Nernst-Planck equation can be applied as an approximately valid formula in cases with an external field which is large compared to the diffusion potentials, it is by no means applicable to simple diffusion processes, as has been done on occasion. Studies carried out on this basis therefore are excluded from the following considerations.

Accordingly, diffusion phenomena on an ion exchanger are considered as transport phenomena in a (concentrated) electrolyte with the use of the chemical potential gradients as the driving force. In principle, these can be calculated if the activity coefficients have been determined by the thermodynamic method described in the first part of this chapter. In general, however, we are dealing with polymeric mixtures, so that a single transport coefficient L_{kj} is not sufficient in the expression for the sum in the center of Equation (1). According to Hirschfelder [275], there are as many independent transport coefficients as there are pairs of components in a mixture. These express the interrelationship between the individual components. If the number of the latter is n, we have $[n(n-1)]/2$ independent transport coefficients. Thus, a single transport coefficient is sufficient only in the case when two components are present; if three components are involved, we already need three such coefficients. Therefore, one will generally attempt to reduce the transport coefficients by ignoring those which are small in absolute value.

Since the transport coefficients of Equation (1) generally are not known nor are the chemical potential gradients, these quantities are summarized by the diffusion coefficients which are first determined empirically. It must be emphasized here that every diffusion coefficient refers to a given reference system [255]. In the case of a binary system, we then arrive at the Fick law with a single diffusion constant.

Since the first treatment of the kinetics of ion exchange adsorption by Boyd, Adamson and Myers [79], the concept has generally been recognized that in principle two different diffusion mechanisms exist during ion exchange. In the region of low external concentrations (about 0.001 N), the rate is determined by the diffusion of ions through the liquid film enveloping the exchanger particle, while at high external concentrations (about 0.3 N), diffusion of the ions through the exchanger particle itself represents the rate-controlling step. On the basis of this division, a distinction is made between film diffusion and particle diffusion.

As an illustration, these relations are shown in Figure 100. The outer boundary of the spherical exchanger particle represents the interface between the adhering Nernst film with film diffusion and the exchanger particle with particle diffusion. This particle can be enveloped by an external solution of different concentrations. If the external concentration is high, the exchanging ions are rapidly resupplied and only then does the resin phase become rate-controlling, and vice versa.

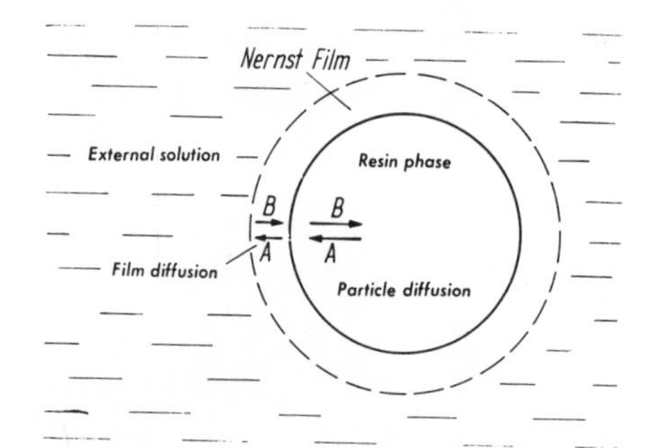

Figure 100. Film diffusion and particle diffusion in ion exchange kinetics.

For a mathematical treatment of ion exchange kinetics, a theory was developed by Adamson and Grossmann [5] in which the different reaction rates of two ions, A and B, in their diffusion currents according to Fick's law are determined by the equations:

$$P_A = -D_A \left(\frac{\partial C_A}{\partial l}\right) = D_A \frac{C_A - \overline{C}_A}{l} \tag{3}$$

$$P_B = -D_B \left(\frac{\partial C_B}{\partial l}\right) = D_B \frac{C_B - \overline{C}_B}{l} \tag{4}$$

The diffusion coefficient D_A expresses the interaction of ion A with water and D_B has the analogous meaning for ion B. The third diffusion coefficient, which is to express the interaction between A and B, can be ignored at low concentrations. At the same time, the concentrations of ions A and B in a film of thickness 1 vary linearly from concentration C_A and C_B to concentration $\overline{C}_A$ and $\overline{C}_B$ of the exchanger surface. If we assume further that concentrations $\overline{C}_A$ and $\overline{C}_B$ are in equilibrium with the ions contained in the exchanger, which are characterized by the mole fractions N according to the law of mass action:

$$K = \frac{\overline{C}_B N_A}{\overline{C}_A N_B} \tag{5}$$

then, with consideration of the continuity condition for the case that $C_B = O$ and $C_A = C_{AO}$ at the time of start of the reaction and that all A-ions accordingly are in the solution, we obtain a differential equation for the increase of N_A with time:

$$\frac{dN_A}{dt} = \left[\frac{OWD_A}{VL}\right]\left[\frac{N^2{}_A(K-1) - N_A(1+\alpha) + \alpha}{1 + (D_A/KD_B - 1)N_A}\right] \tag{6}$$

where O is the total surface of the exchange particles, V is the volume of solution and W the weight of the exchanger; $\alpha = C_A V/WE$ with E as the loading capacity.

Dickel [154] has offered an exact solution of this differential equation as a relation for film diffusion:

$$f(N_A) \equiv (1 + \beta N_A^{\infty}) \ln \left(1 - \frac{N_A}{N_A^{\infty}}\right) - \left(1 + \beta' \frac{N_A}{2N_A^{\infty} - 1}\right) \ln \left[1 - \frac{N_A}{N_A^{\infty}} (2N_A^{\infty} - 1)\right] = \alpha \frac{D_A}{\sqrt{K}} t \qquad (7)$$

where $\alpha = OW/Vl$, $\beta = D_A/(D_B K) - 1$, and N_A is the equilibrium concentration; the latter is attained asymptotically near the end of the exchange process and is identical with the expression:

$$N_A^{\infty} = \frac{K}{K-1} \left[\frac{(\alpha + 1)}{2 \pm \sqrt{\left[\frac{(\alpha+1)}{2}\right]^2 - \alpha \frac{K-1}{K}}} \right] \qquad (8)$$

For the inverse reaction with the initial conditions $C_A = O$, $C_B = C_{BO}$ for $t = O$, we have analogously:

$$f(N_B) \equiv 1 + (\beta' N_B^{\infty}) \ln \left(1 - \frac{N_B}{N_B^{\infty}}\right) - \left(1 + \beta' \frac{N_B}{2N_B^{\infty} - 1}\right) \ln \left[1 - \frac{N_B}{N_B^{\infty}} (2N_B^{\infty} - 1)\right] \qquad (9)$$

with

$$\beta' = \left(\frac{D_B}{D_A}\right) K - 1$$

In the evaluation of test results by these formulas, it must be taken into consideration that the law of mass action does not have precise validity in the simple form given here. Dickel calculated the K-values according to the equation

$$K = \frac{(N_A^{\infty})^2}{(1 - N_A^{\infty})^2} \qquad (10)$$

for the exchange of alkali ions for H^+-ions. The fact that considerable discrepancies result between calculated and directly determined values demonstrates the considerable degree to which the activity coefficients are influenced by the presence of hydrogen ions. Moreover, it must be kept in mind that when the measured values of N_A/N_A^{∞} and K_B/K_B^{∞} are substituted into Equations (5) and (7), the values of the functions on the right side plotted over the corresponding time values must fall on a straight line with the slope $aD_A/\sqrt{VK}$ and $aD_B/\sqrt{K}$. The as yet unknown D_A/D_B-value contained in the expression for β can be determined graphically by plotting the two families of curves resulting from Equations (5) and (7) with different D_A/D_B-values as the parameter and then determining those equal-parameter curves for which the ratio of slopes has the required value $D_A/(D_B K)$. It can be calculated more rapidly on the basis of the relation furnished by Dickel:

$$\frac{D_A}{D_B} = \sqrt{K} \times$$

$$\frac{\ln\left(1-\frac{N_A}{N_A^\infty}\right) - \ln\left(1-\frac{N_B}{N_B^\infty}\right) - \left\{\ln\left[1+\left(\frac{N_A}{N_A^\infty}\right)^b\right] - \ln\left[1-\left(\frac{N_B}{N_B^\infty}\right)^b\right]\right\} \Big/ b}{\ln\left(1-\frac{N_B}{N_B^\infty}\right) - \ln\left(1-\frac{N_A}{N_A^\infty}\right) - \left\{\ln\left[1+\left(\frac{N_A}{N_A^\infty}\right)^b\right] - \ln\left[1-\left(\frac{N_B}{N_B^\infty}\right)^b\right]\right\} \Big/ b} \tag{11}$$

It must be pointed out in this connection that the choice of time t is arbitrary in this calculation; in other words, the equation must be satisfied at any time.

Dickel and Grimmeiss [152] demonstrated the validity of Equations (5) and (7) with Lewatit S 100 during the exchange of $RNa + H^+$ and $RH + Na^+$ in the concentration range of 0.005, while Dickel and Hübner [153] used Lewatit CNO in the exchange of potassium, sodium and lithium in a concentration range of 0.01 N.

Bunzl and Dickel [106] recently published a detailed study of the theory of film diffusion. The measurements were made in small concentration intervals in which the equilibrium constant K can actually be considered constant. The diffusion coefficients observed by Dunlop [166] in the ternary electrolyte for the system Li^+-K^+ were used as a basis for the calculation and led to a quantitative agreement between calculated and measured values.

With the assumption that the exchanger is spherical and that the initial concentration of the exchanged ion in the solution is constant, the relation

$$F = 1 - \frac{6}{\pi^2} \sum_{1=n}^{\infty} \frac{1}{n^2} l^{-\frac{Di\pi^2 n^2 i}{r_0^2}} \tag{12}$$

according to Boyd *et al.* applies to the range of particle diffusion. In this case, only one diffusion coefficient occurs, corresponding to a binary system. The "hydrates" of the ions or resins function as components. Further, $F = Q/Q_0$, r_0 is the particle radius, and Q is the total quantity adsorbed by the exchanger at time t. For an evaluation of experiments, Equation (10) is calculated as a function of Bt with $B = D^2\pi^2/r$, and the calculated F-values are tabulated against the Bt-values corresponding to them. If these experimental results obey Equation (10), the Bt-values corresponding to the experimentally found F-values, plotted versus the corresponding times t, must result in a straight line of slope B. It can then be used to calculate the desired diffusion coefficient for the given radius of the exchanger particles.

A constancy of the external solution is an important prerequisite for the solution of the diffusion problem according to particle diffusion. Boyd *et al.* achieved this by the use of tracers. In the experimental design of Dickel *et al.*, the samples reverse charges in the course of exchange, while the outer concentrations change such a relatively small amount that they can be considered as practically constant. It was found that the test points fall on the required straight line only when the outer solution has concentrations higher than 1 N. At lower concentrations, deviations occur increasingly, while at higher concentrations, all points fall on one and the same line, *i.e.*, no concentration dependence exists, as demanded by theory. At the same time, a different diffusion constant is determined for the forward-and-back reaction. No theoretical interpretation of these findings was offered.

In all of these experiments, the course of diffusion was measured in the nonsteady state, and thus the second law of Fick was used as a basis. In principle, such measurements can also be carried out in the steady state with the use of membranes and the evaluation made according to the first Fick law. This has been discussed by Dickel [149]. Surprisingly, he found in the system of Li^+-K^+ that the concentration dependence of the diffusion coefficient determined by the first Fick law exhibits precisely the inverse course to that obtained on the basis of the second Fick law. It must be concluded that particle diffusion represents a reaction-coupled rather than an ordinary diffusion process. Thus, D in Equation (12) actually represents an apparent diffusion constant. The explanation for the expression of the forward-and-back reaction by different D-values therefore can be sought not in simple diffusion theory but in a more profound reaction theory.

The activation energy follows from the temperature dependence of the diffusion coefficients for film diffusion as well as particle diffusion according to the Arrhenius equation. Dickel and Nieciecki [155] found values of about 5 kcal/mole for film diffusion. For particle diffusion, the values are considerably lower, between 0.595 and 2.080 kcal/mole, depending on the exchange process in the alkali metals and the exchanger types utilized. This difference of activation energies was also confirmed by Sugai and Furuichi [573], who found 1.4 kcal/mole for calcium on Dowex 50. It follows that ions in aqueous solution are located in a considerably deeper potential well than in the exchanger; this must be attributed to the more fully developed hydration envelopes in the solution. The differences between the values for the forward-and-back reaction lead to a relatively low heat of exchange.

To the extent to which ion exchange kinetics are to be considered for practical purposes, it should always be kept in mind that the diffusion coefficients depend not only on the ionic species but particularly also on the exchangers themselves. They increase with increasing particle size and decreasing cross-linking. If the degree of cross-linking decreases, swelling increases, and with an increasing water content of the exchanger phase the diffusion coefficients become increasingly similar to the values of ordinary dilute solutions. In most commercial ion exchange resins, the values of particle diffusion amount approximately to $1/5$–$1/10$ of the diffusion values in dilute aqueous solutions. With regard to the ionic species, it should still be pointed out that large organic ions in particular are exchanged very slowly, and the more slowly as their size becomes commensurable to the distances between tie-points of the exchanger network.

For the present, only the concentration gradients and the chemical potential gradients have been used fundamentally as the driving force in exchanger kinetics, while osmotic pressure diffusion $\Delta\pi$ has been neglected. However, in the investigation of transport processes in membranes, the latter quantity is generally always taken into consideration, so that we now have not only a "chemical pump" but also an "osmotic pump." Finally, if a current is allowed to flow through the membrane, an "electrical pump" is added. However, a discussion of further details of this subject would lead beyond our prescribed scope.

Appendix

The following appendix contains a number of tables listing commercial ion exchangers and their sources of supply as well as computing charts to facilitate practical application.

Table I. Merck
E. Merck AG,

Trade name	*Type**	*Active Group*		*Particle size (mm)*	*Bulk density (g/liter)*
Ion exchanger I	C	$-SO_3^-$	strong acid	0.3–0.9	~650
Ion exchanger II	A	$-NH_2$	weak base	0.3–0.9	~500
Ion exchanger III	A	$-NR_3$	strong base	0.3–0.9	~500
Ion exchanger IV	C	$-COO^-$	weak acid	0.3–0.9	~500
Ion exchanger V	MB	$-SO_3^-$ and $-NR_3$		0.3–0.9	~500

*C = cation exchanger; A = anion exchanger; MB = mixed bed exchanger

Table II. Lewatit
Farbenfabriken Bayer,

Designation	*Type**	*Active group*		*Particle shape*	*Particle size (mm)*
Lewatit S 100	C	$-SO_3^-$	strong acid	beads	0.3–1.2
Lewatit S 115	C	$-SO_3^-$	strong acid	beads	0.3–1
Lewatit SP 120	C	$-SO_3^-$	strong acid	beads	0.3–2
Lewatit CNP	C	$-COO^-$	weak acid	beads	0.3–1.5
Lewatit MP 62	A		weak base	beads	0.3–1.5
Lewatit MP 64	A		medium base	beads	0.3–1.2
Lewatit M 600	A		strong base	beads	0.3–1.2
Lewatit MP 600	A		strong base	beads	0.3–1.5
Lewatit M 500	A		strong base	beads	0.3–1.2
Lewatit MP 500	A		strong base	beads	0.3–1.5
Lewatit M 504	A		strong base	beads	0.3–1.2

*C = cation exchanger; A = anion exchanger

Remarks: Lewatit SP 120, CNP as well as all MP resins are ion for exchanging large organic molecules. Special types for

Ion exchangers
Darmstadt, W. Germany

pH range	*Per-missible temp. (°C)*	*Capacity (meq/g)*	*Moisture content (%)*	*Swelling during exchange (%)*	*Delivered form*
4–14	~ 90	~ 4.5	45–55	~ 5	acid
0– 7	~100	~ 5	40–50	~15	base
1–10	~ 40	~ 3	50–60	~ 5	base
6–14	~ 90	~10	40–50	~55	acid
4–12	~ 40	as I & III	50–60	as I & III	acid & base

Ion Exchanger
Leverkusen, W. Germany

Bulk density (g/liter)	*pH range*	*Permissible temperature (°C)*	*Total capacity, wet resin (eq/liter)*	*Effective capacity ($gCaCO_3$/liter) up to:*	*Specific load ($m^3/h/m^3$) up to:*
800–900	no	120	2.2	90	40
800–900	limit	120	2.0	71	40
700–800		120	1.9	54	40
700–800		80	3.8	125	40
600–700		100	1.9	75	40
650–750		70	1.5	62	40
670–750		40	1.6	43	40
700–780		40	1.2	32	40
700–780		70	1.6	28	40
675–750		70	1.2	25	40
700–750		70	1.3	32	40

exchangers with a macroreticular matrix, particularly suited
selective exchange and adsorber resins on request.

Table III. Diaion

Manufacturer: Mitsubishi Chemical Industries Limited,
U.S. Distributor: Mitsubishi International Corporation,

Trade name	*Type*	*Functional Group*
DIAION SK 1A	gel-type, strong acid	$-SO_3Na$
DIAION SK 1B	gel-type, strong acid	$-SO_3Na$
DIAION SK 102	gel-type, strong acid	$-SO_3Na$
DIAION SK 103	gel-type, strong acid	$-SO_3Na$
DIAION SK 104	gel-type, strong acid	$-SO_3Na$
DIAION SK 106	gel-type, strong acid	$-SO_3Na$
DIAION SK 110	gel-type, strong acid	$-SO_3Na$
DIAION SK 112	gel-type, strong acid	$-SO_3Na$
DIAION SK 116	gel-type, strong acid	$-SO_3Na$
DIAION SA 10A	gel-type, strong base 1	$-CH_2N(CH_3)_3Cl$
DIAION SA 11A	gel-type, strong base 1	$-CH_2N(CH_3)_3Cl$
DIAION SA 20A	gel-type, strong base 2	$-CH_2N(CH_3)_2(C_2H_4OH)Cl$
DIAION SA 21A	gel-type, strong base 2	$-CH_2N(CH_3)_2(C_2H_4OH)Cl$
DIAION SK 1	gel-type strong acid (analytical grade)	$-SO_3Na$
DIAION SA 100	gel-type, strong base (analytical grade)	$-CH_2N(CH_3)_3Cl$
DIAION PK 204	porous-type, strong acid	$-SO_3Na$
DIAION PK 208	porous-type, strong acid	$-SO_3Na$
DIAION PK 212	porous-type, strong acid	$-SO_3Na$
DIAION PK 216	porous-type, strong acid	$-SO_3Na$
DIAION PK 220	porous-type, strong acid	$-SO_3Na$
DIAION PK 224	porous-type, strong acid	$-SO_3Na$
DIAION PK 228	porous-type, strong acid	$-SO_3Na$
DIAION PA 304	porous-type, strong base 1	$-CH_2N(CH_3)_3Cl$
DIAION PA 306	porous-type, strong base 1	$-CH_2N(CH_3)_3Cl$
DIAION PA 308	porous-type, strong base 1	$-CH_2N(CH_3)_3Cl$
DIAION PA 310	porous-type, strong base 1	$-CH_2N(CH_3)_3Cl$
DIAION PA 312	porous-type, strong base 1	$-CH_2N(CH_3)_3Cl$
DIAION PA 314	porous-type, strong base 1	$-CH_2N(CH_3)_3Cl$
DIAION PA 316	porous-type, strong base 1	$-CH_2N(CH_3)_3Cl$
DIAION PA 318	porous-type, strong base 1	$-CH_2N(CH_3)_3Cl$
DIAION PA 320	porous-type, strong base 1	$-CH_2N(CH_3)_3Cl$
DIAION PA 404	porous-type, strong base 2	$-CH_2N(CH_3)_2(C_2H_4OH)Cl$
DIAION PA 406	porous-type, strong base 2	$-CH_2N(CH_3)_2(C_2H_4OH)Cl$
DIAION PA 408	porous-type, strong base 2	$-CH_2N(CH_3)_2(C_2H_4OH)Cl$
DIAION PA 410	porous-type, strong base 2	$-CH_2N(CH_3)_2(C_2H_4OH)Cl$
DIAION PA 412	porous-type, strong base 2	$-CH_2N(CH_3)_2(C_2H_4OH)Cl$
DIAION PA 414	porous-type, strong base 2	$-CH_2N(CH_3)_2(C_2H_4OH)Cl$
DIAION PA 416	porous-type, strong base 2	$-CH_2N(CH_3)_2(C_2H_4OH)Cl$
DIAION PA 418	porous-type, strong base 2	$-CH_2N(CH_3)_2(C_2H_4OH)Cl$
DIAION PA 420	porous-type, strong base 2	$-CH_2N(CH_3)_2(C_2H_4OH)Cl$

ion exchanger

No. 4 2-Chome, Marunouchi, Chiyoda-ky, Tokyo
277 Park Avenue, New York, New York 10017

Particle shape	*Particle size (mesh)*	*Capacity (meq/g)*	*pH range*	*Permissible temperature (°C)*	*Density (g/liter)*
beads	48–16	1.90	0–14	120 (Na-, H-)	800–840
beads	48–16	1.90	0–14	120 (Na-, H-)	800–840
beads	48–16	0.60	0–14	120 (Na-, H-)	690–730
beads	48–16	0.90	0–14	120 (Na-, H-)	710–750
beads	48–16	1.20	0–14	120 (Na-, H-)	740–780
beads	48–16	1.60	0–14	120 (Na-, H-)	780–820
beads	48–16	2.00	0–14	120 (Na-, H-)	810–860
beads	48–16	2.10	0–14	120 (Na-, H-)	820–870
beads	48–16	2.10	0–14	120 (Na-, H-)	890–900
beads	48–16	1.20	0–12	60 (OH-), 80 (Cl-)	630–670
beads	48–16	0.85	0–12	60 (OH-), 80 (Cl-)	600–640
beads	48–16	1.30	0–12	40 (OH-)	650–690
beads	48–16	0.80	0–12	40 (OH-)	600–650
beads	200–100	2.00–2.20	0–14	120 (Na-, H-)	860–880
beads	200–100	1.00–1.30	0–12	60 (OH-), 80 (Cl-)	820–840
beads	48–16	0.65	0–14	120 (Na-, H-)	660–710
beads	48–16	1.20	0–14	120 (Na-, H-)	710–760
beads	48–16	1.50	0–14	120 (Na-, H-)	740–790
beads	48–16	1.75	0–14	120 (Na-, H-)	760–810
beads	48–16	1.90	0–14	120 (Na-, H-)	770–820
beads	48–16	2.00	0–14	120 (Na-, H-)	780–830
beads	48–16	2.05	0–14	120 (Na-, H-)	790–840
beads	48–16	0.45	0–12	60 (OH-), 80 (Cl-)	590–640
beads	48–16	0.70	0–12	60 (OH-), 80 (Cl-)	600–650
beads	48–16	0.90	0–12	60 (OH-), 80 (Cl-)	610–660
beads	48–16	1.00	0–12	60 (OH-), 80 (Cl-)	620–670
beads	48–16	1.05	0–12	60 (OH-), 80 (Cl-)	630–680
beads	48–16	1.05	0–12	60 (OH-), 80 (Cl-)	630–680
beads	48–16	1.05	0–12	60 (OH-), 80 (Cl-)	630–680
beads	48–16	1.05	0–12	60 (OH-), 80 (Cl-)	640–690
beads	48–16	1.05	0–12	60 (OH-), 80 (Cl-)	640–690
beads	48–16	0.55	0–12	40 (OH-), 60 (Cl-)	600–650
beads	48–16	0.70	0–12	40 (OH-), 60 (Cl-)	610–660
beads	48–16	0.90	0–12	40 (OH-), 60 (Cl-)	620–670
beads	48–16	1.00	0–12	40 (OH-), 60 (Cl-)	630–680
beads	48–16	1.10	0–12	40 (OH-), 60 (Cl-)	640–690
beads	48–16	1.15	0–12	40 (OH-), 60 (Cl-)	650–700
beads	48–16	1.20	0–12	40 (OH-), 60 (Cl-)	650–700
beads	48–16	1.20	0–12	40 (OH-), 60 (Cl-)	650–700
beads	48–16	1.20	0–12	40 (OH-), 60 (Cl-)	650–700

Table IV. Wofatit
VEB Farbenfabrik Wolfen, Wolfen

Trade name	*Type**	*Active group*		*Particle shape*
Wofatit KPS 200	C	$-SO_3^-$	strong acid	Beads
Wofatit F	C	$-SO_3^-$	strong acid	Granulate
Wofatit P	C	$-SO_3^-$	strong acid	Granulate
Wofatit CP 300	C	$-COO^-$	weak acid	
Wofatit CN	C		weak acid	Granulate
Wofatit N	A	Aromat. amine	weak acid	Granulate
Wofatit MD	A	Aromat. amine	medium base	Granulate
Wofatit L 150	A	Alkylimine	medium base	Granulate
Wofatit L 165	A	Alkylimine	strong base	Granulate

*C = cation exchanger, A = anion exchanger
Remarks: The Wofatit types are available for analytical and chromatographic (0.3–0.8 mm). Special resins: Wofatit E for decolorizing of solutions;

Table V. Dowex
Dow Chemical Company,

Trade name	*Type**	*Active Group*		*Particle shape*	*Particle size (US mesh)*
Dowex 50	C	$-SO_3^-$	strong acid	Beads	20–50
Dowex 50W	C	$-SO_3^-$	strong acid	Beads	20–50
Dowex 1	A		strong base	Beads	20–50
Dowex 21K	A		strong base	Beads	20–50
Dowex 2	A		strong base	Beads	20–50
Dowex 4	A		weak base	Granulate	16–40

*c = cation exchanger, A = anion exchanger
Remarks: Dowex 50 W is identical to Dowex 50 in chemical properties, but differs

Ion Exchangers
Kr. Bitterfeld, East Germany

Particle size (mm)	*Bulk density (g/liter)*	*Permissible temperature (°C)*	*Capacity (meq/g)*	*Effective volume capacity 100cm³ gCaO/resin*
0.3–1.0	750–800	115	4.5	4.5–5.0
0.3–1.5	700–750	50	2.9	1.8–2.0
0.3–1.5	650–750	97	1.9	0.9–1.0
0.3–1.0	750–800	30	10.0	4.0–4.5
0.3–1.5	800–850	30	2.0	0.8–1.8
0.3–1.5	600–700	30	4.3	1.2
0.3–1.5	700–750	30	5.7	2.0
0.3–1.5	600–700	50	10.0	3.5–4.0
0.3–1.5	600–700	50	8.0	5000 mg SiO_2/l.resin

purposes in the form of p.a.-Wofatits of high purity and finer granulometry
Wofatit CP 300 for the isolation of streptomycin.

Ion exchangers
Midland, Michigan

pH range	*Permissible temperature (°C)*	*Capacity (meq/g)*	*Cross-linking (%DVB)*	
			standard	*special products*
0–14	150	4.8	8	1–16
0–14	150	4.8	8	1–16
0–14	50 (OH-form), 150 (Cl-form)	3.5	8	1–10 and 16
0–14	50 (OH-form), 150 (Cl-form)	4.5		
0–14	30 (OH-form), 150 (Cl-form)	3.5	8	1–10
0– 7	65			porous

substantially in physical properties: higher resistance to attrition, white color.

Table VI. Russian cationites and anionites

Type	Active basic groups	Raw materials	Size in mm	Statistical exchange capacity (meq/g): maximum	Statistical exchange capacity (meq/g): with ions of neutral salts
STRONG ACID CATIONITES (CLASS KU)					
MSF (MSF-3)	SO_3H, OH	phenol, phenol–sulfonic acid	0.3–2.0	4.5	1.8
SDB-3	SO_3H	styrene	0.3–1.5	4.5	4.2
DB	SO_3H		0.3–0.2	3.5	3.0
NSF	SO_3H	naphthalene	0.3–2.0	3.0	2.4
RF (FB, FS)	PO_3H_2		0.3–2.0	5.0	3.2
AR	AsO_3H_2		0.3–2.0	5.0	3.2
KU–3	SO_3H	vinylnaphthalene	0.3–1.5	5.7	5.3
KU–4	SO_3H	acenaphthalene	0.3–1.5	5.7	5.3
KU–5	SO_3H	naphthalene	0.3–2.0	2.5	1.8
KU–6	SO_3H, COOH	acenaphthene	0.3–2.0	5.0	3.0
KU–6 F	SO_3H, COOH, OH	acenaphthene, phenol	0.3–1.5	6.2	3.5
KU–6, KU–8, KU–9	SO_3H, OH	phenol, resorcinol, sulfonic acids, aldehydes, and ketones	0.3–2.0	5.0	3.0
KBU–1 (KBU–2)	SO_3H, COOH	styrene, methacrylic acid	0.3–1.5	7.0	3.5
KF–1, KF–2, KF–3, KF–4	PO_3H_2	styrene, vinylnaphthalene, acenaphthalene	0.3–2.0	6.0	1.5
SN, SNF	SO_3H, OH	phenolic novolaks	0.3–2.0	5.2	3.0

WEAK ACID CATIONITES (CLASS KB)					
KB–1, KB–2, KB–3	COOH	methacrylic acid, methacrylate, nitriloacrylic acid	0.3–1.5	11.0	–
KC–1	COOH	maleic anhydride, methacrylic acid	0.3–2.0	6.5	–
KMD, KMG, KM			0.3–2.0	10.0	–
WEAK BASE ANIONITES (CLASS AN)					
AN–4k, AN–7k	N, HN_2, NH	polyvinyl chloride	0.3–2.0	6.0	–
AN–9	NH, N	phenol	0.3–2.0	5.0	–
AN–10	NH_2, N	allylamines	0.3–2.0	12.0	–
AN–15	NH_2	styrene	0.3–2.0	2.0	–
AN–17	NH	styrene, monomethylamine	0.3–2.0	5.0	–
AN–18	N	dimethylamine	0.3–2.0	4.5	–
AN–19	NH, N	styrene, polyethylene–polyamine	0.3–2.0	4.0	–
AN–20	NH_2, NH	styrene, ammonia	0.3–2.0	7.0	–
AN–21	NH, NH_2	styrene, hexamethylenediamine	0.3–2.0	7.0	–
AN–22	NH, NH_2	styrene, ethylenediamine	0.3–2.0	7.0	–
AN–23	N	vinylpyridine	0.3–2.0	5.0	–
STRONG BASE ANIONITES (CLASS AW)					
AW–15	$N(R)_3$	styrene	0.3–2.0	2.0	1.5
AW–17	$N(R)_3$	styrene, trimethylamine	0.3–2.0	3.8	3.5
AW–18	$N(R)_3$	styrene, pyridine	0.3–2.0	3.0	1.5
AW–19	$N(R)_3$	vinylnaphthalene, trimethylamine	0.3–2.0	3.0	2.5
AW–20	$N(R)_3$	vinylpyridine, epichlorohydrin	0.3–2.0	2.0	1.5
PEK	$N(R)_3$	ethylenediamine	0.3–2.0	6.0	1.8

Table VII. Inorganic Bio-Rad ion-exchangers

BioRad Laboratories, 32nd & Griffin Ave., Richmond, California

Trade name	*Type**	*Composition*	*Capacity (meq/g)*	*Particle size (US mesh)*	*Stability range*
Bio-Rad ZP–1	C	zirconium phosphate	1.5	20– 50 50–100 100–200	strong acid to pH 13
Bio-Rad ZT–1	C	zirconium tungstate	0.8	50–100 100–200	pH 1–pH 6
Bio-Rad ZM–1	C	zirconium molybdate	1.0	50–100 100–200	pH 1–pH 5
Bio-Rad AMP–1	C	ammonium molybdophosphate	1.2	micro-crys-talline	strong acid to pH 6
Bio-Rad KCF–1	C	potassium hexacyanoco-balt(II) ferrate(II)	0.5	20– 50 50–100	strong acid to pH 12
Bio-Rad HZO–1	A, C	zirconium oxide	1.5	20– 50 50–100 100–200	pH 1–5 *N* base
Bio-Rad HTO–1	A	titanium oxide	1.0	20– 50 50–100	–

*C = cation exchanger, A = anion exchanger

Table VIII. MN-Cellulose Ion Exchangers

Manufacturer: Macherey, Nagel & Co., 516 Duren, Werkstrabe 6–8, Postfach 307
Distributors: US: Brinkmann Instruments, Cantiague Road, Westbury, N. Y. 11590
UK: CAMLAB (Glass) LTD., Cambridge/Great Britain

Name	*Designation of the CC grade*[1]	*Exchange group*	*Type*[2]	*Capacity meq/g*	*Designation of the TLC grade*[3]	*Designation of the exchanger-coated Polygram ready-to-use film*
Aminoethylcellulose	MN 2100 AE	$—O \cdot C_2H_4 \cdot NH_2$	A	0.35	–	–
p-aminobenzyl cellulose	MN 2100 PAB	$—O \cdot CH_2 \cdot C_6H_4 \cdot NH_2$	A	0.15	–	–
Carboxymethylcellulose	MN 2100 CM	$—O \cdot CH_2 \cdot COO^-$	C	0.7	MN 300 CM	CEL 300 CM
Diethylaminoethylcellulose	MN 2100 DEAE	$—O \cdot C_2H_4 \cdot N(C_2H_5)_2$	A	0.7	MN 300 DEAE	CEL 300 DEAE
ECTEOLA cellulose	MN 2100 ECTEOLA	unknown	A	0.35	MN 300 ECTEOLA	CEL 300 ECTEOLA
Guanidoethylcellulose	MN 2100 GE	$—O \cdot C_2H_4 \cdot NH \cdot C{<}^{NH}_{NH_2}$	A	0.35	–	–
Phosphorylated cellulose	MN 2100 P	$—O \cdot PO_3H_2$	C	0.7	–	–
Polyethylenimine cellulose	MN 2100 PEI	$(—NH—CH_2—CH_2—)n$	A	1.0	MN 300 PEI	CEL 300 PEI
Polyphosphate cellulose	MN 2100 Poly-P	phosphate	C	0.7	MN 300 Poly-P	–
Sulfoethylcellulose	MN 2100 SE	$—O \cdot C_2H_4 \cdot SO_3^-$	C	0.2	–	–
Triethylaminoethylcellulose	MN 2100 TEAE	$—O \cdot C_2H_4 \cdot N(C_2H_2)_3$	A	0.7	–	–

[1]CC = column chromatography.
[2]A = anion exchanger; C = cation exchanger.
[3]TLC = thin-layer chromatography.

Table IX. Whatman advanced ion-exchange celluloses

Manufacturer: W & R Balston Ltd., Maidstone, Kent, England

Distributor: H. Reeve Angel & Co., 14, New Bridge St., London E.C. 4; 9 Bridewell Pl., Clifton, N.J.

Trade name	*Literature designation*	*Functional group*	*Type**	*Special description*
WHATMAN	carboxymethylcellulose	$-O-CH_2-COO^-$	C	
CM 22				new fibrous
CM 23				new fibrous, fines reduced
CM 32				microgranular
CM 52				microgranular, preswollen
WHATMAN	diethylaminoethylcellulose	$-O-C_2H_5-\overset{+}{N}H{<}^{C_2H_2}_{C_2H_2}$	A	
DE 22				new fibrous
DE 23				new fibrous, fines reduced
DE 32				microgranular
DE 52				microgranular, preswollen

*C = cation exchanger; A = anion exchanger

Table X. SELECTA cellulose ion exchangers

Manufacturer: Carl Schleicher & Schüll, 3354 Dassel, W. Germany

Distributors: US: Mssrs. Carl Schleicher & Schüll Co., 543 Washington Street, Keene, N.H.
UK: Anderman & Co. Ltd., Battlebridge House, 87–95 Tooley Street, London S.E. 1/England

Name	*Exchange group*	*Type**	*Capacity, meq/g*
CAM-cellulose	$-O-CH_2-COOH$	C, weak acid	0.7–0.9
CAM-cellulose, coarse	$-O-CH_2-COOH$	C, weak acid	1.0–1.2
CAS-cellulose	$-O-CO \cdot (CH_2)_2 \cdot COOH$	C, weak acid	0.5
P-cellulose	$-O \cdot PO_3H_2$	C, strong acid	0.8–1.0
DEAE-cellulose	$-O \cdot C_2H_4 \cdot N(C_2H_5)_2$	A	0.5–0.7
DEAE-cellulose, coarse	$-O \cdot C_2H_4 \cdot N(C_2H_2)_2$	A	0.8–1.0
Ecteola-cellulose	quaternary ammonium–	A, weak base	0.3–0.5
TEAE-cellulose	$[-O \cdot C_2H_4 \cdot N(C_2H_5)_3]^+$	A	0.5–0.7
PEI-cellulose	imino–	A, strong base	1.0

C = cation exchanger; A = anion exchanger

Table XI. Conversion table for water hardness units

	US hardness, degree	*British hardness, degree*	*French hardness, degree*	*German hardness, degree*	*Russian hardness, degree*	*ppm as $CaCO_3$*
1° US	1	1.201	1.716	0.961	6.864	17.16
1° British	0.8324	1	1.429	0.7999	5.714	14.29
1° French	0.5828	0.700	1	0.5599	4.000	10.00
1° German	1.041	1.250	1.786	1	7.144	17.85
1° Russian	0.1457	0.175	0.2500	0.1400	1	2.500
1 ppm as $CaCO_3$	0.05828	0.070	0.1000	0.4000	0.4000	1

Table XII. IMAC ion exchangers

Imacti, Postfach 240, Amsterdam-C, Netherlands

	IMAC Cation Exchangers					
	Dusarit S	*Imac C 19*	*Imac Z 5*	*Imac C 12*	*Imac C 8 P*	*Imac C 16 P*
Type	strongly acidic and weakly acidic	strongly acidic and weakly acidic	weakly acidic	strongly acidic	strongly acidic	strongly acidic
Chemical composition	Coal	Coal	Polyacrylic	Polystyrene	Polystyrene porous	Polystyrene macroporous
Active groups	polyfunctional	polyfunctional	COOH	SO_3H	SO_3H	SO_3H
Physical form	granules	granules	beads	beads	beads	beads
Approx. screen grading:						
mm	0.3–1.2	0.3–1.2	0.3–1.2	0.3–1.2	0.3–1.2	0.3–1.2
mesh U.S.S.	16–50	16–50	16–50	16–50	16–50	16–50
Moisture content, in %, approx.		30	48	50	57	48
Total capacity:	strong weak	strong weak				
meq/l resin	700 350	300 1000	3000	2000	1700	2100
kgrs $CaCO_3$/cu.ft. resin	15 8	6 22	66	44	37	46
pH range	0–14	4–14	4–14	0–14	0–14	0–14
Maximum temp.:						
°C	80	80	100	120	120	120
°F	175	175	212	250	250	250
Variation of volume, maximum	H → Na + 15%	H → Na + 15%	H → Na + 60%	H → Na − 5%	H → Na − 5%	H → Na − 5%

Table XII. IMAC ion exchangers—Continued

	IMAC Cation Exchangers					
	Dusarit S	*Imac C 19*	*Imac Z 5*	*Imac C 12*	*Imac C 8 P*	*Imac C 16 P*
Regeneration: meq/l resin	600–1600	500–800	1000–2300	H 1000–3000 Na 2000–6000	H 1000–3000 Na 2000–6000	H 1000–3000 Na 2000–6000
lbs/cu.ft. resin	HCl 1.5-4 H_2SO_4 2 -5	HCl 1 -2 H_2SO_4 1.5-2.5	HCl 2.3-5.2 H_2SO_4 3 -7	HCl 2.3-7 H_2SO_4 3 -9.5 NaCl 7.3-22	HCl 2.3-7 H_2SO_4 3 -9.5 NaCl 7.3-22	HCl 2.3-7 H_2SO_4 3 -9.5 NaCl 7.3-22
Regenerant concentration, in%	HCl 3-6 H_2SO_4 1-5	HCl 3 -6 H_2SO_4 0.7-1.5	HCl 3 -6 H_2SO_4 0.7-1.5	HCl 6-12 H_2SO_4 1-10 NaCl 6-12	HCl 6-12 H_2SO_4 1-10 NaCl 6-12	HCl 6-12 H_2SO_4 1-10 NaCl 6-12
Operating capacity*: meq/l resin	400–800	500–750	1000–2000	800–1400	700–1300	800–1600
kgrs $CaCO_3$/cu.ft. resin	9– 18	11– 16	22– 44	17– 31	15– 28	17– 35

	IMAC Anion Exchangers				
	Imac A 20 *Imac A 21*	*Imac A 27*	*Imac S 5-40*	*Imac S 5-50*	*Imac S 5-42*
Type	weakly basic	mainly weakly basic strongly basic	very strongly basic (type I)	very strongly basic (type I)	strongly basic (type II)
Chemical composition	Polystyrene porous	Polyamine condensate	Polystyrene porous	Polystyrene porous	Polystyrene highly porous
Active groups	A20: I, II, III† A21: III†	(I, II) III, IV†	$R \cdot N(CH_3)_3OH$	$R \cdot N(CH_3)_3OH$	$R \cdot [N(CH_3)_2CH_2CH_2OH]OH$
Physical form	beads	beads	beads	beads	beads

Property					
Approx. screen grading:					
mm	0.4–0.85	0.3–2.0	0.4–0.85	0.4–0.85	0.4–0.85
mesh U.S.S.	20–40	10–50	20–40	20–40	20–40
Moisture content, in %, approx.	A20: 55 A21: 50	60	58	53	53
Total capacity:		strong weak			
meq/l resin	A20: 2050 57	300 1700	1000	1200	1100
kgrs $CaCO_3$/cu.ft. resin	A21: 2100 46	6 38	22	26	23
pH range	0–8	0–8	0–14	0–14	0–14
Maximum temp.:			ROH 50° C 120° F	ROH 50° C 120° F	ROH 40° C 105° F
°C	100	35			
°F	212	95	RCl 80° C 175° F	RCl 80° C 175° F	RCl 80° C 175° F
Variation of volume, maximum	OH → Cl A20: + 25% A21: + 20%	OH → Cl + 15%	OH → Cl – 20%	OH → Cl – 15%	OH → Cl – 15%
Regeneration:					
meq/l resin	1400–2100	1100–2400	1500–3500	1500–3500	1000–2000
lbs/cu.ft. resin	NaOH 3.5-5.3 NH_3 1.5-2.3 Na_2CO_3 4.5-7	NaOH 2.8-6 NH_3 1.2-2.5 Na_2CO_3 3.5-8	NaOH 4-10	NaOH 4-10	NaOH 2.5-5
Regenerant concentration, in %	NaOH 3-8 NH_3 2-5 Na_2CO_3 5-8	NaOH 3-8 NH_3 2-5 Na_2CO_3 5-8	NaOH 4-6	NaOH 4-6	NaOH 3-6
Operating capacity*:					
meq/l resin	1300–1800	900–1900	500–800	600–900	600–900
kgrs $CaCO_3$/cu.ft. resin	28– 39	20– 42	11– 17	13– 20	13– 20

Table XII. IMAC ion exchangers—Continued

	ASMIT Decolorizing Resins		
	Asmit 173 N *Asmit 173 NP*	*Asmit 259 N*	*Asmit 261*
Type	—	strongly basic scavenger	strongly basic
Chemical composition	—	Polystyrene porous	Polystyrene porous
Active groups	—	$R \cdot N(CH_3)_3OH$	$R \cdot N(CH_3)_3OH$
Physical form	granules	beads	beads
Approx. screen grading: mm	0.3–2.0	0.5–1.1	0.4–0.85
mesh U.S.S.	10–50	16–30	20–40
Moisture content, in %, approx.			
Total capacity: meq/l resin	—	500	800
kgrs $CaCO_3$/cu.ft. resin		11	17
pH range	1–7 preferably 3–5	3–9	3–9
Maximum temp.: °C	100	80	80
°F	212	175	175

*Depending on the amount of regenerant, the pH and the composition of the solution to be treated; for weak base anion exchangers especially on the ratio Cl : SO_4 and the temperature.

†I primary amines; II secondary amines; III tertiary amines; IV quaternary ammonium.

Table XIII. Conversion of weights

	grain *gr*	*gram* *g*	*pound* *lb*	*kilogram* *kg*	*ton (metric)* *t*	*ton (long)* *long ton*
1 gr	1	0.064799	$1.4286 \cdot 10^{-4}$	$6.4799 \cdot 10^{-4}$	$6.4799 \cdot 10^{-8}$	$6.378 \cdot 10^{-8}$
1 g	15.4324	1	$2.2046 \cdot 10^{-3}$	10^{-3}	10^{-6}	$9.842 \cdot 10^{-7}$
1 lb	7,000	453.592	1	0.45359	$4.5359 \cdot 10^{-4}$	$4.464 \cdot 10^{-4}$
1 kg	$1.5432 \cdot 10^{4}$	10^{3}	2.20462	1	10^{-3}	$9.842 \cdot 10^{-4}$
1 t	$1.5432 \cdot 10^{7}$	10^{6}	2,204.62	10^{3}	1	0.9842
1 long ton	$1.5679 \cdot 10^{7}$	$1.01605 \cdot 10^{6}$	2,240	1,016.05	1.01605	1

Table XIV. Conversion of Volumes

	Milliliter (*ml*)	*Liter* (*l*)	*Gallon* (*US*) *gal* (*US*)	*Gallon* (*imp*) *gal* (*imp*)	*Cubic foot* *cu ft*	*Cubic meter* m^3
1 ml	1	10^{-3}	$264.178 \cdot 10^{-6}$	$219.976 \cdot 10^{-6}$	$3.53157 \cdot 10^{-5}$	$1.000028 \cdot 10^{-6}$
1 l	10^3	1	$264.178 \cdot 10^{-3}$	$219.976 \cdot 10^{-3}$	0.0353157	$1.000028 \cdot 10^{-3}$
1 gal (US)	3,785.33	3.78533	1	0.832680	0.133681	$3.78543 \cdot 10^{-3}$
1 gal (imp)	4,545.96	4.54596	1.20094	1	0.160544	$4.54609 \cdot 10^{-3}$
1 cu ft	28,316.1	28.3161	7.48047	6.22884	1	$28.3168 \cdot 10^{-3}$
1 m^3	999,972	999.972	264.170	219.969	35.3147	1

Table XV. Conversion of densities and concentrations

	Parts per million (ppm)	*Grain per gal (imp) [gr/gal (imp)]*	*Grain per gal (US) [gr/gal (US)]*	*Gram per liter (g/l)*	*Pound per cubic foot (lb/cu ft)*	*Gram per cubic centimeter (g/cm³)*
1 ppm	1	0.070157	0.058418	0.001000	$6.2426 \cdot 10^{-5}$	10^{-6}
1 gr/gal (imp)	14.2538	1	0.832680	0.0142538	$8.8983 \cdot 10^{-4}$	$1.42538 \cdot 10^{-5}$
1 gr/gal (US)	17.118	1.20094	1	0.017118	0.0010686	$1.7118 \cdot 10^{-5}$
1 g/l ($\equiv$ kg/m³)	10^{3}	70.157	58.418	1	0.062428	10^{-3}
1 lb/cu ft	$1.60189 \cdot 10^{4}$	$1.1238 \cdot 10^{3}$	935.767	16.0189	1	0.0160189
1 g/cm³	10^{6}	$70.157 \cdot 10^{3}$	$5.8418 \cdot 10^{4}$	10^{3}	62.426	1

Table XVI. Conversion of flow rates

	Gallon (imp) per hour [*gal (imp)/h*]	*Gallon (US) per minute* [*gal (US)/min*]	*Gallon (imp) per minute* [*gal (imp)/min*]	*Cubic meter per hour* (m^3/h)	*Cubic foot per minute* (*cu ft/min*)	*Liter per second* (*l/s*)	*Cubic foot per second* (*cu ft/sec*)
1 gal (imp)/h	1	0.020016	0.016667	0.004546	0.0026757	0.0012628	0.0000446
1 gal (US)/min	49.9610	1	0.83268	0.22712	0.13368	0.06309	0.002228
1 gal (imp)/min	60.00	1.20094	1	0.272765	0.16054	0.75766	0.0026757
1 m^3/h	219.969	4.4029	3.66615	1	0.588578	0.27777	0.0098096
1 cu ft/min	373.731	7.4806	6.22883	1.69901	1	0.47193	0.016667
1 l/s	791.912	15.8509	13.1985	3.600	2.11894	1	0.0353156
1 cu ft/sec	22,423.8	448.83	373.731	101.941	60.00	28.316	1

Table XVII. Chemical equivalents

1 lb as $CaCO_3$	= 9.08 geq	1 geq	= 0.11 lb as $CaCO_3$
1 kg as $CaCO_3$	= 1.3 geq	1 geq	= 0.77 kg as $CaCO_3$
1 lb/ft^3 as $CaCO_3$	= 0.32 geq/l	1 geq/l	= 3.12 kg/ft^3 as $CaCO_3$
1 g/gal as $CaCO_3$	= 0.286 meq/l	1 meq/l	= 35.0 g/gal as $CaCO_3$

Bibliography

1. van Abbé, N. J., and J. T. Rees. "Amberlite XE-88 as a Tablet-disintegrating Agent," J. Am. Pharmacol. Assoc. Sci. Sd. **47,** 487 (1958).
2. Abrams, I. M., and S. M. Lewon. "Removal of ABS from Water by Chloride Cycle Anion Exchange," J. Am. Water Works Assoc. **54,** 537 (1962).
3. Adams, B. A., and E. L. Homes. Chem. Age **38,** 117 (1938).
4. Adams, B. A., and E. L. Homes. J. Soc. Chem. Ind. **T54,** 1 (1935).
5. Adamson, A. W., and J. J. Grossmann. "A Kinetic Mechanism for Ion-Exchange," J. Chem. Physiol. **17,** 1002 (1949).
6. Ahrland, S., J. Grenthe, and B. Noren. "Ion Exchange Properties of Silica Gel," Acta Chem. Scand. **14,** 1059, 1077 (1960).
7. Alberti, G., and A. Conte. "Effects of the Drying Temperature on the Ion Exchange Properties of Zirconium Phosphate," J. Chromatog. **5,** 244 (1961).
8. Alberti, G., *et al.* "Comparison of Ion Exchange Properties of Amorphous and Crystalline Insoluble Acid Salts of Tetravalent Metals and Polybasic Acids," Conference on Ion Exchange, London, 1969.
9. Allen, K. A. J. Phys. Chem. **60,** 943 (1956).
10. Alm, R. S., R. J. P. Williams, and A. Tiselius. "Gradient Elution Analysis: I. A General Treatment," Acta Chem. Scand. **6,** 826 (1952).
11. Alstad, J., and A. O. Brunfelt. "Adsorption of Rare Earth Elements on an Anion Exchange Resin from Nitric Acid–Acetone Mixtures," Anal. Chim. Acta **38,** 185 (1967).
12. Amphlett, C. B., and L. A. McDonald. "Equilibrium Studies on Natural Ion Exchange Minerals. I. Cesium and Strontium," J. Inorg. Nucl. Chem. **2,** 403 (1956).
13. Amphlett, C. B., and L. A. McDonald. "Equilibrium Studies on Natural Ion Exchange Minerals: II. Cesium, Sodium and Ammonium Ions," J. Inorg. Nucl. Chem. **6,** 145 (1958).
14. Amphlett, C. B., L. A. McDonald, and M. J. Redman. "Ion Exchange Properties of Hydrous Zirconium Oxide," Chem. Ind. **1957,** 365.
15. Amphlett, C. B., L. A. McDonald, and M. J. Redman. "Synthetic Inorganic Ion Exchange Materials: II. Hydrous Zirconium Oxide and other Oxides," J. Inorg. Nucl. Chem. **6,** 220, 236 (1958).
16. Amphlett, C. B., *et al.* Chem. Ind. **1956,** 1314.
17. Anders, J. "Ion Exchangers for Pickling Effluents," Städtehygiene **14,** 47 (1963).
18. Anderson, R. E. "A Contour Map of Anion Exchange Resin Properties," Ind. Eng. Chem. Prod. Res. Dev. **3,** 85 (1964).

19. Anderson, R. E., and R. D. Hansen. "Phenol Sorption on Ion Exchange Resins," Ind. Eng. Chem. **47,** 71 (1955).
20. Anon. "Process uses Powdered Ion Exchange Resins," Chem. Eng. News **40** (3) 61 (1962).
21. Aoki, F. Bull. Chem. Soc. Japan **26,** 480 (1953).
22. Applezweig, N. "Cinchona Alkaloids Prepared by Ion Exchange," J. Am. Chem. Soc. **66,** 1990 (1944).
23. Arden, T. V., and M. Giddings. "Anion Exchange in Chromate Solution." J. Appl. Chem. **11,** 229 (1961).
24. Argonne National Laboratory. *Reactor Fuel Processing,* Vol. 2 (U.S.A.E.C.: Jan. 1959), No. 1, p. 19.
25. Assalini, G., and G. Brandoli. "Ion Exchangers for the Treatment of Beet Sugar Juices," J. Am. Soc. Sugar Beet Technologists **11,** 341, 349 (1960).
26. Ass. Sci. and Techn. Soc. of South Africa. *Uranium in South Africa 1946–56,* Vol. I and II (Johannesburg, 1957).
27. Astle, M. J., and R. W. Etherington. Ind. Eng. Chem. **44,** 2871 (1952).
28. Astle, M. J., and J. A. Zaslowsky. Ind. Eng. Chem. **44,** 2867 (1952).
29. Atteberry, R. W., and G. E. Boyd. J. Am. Chem. Soc. **72,** 4805 (1950).
30. Austrian Patent No. 216,976. International Atomic Energy Organization, 1960 (O. Bobleter and K. Buchtela).
31. Baestlè, L. H., and J. Pelsmaekers. "Ion Exchange Properties of Zirconyl Phosphates: I. Contribution to the Structure of Zirconyl Phosphates," J. Inorg. Nucl. Chem. **21,** 124 (1961).
32. Baestlè, L. H., *et al.* "Ion Exchange Properties of Zirconium and Titanium Phosphates and their Use in the Separation of Nuclear Reaction Products," Conference on Ion Exchange, London, 1969.
33. Baestlé, L. H., *et al.* Proc. 3rd U. N. Intern. Conf. on the Peaceful Uses of Atomic Energy, Geneva, 1964 **10,** 580 (1965).
34. Bähr, G. "Voltage-Time Measurements of Polar Double Membranes: I. Qualitative Description of Experiments," Ber. Bunsenges. Physik. Chem. **71,** 883 (1967).
35. Baldwin, P. La Tribune de CEBEDEAU **17,** 248 (1964).
36. Baldwin, P. "A New Method for Saving Regenerants in Water Treatment by Ion exchangers," Techn. Ueberwach. **5,** 415 (1964).
37. Barrer, R. M. "Some Features of Ion Exchange in Crystals," Chem. Ind. **1962,** 1258.
38. Barrer, R. M., and L. Hinds. "Ion Exchange in Crystals of Analcite and Leucite," J. Chem. Soc. (London) **1953,** 1879.
39. Barrer, R. M., and D. M. MacLeod. "Trans. Faraday Soc. **50,** 980 (1954).
40. Barrer, R. M., *et al.* "Hydrothermal Chemistry of the Silicates: VIII. Low Temperature Crystal Growth of Aluminosilicates and of some Gallium and Germanium Analogues," J. Chem. Soc. (London) **1959,** 195.

41. Barrer, R. M., *et al.* Trans. Brit. Ceram. Soc. **56,** 155 (1957).
42. Basinski, A., and M. Sierocka. "Course of Brine Formation from Fe(III)-Aluminum Hydroxide and Chromium Hydroxide with Ion Exchangers," Ann. Soc. Chim. Polonorum **29,** 656 (1955).
43. Baxter, G. P., and R. G. Griffen. Am. Chem. J. **34,** 204 (1905).
44. Becker-Boost, E. H. "Decisive Factors for the Design of Ion Exchangers," Chem. Eng. Sci. **14,** 331 (1961).
45. Becker-Boost, E. H. "Process Technology of Ion Exchange: I. Designing by Empirical Principles," Chem. Ingr.-Tech. **27,** 579 (1955).
46. Becker-Boost, E. H. "Process Technology of Ion Exchange: II. Static and Industrial Reaction Kinetics," Chem. Ingr.-Tech. **28,** 411 (1956).
47. Becker-Boost, E. H. "Process Technology of Ion Exchange: III. Designing by Theoretical Principles." Chem. Ingr.-Tech. **28,** 532 (1956).
48. Belgian Patent No. 633,826, Pintsch Bamag AG, 1963.
49. Bendich, A., *et al.* J. Am. Chem. Soc. **77,** 3671 (1955).
50. Bendich, A., *et al.* J. Am. Chem. Soc. **80,** 3949 (1958).
51. Benerito, R. R., B. B. Woodward, and J. D. Guthrie. "Preparation and Properties of Quaternary Cellulose Anion Exchangers," Anal. Chem. **37,** 1963 (1965).
52. Benesi, H. A., *et al.* Anal. Chem. **27,** 1963 (1955).
53. Bennett, B. A., F. L. D. Cloete, and M. Streat. "A Systematic Analysis of the Performance of a New Continuous Ion-Exchange Technique," Conference on Ion Exchange, London, 1969.
54. Beukenkamp, J., and W. Rieman. Anal. Chem. **22,** 582 (1950).
55. Beukenkamp, J., *et al.* Anal. Chem. **26,** 505 (1954).
56. Beyer, W. A., and D. B. James. "Independence of the Performance of an Ion-Exchange Column on its Shape," Ind. Eng. Chem. Fundamentals **5,** 433 (1966).
57. Bhakuni, T. S., and C. S. Sastry. Envir. Health, Nagpur **5,** 61 (1963).
58. Bhakuni, T. S., and C. S. Sastry. Envir. Health, Nagpur **6,** 246 (1964).
59. Blaedel, W. J., and T. J. Haupert. "Exchange Equilibrium though Ion Exchange Membranes: Analytical Applications," Anal. Chem. **38,** 1305 (1966).
60. Blaedel, W. J., and E. D. Olsen. "Application of Chlorine in Cation Exchange Separations," Anal. Chem. **33,** 531 (1961).
61. Blake, W. E., and J. Randle. "Removal of Zn^{2+} from the Ternary System Zn^{2+}-Na^{+}-H^{+} by Cation Exchange Column." J. Appl. Chem. **17,** 358 (1967).
62. Blanchard, J., and J. G. Nairn. "The Binding of Cholate and Glycholate Anions by Anion-Exchanger Resins," J. Phys. Chem. **72,** 1204 (1968).
63. Blasius, E., and B. Brozio. *Chelating Ion Exchange Resins in Chelates in Analytical Chemistry* (New York: M. Dekker Publishing Co., 1969).
64. Blasius, E., and H. Pittack. Angew. Chem. **71,** 445 (1959).

65. Blasius, E., and R. Schmitt. "On the Sorption Properties of Cation Exchangers based on Polystyrene in Aqueous Dioxane and dimethylsulfoxide," J. Chromatog. **42,** 53 (1969).
66. Blasius, E., and R. Schmitt. "Residual Water Analysis in Ion Exchangers with the Aid of Tritium-labeled Water and Comparison of this method with the Karl-Fischer Titration. Z. Anal. Chem. **241,** 4 (1968).
67. Blasius, E., *et al.* "Ionic Sieves: Capillary Properties of Differently Cross-linked Anion Exchanger Based on Synthetic Resins," Angew. Chem. **68,** 671 (1956).
68. Blesing, N. V. *et al.* "Some Ion Exchange Processes for Partial Demineralization," Conference on Ion Exchange, London, 1969.
69. Bock, K. J., and W. Lukaschek. "Effect of the Plant Effluent Treatment Installations of Chemische Werke Hüls AG on Central Treatment Plants and Sewer Systems," Chem. Ingr.-Tech. **40,** 286 (1968).
70. Bodamer, G. W., and R. Kunin. "Behavior of Ion Exchange Resins in Solvents other than Water," Ind. Eng. Chem. **45,** 2577 (1953).
71. Boehm, H.-P., and K. H. Lieser. "Radiomechanical Determination of the Capacity of Clay Minerals for Cation Exchange," Z. anorg. allgem. Chem. **304,** 207 (1960).
72. Boirie, Ch. Bull. Soc. Chim. France **1958,** 1088.
73. Bolto, B. A., *et al.* "Some Fundamental Aspects of the Thermal Regeneration of Weak Electrolyte Resins," Conference on Ion Exchange, London, 1969.
74. Bonner, O. D., and J. C. Moorefield. "Ion Exchange in Mixed Solvents," J. Phys. Chem. **58,** 555 (1954).
75. Bonner, O. D., and R. R. Pruett. "Variations in the Structure of Sulfonic Acid Type Cation Exchanger Resins and the Effect of these Variations on their Properties," Z. phys. Chem. N.S. **25,** 75 (1960).
76. Borgolte, Th. "Economical Methods for the Treatment of Hydrochloric Acid Iron Pickles by Ion Exchange." Chem.-Ztg. **92,** 621 (1968).
77. Bors, G., *et al.* "The Isolation of Atropine, Morphine and Strychnine by Means of Ion Exchange Resin in Forensic Chemistry," Farmacia **12,** 479 (1964).
78. Bouchard, J. "The Development of the Degrémont-Asahi Continuous Ion Exchange Process," Conference on Ion Exchange, London, 1969.
79. Boyd, G. E., A. W. Adamson, and L. S. Myers, Jr. "The Exchange Adsorption of Ions from Aqueous Solutions by Organic Zeolites." II. Kinetics. J. Am. Chem. Soc. **69,** 2836 (1947).
80. Brantner. "Bacteriological Investigations of Ion Exchange Installations," Gas-Wasserfach **109,** 241 (1968).
81. Braun, D., and A. Y. Kim. "On the Structure of Cross-linked Polymers of Styrene, P-Iodostyrene and Divinylbenzene," Kolloid-Z. **216-217,** 321 (1967).
82. Breck, D. W., W. G. Eversole, R. M. Milton, T. B. Reed, and T. L. Thomas. "Crystalline Zeolites. I. The Properties of a New Synthetic Zeolite Type A," J. Am. Chem. Soc. **78,** 5963 (1956).

83. Breck, D. W., and J. V. Smith. "Sci. Am." **1,** (1959).
84. Breitling, V. "Economy in Water Treatment," Techn. Mitt. Krupp **59,** (8) 413 (1966).
85. British Patent No. 450,308/9 of 11–13–1934 (B. A. Adams and E. L. Homes).
86. British Patent No. 553,233 of 5–13–1943 (B. A. Adams and E. L. Homes).
87. British Patent No. 806,107, 1956 (Priority 1955 to Illinois Water Treatment (Co.).
88. British Patent No. 831,206 1956 (R. R. Porter and T. V. Arden).
89. British Patent No. 958,048, 1964. Process for removal of radionuclides from milk.
90. British Patent No. 1,013,069, 1967 (International Analyser Co.).
91. Broadbank, R. W. C., *et al.* "A Possible Use of Ammonium-12-Molybdateophosphate for assaying Certain Radioactive Fission Products in Water," Analyst **85,** 365 (1960).
92. Broadbank, R. W. C., *et al.* "The Ion Exchange and Other Properties of 12-Molybdateophosphates," J. Inorg. Nucl. Chem. **23,** 311 (1961).
93. Brodsky, A., *et al.* "Harmful Effects to Organic Materials in Full Desalination of Water." Mitt. Ver. Grosskesselbesitzer **1964,** No. 90, 190.
94. Brooks, R. R. "The Use of Ion Exchange Enrichment in the Determination of Trace Elements in Sea water," Analyst **85,** 745 (1960).
95. Brown, K. B., C. F. Coleman, and J. C. White. Proc. 2nd U. N. Intern. Conf. on the Peaceful Uses of Atomic Energy, Geneva, 1958 **3,** 472 (P/509).
96. Brown, W. E., and W. Rieman. J. Am. Chem. Soc. **74,** 1278 (1952).
97. Brunner, R. Personal communication.
98. Brunner, R. "Economical Design Aspects of Ion Exchange Installations," Thirty Years of Synthetic Ion Exchange Resins, Symposium. Leipzig, 1968.
99. Brunner, R., and B. Vassiliou. "The Desal Process: an Ion Exchange Process for Highly Saline Raw Water," VGB-Potable Water Conference, 1968, p. 61.
100. Brutovsky, M., and M. Zaduban. "Application of the Anion Exchanger Wofatit SBW for the Concentration of I^{131}," Collection Czech. Chem. Commun. **32,** 505 (1967).
101. Büchi, J. "Ion Exchangers in Pharmacy and Medicine," Arzneimittel-Forsch. **1,** 247 (1951).
102. Büchi, J. "Technical Applications of Ion Exchange," J. Pharm. Pharmacol. **8,** 369 (1956).
103. Büchi, J., and F. Furrer. Arzneimittel-Forsch. **3,** 1 (1953).
104. Büchi, J., and M. Soliva. "Demineralized Water by Ion Exchange," Pharm. Acta Helv. **29,** 221 (1954).
105. Buchwald, H., and W. P. Thistlewhaite. "Some Cation Exchange Properties of Ammonium 12-Molybdophosphate," J. Inorg. Nucl. Chem. **5,** 341 (1958).
106. Bunzl, K., and G. Dickel. Z. Naturforsch. **24a,** 110 (1969).

107. Burstall, F. H., *et al.* Ind. Eng. Chem. **45,** 1648 (1953).
108. Busch, L. "Sodium Loss with Ion Exchangers," Therap. Berichte **1954,** 163.
109. Buser, W. Helv. Chim. Acta **34,** 1635 (1951).
110. C. A. **54:** 11,415 (1960).
111. C.A. **57:** 10747 g. (H. J. Riedel).
112. C.A. **64:** 6305c.
113. Calman, C., and G. P. Simon. "Behavior of Ion Exchangers in Ultra-Pure Water Systems," Conference on Ion Exchange, London, 1969.
114. Carlsson, B., T. Isaksson, and O. Samuelson. "Automatic Chromatography of Hydroxy Acids on Anion-Exchange Resins," Anal. Chim. Acta **43,** 47 (1968).
115. Cassidy, H. G. "Electron Exchange Polymers," J. Am. Chem. Soc. **71,** 402 (1949).
116. Cassidy, H. G., and A. K. Kun. "Oxidation-Reduction Polymers (Redox-polymers)," Polymer Rev. **1965,** 11.
117. Chapman, C., and J. A. S. Wyatt. "Electrodialysis: the Interaction between Costs and Design," Conference on Ion Exchange, London, 1969.
118. Chatterjee, A., and J. A. Marinsky. "Dissociation of Methacrylic Acid Resins," J. Phys. Chem. **67,** 41 (1963).
119. Chatterjee, A., and J. A. Marinsky. "A Thermodynamic Interpretation of the Osmotic Properties of Cross-Linked Polymethacrylic Acid," J. Phys. Chem. **67,** 47 (1963).
120. Chu, B., D. C. Whitney, and R. M. Diamond. "On Anion-Exchange Selectivities," J. Inorg. Nucl. Chem. **24,** 1405 (1962).
121. Clayton, R. C. "Systematic Analysis of Continuous and Semi-Continuous Ion-Exchange Techniques and the Development of a Continuous System," Conference on Ion Exchange, London, 1969.
122. Clearfield, A. "Structure and Properties of Zirconium Phosphate-Type Ion Exchangers," Conference on Ion Exchange, London, 1969.
123. Cloete, F. L. D., C. R. Frost, and M. Streat. "Fractional Separations by Continuous Ion Exchange," Conference on Ion Exchange, London, 1969.
124. Cohn, W. E. J. Am. Chem. Soc. **72,** 1471 (1950).
125. Cohn, W. E., and H. W. Kohn. J. Am. Chem. Soc. **70,** 1986 (1948).
126. Cole, R., and S. L. Shulman. "Adsorbing Sulfur Dioxide on Dry Ion Exchange Resins," Ind. Eng. Chem. **52,** 859 (1960).
127. Coleman, C. F., *et al.* Ind. Eng. Chem. **50,** 1756 (1958).
128. Commoner, B., *et al.* Nature **178,** 767 (1956).
129. Cookson, D., *et al.* "Use of Anion-Exchange Resins for Syrup Decolorization in Sugar Refining," Conference on Ion Exchange, London, 1969.
130. Cooney, D. O. "Column Configuration Performance in Fixed-Bed Ion-Exchange Systems," Ind. Eng. Chem. Fundamentals **6,** 159 (1967).
131. Cooper, R. S. "Slow Particle Diffusion in Ion-Exchange Columns," Ind. Eng. Chem. Fundamentals **4,** 308 (1965).

132. Coursier, J., and J. Huré. Anal. Chim. Acta **18,** 272 (1958).
133. Crouch *et al.* AERE C/R 2325.
134. Crouse, D. J., and K. B. Brown. "The Amex Process for Extracting Thorium Ores with Alkyl Amines," Ind. Eng. Chem. **51,** 1461 (1959).
135. D'Ans, J., E. Blasius, H. Guzatis, and U. Wachtel. "The Application of Anion Exchangers in Analytical and Preparative Chemistry," Chem.-Ztg. **76,** 811 (1952).
136. Davies, C. W., and Owen, B. D. R. J. Chem. Soc. **1956,** 1676.
137. Davison, J., F. O. Read, F. D. L. Noakes, and T. V. Arden. "Ion Exchange for Gold Recovery," Bull. Inst. Mining Met. **651,** 247 (1961).
138. Dawson, J., and R. J. Magee. Mikrochim. Acta **1958,** 325.
139. De, A. K., and Sen. Talanta **13,** 1313 (1966).
140. De, A. K., and Sen. Z. Anal. Chem. **211,** 243 (1965).
141. Dehner, J. "Stabilization of Effervescent Wines by Cation Exchangers," Weinberg Keller **12,** 403 (1965).
142. Deuel, H., and F. Hostettler. "One Hundred Years of Ion Exchange," Experienta **6,** 445 (1950).
143. Deuel, H., and K. Hutschneker. "The Structure and Mechanism of Action of Ion Exchangers," Chimia **9,** 49 (1955) (Table 4).
144. Deuel, H., *et al.* Helv. Chim. Acta **34,** 1849 (1951).
145. Deutsche Pharmacia, Frankfurt/Main.
146. Diamond, R. M. J. Am. Chem. Soc. **77,** 2978 (1955).
147. Dickel, G. J. Chem. Physiol. **60,** 73 (1963).
148. Dickel, G. Z. Elektrochem. **54,** 353 (1950).
149. Dickel, G. "Kinetics of the Ion Exchangers. Coupling of Diffusion and Reaction in the Ion Exchanger," Z. Naturforsch. **23a,** 2077 (1968).
150. Dickel, G. "Thermodynamic Treatment of Ion Exchange Equilibria according to the Gibbs-Donnan-Guggenheim Membrane Model," Z. Physik, Chem. **25,** 233 (1960).
151. Dickel, G., and K. Bunzl. "The Swelling Behavior of Ion Exchangers. I. Swelling Pressure According to the Gibbs-Kelvin Concept," Z. Physik. **39,** 198 (1963).
152. Dickel, G., and H. Grimmeiss. "Kinetics of Ion Exchange," J. Chim. Phys. **1958,** 269.
153. Dickel, G., and E. Hübner. "Kinetic Studies on Strong and Weak Ion Exchangers," Kolloid-Z. **179,** 60 (1961).
154. Dickel, G., and A. Meyer. "On the Kinetics of Ion Exchange on Exchange Resins," Z. Elektrochem. **57,** 901 (1953).
155. Dickel, G., and L. V. Nieciecki. "Kinetics of Ion Exchange on Exchange Resins: II," Z. Elektrochem. **59,** 913 (1955).
156. Diemaier, W., and G. Maier. "Z. Lebensm." Untersuch.-Forsch. **119,** 123 (1963).
157. Donnan, F. G. "Theory of Membrane Equilibria and Membrane Potentials in the Presence of Non-Dialyzing Electrolytes. Contribution to Physicochemical Physiology," Z. Elektrochem. Angew. physik. Chem. **17,** 572 (1911).

158. Donnan, F. G. "Precise Thermodynamics of Membrane Equilibria. II," Z. physik. Chem. Dept. A **168,** 369 (1934).
159. Donnan, F. G., and E. A. Guggenheim. "Precise Thermodynamics of Membrane Equilibria," Z. physik. Chem. Dept. A **162,** 346 (1932).
160. Dorfner, K. *Ion Exchange Chromatography* (Berlin: Akademie-Verlag, 1963), 236 pp., 149 figs., 13 tables.
161. Dorfner, K. "Redox Exchangers," Chem.-Ztg. **85,** 80, 113 (1961).
162. Dorfner, K. "State of the Art and Recent Developments in Gradient Elution," Chem.-Ztg. **87,** 871 (1963).
163. Dorfner, K. "The Use of Ion Exchangers for Preparative Purposes," Medizinal-Markt **1968,** (6) 245.
164. Downing, D. C. "Calculating Minimum-Cost Ion-Exchange Units," Chem. Eng. **1965,** 6.
165. Dubourg, J. "Ion Exchanger in the Sugar Refinery," Allimentation Agricult. Arch. Intern. Droit **76,** 11 (1959).
166. Dunlop, P. J. "Recalculated Values for the Diffusion Coefficients of Several Aqueous Ternary Systems at 25°C," J. Phys. Chem. **68,** 3062 (1964).
167. Dusek, K. "Ion Exchange Matrices: 3. Copolymers of Styrene and Divinylbenzene. Elastic Properties of Toluene-Swollen Copolymers," Collection Czech. Chem. Commun. **27,** 2841 (1962).
168. Dutch Patent Announcement No. 6,413,808, Farbenfabriken Bayer AG, 1964.
169. Dybczynski, R. "Influence of Temperature on Tracer Level Separations by Ion Exchange Chromatography," J. Chromatog. **31,** 155 (1967).
170. Dyer, A., and R. B. Gettins. "Self-Diffusion and Ion-Exchange Processes in Synthetic Zeolites," Conference on Ion Exchange, London, 1969.
171. Edge, R. A. "Anion Exchange Behavior of Some Rare Earths in Dilute Sulfuric Acid Solutions containing Ethanol," J. Chromatog. **6,** 452 (1961).
172. Eisenman, G. *Membrane Transport and Metabolism* (New York: Academic Press, 1961), p. 163.
173. Eisner, U., and H. B. Mark. "Semipermeable Ion-Exchange Membranes as a Preconcentration Matrix for Trace Analysis by Electrochemical and Neutron-Activation Techniques," Talanta **16,** 27 (1969).
174. Erne, K., and T. Canbäck. "The Fluorimetric Determination of Noradrenaline," J. Pharm. Pharmacol. **7,** 248 (1953).
175. Fallon, H. J., and J. W. Woods. "Response of Hyperlipoproteinemia to Cholestyramine Resin," J. Am. Med. Assoc. **204,** 1161 (1968).
176. Faris, J. P. "Adsorption of the Elements from Hydrofluoric Acid by Anion Exchange," Anal. Chem. **32,** 520 (1960).
177. Farrell, J. B., and R. N. Smith. "Process Applications of Electrodialysis," Ind. Eng. Chem. **54,** (6) 29 (1962).
178. Faucher, J. A., R. W. Sothwort, and H. C. Thomas. "Adsorption Studies on Clay Minerals: I. Chromatography on Clays," J. Chem. Phys. **20,** 157 (1952).

179. Faucher, J. A., and H. C. Thomas. "IV. The System Montmorillonite-Cesium-Potassium," J. Chem. Phys. **22,** 258 (1954).
180. Fisher, S., and R. Kunin. "Routine Exchange Capacity Determination of Ion Exchange Resins," Anal. Chem. **27,** 1191 (1955).
181. Fisher, S., and R. Kunin. "Ion Exchange Preparation of Low-Silica Hydroxide Solutions for Colorimetric Determination of Total Silica," Nature **177,** 1125 (1956).
182. Fisher, J. W., and A. J. Vivyurka. "Combined Ion Exchange-Solvent Extraction Process (Eluex) for Ammonium Diuranate Production," Conference on Ion Exchange, London, 1969.
183. Freeman, D. H., and G. Scatchard. "Volumetric Studies of Ion-Exchange Resin Particles using Microscopy," J. Phys. Chem. **69,** 70 (1965).
184. French Patent No. 1,337,578, 1962 (Mastrorilli Erf.).
185. French Patent No. 1,349,078, 1963 (Asahi Kasei Kogyo).
186. French Patent No. 1,352,176, 1962 (Ets. Degrément).
187. French Patent No. 1,394,306, 1964 (Union Carbide).
188. French Patent No. 1,395,281, 1964 (Farbenfabriken Bayer AG).
189. French Patent No. 1,398,421, 1964 (Farbenfabriken Bayer AG).
190. Fresenius, W., F.-J. Bibo, and W. Schneider. "On the Removal of Nitrate Ions from Potable Water with the Use of anion Exchangers in a Semi-Industrial Installation," Gas-Wasserfach **107,** 306 (1966).
191. Friedland, D. H., *et al.* "Recognition of Cross-Linking in the Infra-Red Spectra of Poly (Styrene–Divinylbenzene)," Conference on Ion Exchange, London, 1969.
192. Fritz, J. S., and Abbink. Anal. Chem. **37,** 1274 (1965).
193. Fritz, J. S., and M. L. Gilette. "Anion Exchange Separation of Metal Ions in Dimethyl Sulphoxide–Methanol–Hydrochloric Acid," Talanta **15,** 287 (1968).
194. Fritz, J. S., and D. J. Pietrzyk. "Non-Aqueous Solvents in Anion-Exchange Separations," Talanta **8,** 143 (1961).
195. Fritz, J. S., and A. Tateda. "Studies on the Anion Exchange Behavior of Carboxylic Acids and Phenols," Anal. Chem. **40,** 2115 (1968).
196. Frost, C. R., and D. Glasser. "An Apparatus for Fractional Ion-Exchange Separation," Conference on Ion Exchange, London, 1969.
197. Fujimoto, M. "Ion Exchange Resins as Reaction Media for Microdetection Tests," Chemist-Analyst **49,** 4 (1960).
198. Fulham, H. T., *et al.* "Thermochemical Instabilities in Anion-Exchange Processing," Conference on Ion Exchange, London, 1969.
199. Furrer, F. "Ion Exchange Installations for the Plant Water Cycle in Electroplating with Consideration of Cyanide-Containing and Acid Effluents," Galvanotechnik **51,** 105 (1960).
200. Gable, R. W., and H. A. Strobel. "Nonaqueous Ion Exchange. I. Some Cation Equilibrium Studies in Methanol," J. Phys. Chem. **60,** 513 (1956).
201. Gabrielson, G., and O. Samuelson. Svensk Kem. Tidskr. **62,** 214 (1950).
202. Gage, T. B., *et al.* Science **113,** 522 (1951).

203. Gaines, G. L., and H. C. Thomas. "V. Montmorillonite–Cesium–Strontium at Several Temperatures," J. Chem. Phys. **23**, 2322 (1955).
204. Galat, A. J. J. Am. Chem. Soc. **70**, 3945 (1948).
205. Galton, V. A., and R. Pitt-Rivers. Biochem. J. **72**, 310 (1959).
206. Gans, R. Chem. Ind. **32**, 197 (1909).
207. Gans, R. "Manganese Removal from Potable Water by Aluminosilicates," Chem.-Ztg. **31**, 355 (1907).
208. Gans, R. Jahrb. Preuss. Geol. Landesanstalt **26**, 179 (1905).
209. Garlanda, T. "Health Regulations for the Use of Ion Exchangers" (German, American, French and Italian Standards for the use of ion exchangers in the foods industry), Materie Plast. Elast. **31**, 719, 786 (1965).
210. Garten, V. A., and D. E. Weiss. "The Ion and Electron Exchange Properties of Activated Carbon in Relation to its Behavior as a Catalyst and Adsorbent," Rev. Pure Appl. Chem. **7**, 69 (1957).
211. Gauvreau, A., and A. Lattes. Compt. Rend. C **266**, (15) 1162 (1968).
212. Gehlsen, H., and E. Fürstenberg. "New Methods for the Application of the Countercurrent Principle to the Treatment of Water from the Saale for Stream Converter Feed Water by Ion Exchangers," Energietechnik **16**, 123 (1966).
213. German Patent Application No. I 75,869 IV b, I 77,574; 1944 (R. Griessbach, H. Lauth, and E. Meier).
214. German Published Patent Application No. 1,221,197 (Farbenfabriken Bayer AG; F. Limbach).
215. German Published Patent Application No. 1,249,220.
216. German Published Patent Application No. 1,275,997 (L. und C. Steinmüller; G. Plura).
217. German Utility Model 1,943,442.
218. Gerstner, F. "The Recovery of Copper in Copper Fiber production," Z. Elektrochem. **57**, 221 (1953).
219. Ghatge, N. D. J. Appl. Polymer Sci. **8**, (3) 1305 (1964).
220. Gibbs, W. Collected Works, Vol. 1, p. 429.
221. Gilwood, M. E. "Saving Capital and Chemicals with Countercurrent Ion Exchange," Chem. Eng. **74** (Dec. 18, 1967).
222. Glass, R. A. J. Am. Chem. Soc. **77**, 807 (1953).
223. Glueckauf, E. "A New Approach to Ion Exchange Polymers," Proc. Royal Soc. Ser. A, **268**, 350 (1962).
224. Glueckauf, E., and G. P. Kitt. "A Theoretical Treatment of Cation Exchangers: III. The Hydration of Cations in Polystyrene Sulfonates," Proc. Royal Soc. **228**, 322 (1955).
225. Glueckauf, E., and R. E. Watts. "The Donnan Law and its Application to Ion Exchange Polymers," Proc. Royal Soc. Ser. A **268**, 339 (1962).
226. Goldberg. Instrument News **18**, 8 (1968).
227. Good, M. L., *et al.* J. Inorg. Nucl. Chem. **6**, 73 (1958).
228. Gordon, M., *et al.* "The Use of Cyanide-Form Ion Exchange Resins in the Preparation of Nitriles," Chem. Ind. **91**, 1019 (1962).

229. Gordon, M., *et al.* "Anion Exchange Resins in the Synthesis of Nitriles," J. Org. Chem. **28,** 698 (1963).
230. Grammont, P. "Ion Exchangers and Continuous Systems," Conference on Ion Exchange, London, 1969.
231. Grande, J. A., and J. Beukenkamp. Anal. Chem. **28,** 1497 (1956).
232. Graul, E. H., and E. K. Reinhart. "Experimental Studies on the decontamination of Water with a View for the Preparation of Potable Water from Radioactively Contaminated Surface Water," Atompraxis **4,** 397 (1958).
233. Graydon, W. F., and R. J. Stewart. "Ion Exchange Membranes: I. Membrane Potentials," J. Phys. Chem. **59,** 86 (1955).
234. Gregor, H. P. "A General Thermodynamic Theory of Ion Exchange Process," J. Am. Chem. Soc. **70,** 1293 (1948).
235. Gregor, H. P. "Gibbs-Donnan Equilibria in Ion Exchange Resin Systems," J. Am. Chem. Soc. **73,** 642 (1951).
236. Griessbach, R. "On the Preparation and Application of New Exchange Adsorbents Particularly Based on Resins," Angew. Chem. **52,** 215 (1939).
237. Griessbach, R. Z. Ver. Deut. Chem. Beiheft [Supplement] **31,** 1 (1939).
238. Griessbach, R., and G. Neumann. "Ion Exchangers and Catalysis," Chem. Tech. **5,** 187 (1953).
239. Grimshaw, R. W. Trans. Brit. Ceram. Soc. **57,** 340 (1958).
240. de Groot, S. R., and P. Mazur. *Non-Equilibrium Thermodynamics* (Amsterdam: North Holland Publishing Co., 1959).
241. Gruber, P. E., and N. Noller. Z. physik. Chem. N.S. **38,** 184, 203 (1963).
242. Guggenheim, E. A. J. Phys. Chem. **33,** 842 (1929).
243. Gustafson, R. L., *et al.* "Adsorption of Organic Species by High Surface Area Styrene–Divinylbenzene Copolymers," Ind. Eng. Chem. Prod. Res. Dev. **7,** 107 (1968).
244. Guthrie, J. D., and A. L. Bullock. "Ion Exchange Celluloses for Chromatographic Separations," Ind. Eng. Chem. **52,** (11) 935 (1960).
245. Haagen, K. Z. Elektrochem. **57,** 178 (1953).
246. Haissinsky, M. J. Chim. Phys. **49,** C, 133 (1952).
247. Hale, D. K., and D. J. McCauley. "Structure and properties of heterogeneous Cation-Exchange Membranes." Trans. Faraday Soc. **57,** 135 (1961).
248. Hall, G. R., and M. Streat. "Radiation-Induced Decomposition of Ion-Exchange Resins: Part I. Anion-Exchange Resins," J. Chem. Soc. (London) **1963,** 5205.
249. Hall, G. R., M. Streat, and G. R. B. Creed. "Ion Exchange in Nuclear Chemical Processes: Part I. The Effect of Heat Ionizing Radiation on Resin Performance," Chem. Eng. No. 316. Trans. Inst. Chem. Engrs. London **46,** 2 (1968).
250. Hall, G. R., *et al.* "Thermal Stability of Ion-Exchange Resins," Conference on Ion Exchange, London, 1969.

251. Hall, N. F., and Bryson. Anal. Chim. Acta. **24,** 138 (1961).
252. Hall, N. F., and D. H. Johns. J. Am. Chem. Soc. **75,** 5787 (1953).
253. Hamilton, P. B. Anal. Chem. **30,** 914 (1958).
254. Harris, F. E., and S. A. Rice. "Model for Ion-Exchange Resins," J. Chem. Phys. **24,** 1258 (1956).
255. Hartley, G. S., and J. Crank. "Some Fundamental Definitions and Concepts in Diffusion Processes." Trans. Faraday Soc. **45,** 801 (1949).
256. Hatch, M. J., J. A. Dillon, and H. B. Smith. "Preparation and Use of Snake-Cage Polyelectrolytes," Ind. Eng. Chem. **49,** 1812 (1957).
257. Haug, A., and O. Šmidsrod. "Strontium–Calcium Selectivity of Alginates," Nature (London) **215,** 757 (1967).
258. Hazan, *et al.* Z. Anal. Chem. **213,** 182 (1965).
259. Hein, F., and H. Lilie. "Preparation of Complex Acids by the Ion Exchange Method," Z. Anorg. Allgem. Chem. **270,** 45 (1952).
260. Hein, W. "Investigation of the Capillary Characteristics of Ion Exchangers," Dissertation, Berlin, 1967.
261. Helfferich, F. "Calculation of Industrial Ion Exchange Installations: a Critical Study of Theoretical Principles and Methods."
262. Helfferich, F. "Ligand Exchange: I. Equilibria," J. Am. Chem. Soc. **84,** 3237 (1962).
263. Helfferich, F. "Ligand Exchange: II. Separation of Ligands having Different Coordinative Valences," J. Am. Chem. Soc. **84,** 3242 (1962).
264. Helfferich, F. Kolloid-Z. **185,** 157 (1962).
265. Hendricks, S. B. "Base Exchange of Crystalline Silicates," Ind. Eng, Chem. **37,** 625 (1945).
266. Henneberg, W., and F. Stohmann. Am. Chem. Pharm. **107,** 152 (1958).
267. Hering, R. "Application of Chelate-Forming Ion Exchange Resins in Analytical Chemistry," Z. Chem. **5,** 402 (1965).
268. Herz, G. P. "Field Experience with Macroreticular Resins," Effluent Water Treat. J. **5,** 453 (1965).
269. Herz, G. P. "The Influence of Organic Substances on Anion Exchangers of Full Deionization Plants," Techn. Ueberwach. **3,** 77 (1962).
270. Hesse, G., and O. Sauter. "On the Independence of Exchange Adsorption and van der Waals Adsorption of Aluminum Oxide," Naturwiss. **34,** 250 (1947).
271. Hetherington, R. "Your Water Supply: How it Affects Processing, What You Can Do to Improve It," Textile World **114,** 86 (Feb. 1964).
272. Higgins, I. R. "Continuous Ion Exchange Equipment," Ind. Eng. Chem. **53,** 635 (1961).
273. Higgins, I. R., and J. T. Roberts. Chem. Eng. Progr. Symp. Ser. **50,** No. 14, 87 (1954).
274. Himmelhoch, S. R., and E. A. Peterson. "Experimental Problems in the Use of Commercially Prepared DEAE-Celluloses for Protein Chromatography." Anal. Biochem. **17,** 383–389 (1966).
275. Hirschfelder, O., Ch. Curtis, and R. Bird. *Molecular Theory of Gases and Liquids* (New York: John Wiley and Sons, 1954).

276. Hofmann, K. "Experiences with Ion Exchangers Operating by the Countercurrent Process," Mitt. Ver. Grosskesselbesitzer **69,** 92 (1960).
277. Hofmann, K. Mitt. Ver. Grosskesselbesitzer **70,** 48 (1961).
278. Hofmann, U. "The Characteristics of Ion Exchange on Clay Minerals," in K. Issleib (ed.). *Anomalies in Ion Exchange Processes* (Berlin: Akademie Verlag, 1962).
279. Hollaway and Nelson. J. Chromatog. **14,** 255 (1964).
280. Hollis, R. F., and C. K. McArthur. Mining Eng. **9,** 442 (1957).
281. Holzapeel, H., and O. Gürtler. "Determination of Relative Affinities of Some Mono-, Bi- and Trivalent Anions for Wofatit SBW," J. Prakt. Chem. **34,** 91 (1966).
282. Honda, M. J. Chem. Soc. Japan **70,** 55 (1949).
283. Horembala, L. E., and C. A. Feldt. "Ion-Exchange Screens Prevent Fouling," Power **112,** 67 (1968).
284. Huff, E. A. "Anion Exchange Study of a Number of Elements in Nitric–Hydrofluoric Acid Mixtures: Analytical Applications of the System," Anal. Chem. **36,** 1921 (1964).
285. Iguchi, A. Bull. Chem. Soc. Japan **31,** 597, 600 (1958).
286. Inczédy, J. "On the Application of Ion Exchangers in Water Analysis," in *Progress in Hydrochemistry* (Berlin: Akademie-Verlag, 1964), p. 161.
287. Inczédy, J., and L. Erdey. "Some Analytical Applications of Ion-Exchange Membranes," Symp. Balatonszelplak **1963,** 207 (1965).
288. Inczédy, J., and I Högye. "II. Swelling and Functioning Properties of Macroporous Ion Exchange Resins," Acta Chim. Acad. Sci. Hung. **56,** 109 (1968).
289. Inczédy, J., and E. Pasztler. "The Analytical Use of Ion Exchangers in Organic Solvents: I. Swelling and Salt Splitting in Non-aqueous Media," Acta Chim. Acad. Sci. Hung. **56,** 9 (1968).
290. Inczédy, J., *et al.* "Application of Ammonium–Sulfosalicylate as a Complexing Agent in Ion Exchange Chromatography," Acta Chim. Acad. Sci. Hung. **50,** 105 (1966).
291. Indusekhar, V. K., and N. Krishnaswamy. "Diffusion Effect during Electro-dialysis with Ion Exchange Membranes."
292. Isagulyants, V. J. Chem. Abstr. **67,** 108 609.
293. Isagulyants, V. J. Chem. Ind. (Moscow) **4,** 258 (1967).
294. Ishida, and R. Kuroda. Anal. Chem. **39,** 212 (1967).
295. Jakubovic, A. O. Nature (London) **184,** 1065 (1959).
296. Jakubovic, A. O., and B. N. Brook. Polymer **2,** 18 (1961).
296a. Jentzsch, D., and I. Pawlik. Z. Anal. Chem. **146,** 88 (1955).
297. Jindra, A. Česk. Farm. **8,** 15 (1959).
298. Jindra, A., and O. Motl. "Ion Exchangers in Pharmacy," Pharmazie **8,** 547 (1953).
299. Jjsseling, F. P., and E. van Dalen. "Potentiometric Titrations with Ion-Exchanging Membrane Electrodes: I. Theoretical Aspects of Simple Precipitation Titrations," Anal. Chem. Acta **36,** 166 (1966).

300. Johnson, R. H., and T. C. Reavey. "Development of Ion Exchange Cartrages for Field Testing of Milk for Iodine-131," Nature (London) **208,** 750 (1965).
301. Johnson, R. H., and T. C. Reavey. Public Health Report **80,** (10) 919 (1965).
302. de Jong, G. J. "Experience in Pretreatment for Full Desalination with Synthetics Resins," Vom Wasser **27,** 306 (1960).
303. Joustra, M., and H. Lundgren. "Preparation of Freeze-Dried, Monomeric and Immunochemically Pure IgG by a Rapid and Reproducible Chromatographic Technique," 17th Ann. Colloquium, Proteides of the Biologicial Fluids, Brügge, Belgium, 1969, p. 98.
304. Kaiser, J. R. Ind. Eng. Chem. Prod. Res. Dev. **1,** 296 (1962).
305. Kauhere, S. S., *et al.* "Ion-Exchange Resins and Cinchona Alkaloids: I. Exchange Equilibria," J. Pharm. Sci. **57,** 342 (1968).
306. Kelso, F. S., *et al.* "Ion Exchange Methods for the Determination of Fluoride in Drinking Water," Anal. Chem. **36,** 577 (1964).
307. Khopkar, S. M., and A. K. De. Anal. Chim. Acta **22,** 153 (1960).
308. Khym, J. X., and L. P. Zill. J. Am. Chem. Soc. **74,** 2090 (1952).
309. Khym, J. X., *et al.* J. Am. Chem. Soc **76,** 5523 (1954).
310. King, P., and J. R. Simmler. "Ion Exchange Separation of Acetamide from Ammonium Acetate and its Determination," Anal. Chem. **36,** 1837 (1964).
311. Klamer, K., and D. W. van Krevelen. "Investigation of Ion Exchange: VI. Construction of Ion Exchange Columns and Conversion to a Larger Scale," Chem. Eng. Sci. **9,** 20 (1958).
312. Klein, G., D. Tondeur, and T. Vermeulen. "Multi-Component Ion Exchange in Fixed Beds: General Properties of Equilibrium Systems," Ind. Eng. Chem. Fundamentals **6,** 339 (1967).
313. Klein, G., M. Villena-Blanco, and T. Vermeulen. "Ion Exchange Equilibria Data in the Design of a Cyclic Sea Water Softening Process," Ind. Eng. Chem. Proc. Des. Dev. **3,** 280 (1964).
314. Klement, R. "Application of Ion Exchange Resins for the Preparative Formation of Free Acids and their Salts," Z. Anorg. Allgem. Chem. **260,** 267 (1949).
315. Klement, R., and A. Kuhn. Z. Anal. Chem. **152,** 146 (1956).
316. Klement, R., and R. Popp. "Preparation and Properties of Some Polyphosphates," Chem. Ber. **93,** 156 (1960).
317. Klement, R., and H. Sandmann. Z. Anal. Chem. **145,** 9 (1955).
318. Kline, L. B., R. P. Sterner, and R. G. Swanson. "Upflow Regeneration: a Practical Way to Reduce Sodium Zeolite Hardness Leakage," Proc. Intern. Water Conf. (Oct. 1965), pp. 158–160.
319. Kloth, L., and W. Poethke. "The Separation and Determination of B-Complex Vitamins by Ion Exchange Resins," Pharm. Zentralhalle (Jena) **103,** 1, 169, 255 (1964).
320. Kloth, L., and W. Poethke. Pharm. Zentralhalle (Jena) **104,** 393 (1965).
321. Klump, K. "Considerations on the Ammoniac Regeneration of Weak Base Anion Exchangers," Energie **19,** 212 (1967).

322. Klump, W. "The Suitability of Ammonia Regeneration of Weak Base Ion Exchangers," VGB-Special Edition, 1965.
323. Knabe, J. "A Method for the Concentration Determination of Salts of Quaternary Ammonium Bases: Report I," Deut. Apotheker-Ztg. **96,** 874 (1956).
324. Knabe, J. "A Method for the Concentration Determination of Salts of Quaternary Ammonium Bases: Report II," Deut. Apotheker-Ztg. **96,** 1243 (1956).
325. Koganowski, A. M., and J. M. Sagrai. Chemiefasern (USSR) **2,** 58 (1963). C 1964, 48 No. 1904.
326. Koganowski, A. M., and J. M. Sagrai. "The Application of a Fluidized Bed of Cation Exchangers for the Removal of Zinc from Effluents," Nichteisenmetalle (USSR) **35,** 35 (1962).
327. Kohlschütter, H. W. "Principles of the Chromatographic Separation Effect of Silica Gel," Ber. Bunsenges. Physik. Chem. **69,** 849 (1965).
328. Kokima, M. Japan Analyst **7,** 177 (1958).
329. Kolf, G. "Measurement and Interpretation of Some Statistical Effects on Polar Double Membranes," Ber. Bunsenges. Physik. Chem. **71,** 877 (1967).
330. Korkisch, J. "Combined Ion Exchange-Solvent Extraction: a New Dimension in Inorganic Separation Chemistry," Nature (London) **210,** 626 (1966).
331. Korkisch, J. "Combined Ion Exchange-Solvent Extraction: a New Separation Principle in Analytical Chemistry," Oesterr. Chemiker-Ztg. **67,** 309 (1966).
332. Korkisch, J., and S. S. Ahluwalia. "Anion Exchange Properties of Different Metal Ions in Sulfuric Acid Solutions Containing Organic Solvents," Z. Anal. Chem. **215,** 86 (1966).
333. Korkisch, J., and H. Gross. "Selective Cation-Exchange Separation of Cobalt in Hydrochloric Acid–Acetone Solutions," Separ. Sci. **2,** 169 (1967).
334. Korkisch, J., and A. Hazan. Anal. Chem. **36,** 2464 (1964).
335. Korkisch, J., and A. Huber. "Cation-Exchange Behavior of Several Elements in Hydrofluoric Acid–organic Solvent Media," Talanta **15,** 119 (1968).
336. Korkisch, J., and K. A. Orlandini. "Cation Exchange Separation of Thorium from Rare Earths and Other Elements in Methanol–Nitric Acid Medium Containing Trioctylphosphine Oxide," Anal. Chem. **40,** 1952 (1968).
337. Korkisch, J., and K. A. Orlandini. Talanta **16,** 45 (1969).
338. Korkisch, J., and S. Urubay. "Ion Exchange in Non-aqueous Solvents: Adsorption Behavior of the Uranium and Other Elements on Strong-Base Anion-Exchange Resins from Organic Acid–organic Solvent Media: Method for the Separation of Uranium," Talanta **11,** 721 (1964).
339. Korn, P. Private communication.
340. de Körösy, F. "Enhanced Salt Diffusion along the Borderline of Mosaic Permselective Membranes." Nature (London) **197,** 685 (1963).

341. de Körösy, F., and J. Shorr. "Mosaic Permselective Membranes," Nature (London) **197,** 685 (1963).
342. Kourim, V., *et al.* J. Inorg. Nucl. Chem. **26,** 1111 (1964).
343. Kovka, F., J. Trojanek, and Z. Cekan. "Isolation of Morphine from Poppy Seeds," Pharmazie **20,** 220, 429, 434 (1965).
344. Krampitz, G., and W. Albersmeyer. Experientia **15,** 375 (1959).
345. Kraus, K. A. ORNL 2159.
346. Kraus, K. A., and G. E. Moore. "Separation of Zirconium and Hafnium with Anion-Exchange Resins." J. Am. Chem. Soc. **71,** 3263 (1949).
347. Kraus, K. A., and G. E. Moore. J. Am. Chem. Soc. **75,** 1460 (1955).
348. Kraus, K. A., and F. Nelson. "Anion Exchange Studies of the Fission Products," Peaceful Atomic Energy **7,** 113 (1955).
349. Kraus, K. A., and H. O. Phillips. "Anion-Exchange Studies: XIX. Anion Exchange Properties of Hydrous Zirconium Oxide," J. Am. Chem. Soc. **78,** 249, 694 (1956).
350. Kraus, K. A., *et al.* J. Am. Chem. Soc. **77,** 3972 (1955).
351. Kraus, K. A., *et al.* J. Phys. Chem. **58,** 11 (1954).
352. Krause, H., and O. Nentwich. "Decontamination of Radioactive Effluents in the Karlsruhe Research Center," Chem. Ingr.-Tech. **40,** 301 (1968).
353. Kreiling, A., W. Ludwig, and G. Pfleiderer. Biochem. Z. **336,** 241 (1962).
354. Krejcar, E. Chem. Prumysl. **15,** 77 (1965).
355. Kressman, T. R. E. "Isopor Resins and the Organic Fouling Problem," Effluent Water Treat. J. **1966,** 119.
356. Kressman, T. R. E. "Properties of Some Modified Polymer Networks and Derived Ion Exchangers," Conference on Ion Exchange, London, 1969.
357. Kressman, T. R. E., and J. A. Kitchener. "Cation Exchange with a Synthetic Phenolsulphonate Resin," J. Chem. Soc. (London) **1949,** 1190, 1201, 1208, 1211.
358. Kromrey, W. "Indirect Acidity Determination of Gastric Juice without Intubation with the Aid of a Quinine Exchanger," Gastroenterology **85,** 20 (1956).
359. Krtil, J. J. Chromatog. **20,** 384 (1965).
360. Krtil, J. "Ion Exchange Characteristics of Vanadium Hexacyanoferrate(II): II. Dyanmic experiments," J. Chromatog. **21,** 85 (1966).
361. Krtil, J., and M. Chavko. "VII. Sorption of Cs^{137} and Rb^{86} on Acid and Normal Ammonium and Thallous Salts of Phosphotungstic and Phosphomolybdic Acid," J. Chromatog. **27,** 460 (1967).
362. Krtil, J., and V. Kourim. "Exchange Properties of Ammonium Salts of 12-Heteropoly Acids: I. Sorption of Cesium on Ammonium Phosphotungstate and Phosphomolybdate," J. Inorg. Nucl. Chem. **12,** 367 (1960).
363. Krug, J. "Ion Exchangers in Galvano-Technology," Galvano **54,** 423 (1963).

364. Krug, J. "Ion Exchangers in Galvanotechnology," Galvano **55,** 295 (1964).
365. Kühne, G. "Past, Present and Future of Ion Exchangers," Chem. Ind. **20,** (9) 621 (1968).
366. Kun, K. A., and R. Kunin. "Probe Structure of Some Macroreticular Ion Exchange Resins," J. Polymer Sci. Part B. Polymer Letters **2,** 587 (1964).
367. Kun, K. A., and R. Kunin. "J. Polymer Sci." **C16,** 1457 (1967).
368. Kunin, R. "A New Ion Exchange Desalination Technique," Brit. Chem. Eng. **11,** 1222 (1966).
369. Kunin, R. "A Critical Examination of the Pore Structure of Macroreticular Ion Exchange Resins," Conference on Ion Exchange, London, 1969.
370. Kunin, R. "Further Studies of the Weak Electrolyte Ion Exchange Resin Desalination Process (Desal process)," Desalination **4,** 38 (1968).
371. Kunin, R. "Chemical Synthesis through Ion Exchange," Ind. Eng. Chem. **56,** 35 (Jan. 1964).
372. Kunin, R. *et al.* Ind. Eng. Chem. Prod. Res. Dev. **1,** 140–44 (June, 1962).
373. Kunin, R., and F. McGarvey. "Ion Exchange," Ind. Eng. Chem. **47,** 565 (1955).
374. Kunin, R., E. Meitzner, and N. Bortnick. "Macroreticular Ion Exchange Resins," J. Am. Chem. Soc. **84,** 305 (1962).
375. Kunin, R., and B. Vassiliou. "Regeneration of Carboxylic Cation Exchange Resins with Carbon Dioxide," Ind. Eng. Chem. Prod. Res. Dev. **2,** 1 (1963).
376. Kunin, R., and B. Vassiliou. "New Deionization Techniques Based on Weak Electrolyte Ion Exchange Resins," Ind. Eng. Chem. Prod. Res. Dev. **3,** 404 (1964).
377. Kunin R., and B. Vassiliou. *Sweet Water from the Sea* (Dechema-Monographs, Vol. **47,** K. Fischbeck, ed., Weinheim/Bergstrasse: Verlag Chemie, 1962), p. 735.
378. Kunin, R., and A. G. Winger. "Technology of Liquid Ion Exchangers," Chem. Ingr.-Tech. **34,** 461 (1962).
379. Kuroda, R., *et al.* "Adsorption Behavior of a Number of Metals in Hydrochloric Acid on a Weakly Basic Anion Exchange Resin," Anal. Chem. **40,** 1502 (1968).
380. Lacy, W. J., and D. C. Lindsten. "Removal of Radioactive Contaminants from Water by Ion Exchange Slurry," Ind. Eng. Chem. **49,** 1725 (1957).
381. Lakshminarayanaiah, L. "Transport Phenomena in Artificial Membranes," Chem. Rev. **65,** 491 (1965).
382. Lakshminarayanaiah, N., and V. Subrahmanyan. "Measurement of Membrane Potentials and Test of Theories," J. Polymer Sci. **A2,** 4491 (1964).
383. Läuger, P. "Transport Phenomena in Membranes," Angew. Chem. **81,** 56 (1969).

384. Lee, and O. Samuelson. Anal. Chim. Acta **37,** 359 (1967).
385. Legradi, L. "Ion Exchange Resins as Indicators," Magy. Kem. Folyoirat **66,** 76 (1960).
386. Lemberg, J. Z. Deut. Geol. Ges. **22,** 335 (1870).
387. Lemberg, J. Z. Deut. Geol. Ges. **28,** 519 (1876).
388. Lengborn, N. "Continuous Ion Exchange," Svensk. Kem. Tidskr. **70,** 255 (1958).
389. Levendusky, J. A., and B. J. Peters. "Application of the Powdex Process for the Purification of Utility Power Plant Condensate," Southeastern Electric Exchange Conf., Tampa, Florida, Oct. 1963.
390. Levendusky, J. A., *et al.* "Continuous Countercurrent Ion Exchange," Ind. Water Eng. **2,** 11 (1965).
391. Levesque, C. L., and A. M. Craig. Ind. Eng. Chem. **40,** 96 (1948).
392. Lewis, D. J. "Liquid Ion-Exchange Processes," Chem. Eng. **72,** (14) 101 (1965).
393. Li, C. H. Acta Endocrinol. **10,** 255 (1952).
394. Li, C. H. Nature **173,** 860 (1954).
395. Li, C. H., *et al.* J. Biol. Chem. **213,** 171 (1955).
396. Lieberman. Clin. Acta **7,** 159 (1962).
397. Lieser, K. H., J. Bastian, and A. B. H. Hecker. "Separation of Cations on Titanium Hexacyanoferrate(II) Columns," Z. Anal. Chem. **228,** 98 (1967).
398. Ling, G. N. *A Physical Theory of the Living State* (New York: Blaisdell, 1962), Ch. 4.
399. Lister, B. A. J. J. Chem. Soc. **1951,** 3123.
400. Lloyd, P. J., and E. A. Mason. "Extraction of Hexavalent Uranium by Trilaurylamine Nitrate," J. Phys. Chem. **68,** 3120 (1964).
401. van Loon and Beamish. Anal. Chem. **36,** 1771 (1964).
402. Luke, C. L. "Ultratrace Analysis of Metals using a Curved Crystal X-Ray Milliprobe," Anal. Chem. **36,** 318 (1964).
403. Lumbroso, T., and M. Reverbori. "Study of the Porosity of a Cation Exchanger of the Sulfonated Styrene–Divinylbenzene Copolymer type," Genie Chim. **92,** (4) 89 (1964).
404. Maeck, W. J., *et al.* "Adsorption of the Elements on Inorganic Ion Exchangers from Nitrate Media," Anal. Chem. **35,** 2086 (1963).
405. Manecke, G. "The Synthesis and Application of Polymeric Redox Systems," Angew. Makromol. Chem. **45,** 26 (1968).
406. Manecke, G. "The Application of Ion Exchange Membranes," Chem. Ingr.-Tech. **30,** 311 (1958).
407. Manecke, G., and P. Gregs. "Amphoteric Cellulose Ion Exchangers," Naturwiss. **50,** 329 (1963).
408. Manecke, G., and H. Heller. "Amphoteric Ion Exchangers I," Makromol. Chem. **55,** 51 (1962).
409. Manecke, G., and H. Heller. "Amphoteric Ion Exchangers II," Makromol. Chem. **59,** 106 (1963).
410. Manecke, G., and H. Heller. "Amphoteric Ion Exchangers III," Makromol. Chem. **82,** 146 (1965).

411. Mann, C. K., and C. L. Swanson. "Cation Exchange Elution of Metallic Chlorides by Hydrochloric Acid," Anal. Chem. **33,** 459 (1961).
412. Marinsky, J. A. "Prediction of Ion-Exchange Selectivity," J. Phys. Chem. **71,** 1572 (1967).
413. Martinola, F. "Macroporous Ion-Exchange Resins for Water Conditioning," Effluent Water Treat. J. (May/June 1966).
414. Martinola, F. "Monitoring of Anion Exchangers in the Production Laboratory," Ber. VGB-Speisewassertagung **1965,** 1.
415. Martinola, F. "Problems of Decontamination with Ion Exchangers," Wasser Luft Betrieb **6,** 457 (1962).
416. Martinola, F., and G. Kühne. "Properties and Application of Powdered Ion-Exchange Resins," Intern. Conference on Ion Exchange, London, 1969.
417. Martinola, F., and L. A. Wegner. "Experiments for the Elution of Radioactive Substances from Ion Exchangers," Atompraxis **7,** 223 (1961).
418. Mastagli, P., *et al.* Compt. Rend. **232,** 1848 (1951).
419. McGarvey, F. X., *et al.* "Ion Exchange Develops as a Process in the Wine Industry," Am. J. Oenol. **9,** 168 (1958).
420. McIntire, F. C., and J. R. Schenck. J. Am. Chem. Soc. **70,** 1193 (1948).
421. McNevin, W. M., and W. B. Crummett. Anal. Chem. **25,** 1628 (1953).
422. Mehls, K. F. H. "Rational Waste Water Planning," Industrie-Anzeiger **1961,** No. 21 (14 March).
423. Mehltretter, C. L., and F. B. Weakley. "Extraction of Morphine from Poppy-Seeds and its Recovery by Ion Exchange," J. Am. Pharm. Assoc. Sci. Ed. **46,** 193 (1957).
424. Menke, K. H. J. Chromatog. **7,** 86 (1962).
425. Menke, K. H. Naturwiss. **45,** 263 (1958).
426. Merciny. Z. Anal. Chem. **236,** 498 (1968).
427. Merriam, C. N., and H. C. Thomas. "VI. Alkali Ions on Attapulgite," J. Chem. Phys. **24,** 993 (1956).
428. Merz, E. "Studies of the Ion Exchange Properties of Inorganic Tin and Zirconium Compounds with Tracer Isotopes," Z. Elektrochem. **63,** 288 (1959).
429. Michalson, A. W. "High Quality Water via Ion Exchange," Chem. Eng. Proc. **64,** (10) 67 (1968).
430. Mikes, J. A. "Pore Structure in Ion-Exchange Materials," Conference on Ion Exchange, London, 1969.
431. Mikes, J. A., and L. I. Kovacs. "Chemical Properties of Bipolar Electrolyte-Exchange Resins," J. Polymer Sci. **59,** (167) 209 (1962).
432. Miller, J. F., F. Bernstein, and H. P. Gregor. "Theory of Selective Uptake of Ions of Different Size by Polyelectrolyte Gels: Experimental Results with Potassium and Quaternary Ammonium Ions and Methacrylic Acid Resins," J. Chem. Phys. **43,** 1783 (1965).

433. Miller, W. E. "Ion Exchange Resins as Indicators," Anal. Chem. **30,** 1462 (1958).
434. Milton, G. M., and W. E. Grummit. Can. J. Chem. **35,** 541 (1957).
435. Minken, J. W. *Desalting of Seawater.* (Oxford-Milan: Pergamon Press-Tamburini Editore, 1965).
436. Modrzejewski, F., and L. Kalinski. "Extraction of Alkaloids from Botanical Raw Materials with the Aid of Ion Exchangers," Acta Polon. Pharm. **16,** 263 (1959).
437. Mongar, I. L., and A. Wassermann. "Adsorption of Electrolyte by Alginate Gells Without and With Cation Exchange," J. Chem. Soc. **1952,** 492.
438. Mongar, I. L., and A. Wassermann. "Influence of Ion Exchange on Optical Properties, Shape and Elasticity of Fully-Swollen Alginate Fibers," J. Chem. Soc. **1952,** 500.
439. Moore, F. L. Anal. Chem. **29,** 1660 (1957).
440. Moore, G. E., and K. A. Kraus. J. Am. Chem. Soc. **72,** 5792 (1950).
441. Moore, S., and W. H. Stein. J. Biol. Chem. **211,** 893 (1954).
442. Moretti, G., *et al.* "Orientation of a Cation Exchange Resin for Trapping Ammonium from Blood in Extracorporeal Circulation," Rev. Franc. d'Etudes Clin. Biol. **11,** 938 (1966).
443. Mottlau, A. Y., and N. E. Fisher. Anal. Chem. **34,** 714 (1962).
444. Müller, H. "Automatic and Recording Control of Conductivity in a Full Deionization Plant," Mitt. VGB **79,** 262 (1962).
445. Murachi *et al.* Biochem. **3,** 48 (1964).
446. Mutter. Tenside **5,** 138 (1968).
447. Naumann, G. "Catalysis by Ion Exchangers," Chem. Tech. **11,** 18 (1959).
448. Nelson, F., and D. C. Michalson. "Cation Exchange in HBr Solutions," J. Chromatog. **25,** 414 (1966).
449. Nelson, F., T. Murase, and K. A. Kraus. "Cation Exchange in Concentrated HCl and $HClO_4$ Solutions," J. Chromatog. **13,** 503 (1964).
450. Nelson, F., *et al.* J. Chromatog. **20,** 107 (1965).
451. Newman, J. "Water Demineralization Benefits from Continuous Ion Exchange Process," Chem. Eng. **74,** 72 (Dec. 18, 1967).
452. Newman, M. S. J. Am. Chem. Soc. **75,** 4740 (1953).
453. Nickless, G., and G. R. Marshall. "Polymeric Coordination Compounds: the Synthesis and Applications of Selective Ion-Exchangers and Polymeric Chelate Compounds," Chromatog. Rev. **6,** 154 (1964).
454. Nitschmann, N., P. Kistler, H. R. Renger, A. Hässig, and A. Joss. "The Isolation of a Thermally Stable Human Plasma Protein Solution by Desalting (PPS)," Vox Sanguinis **1,** 183 (1956).
455. Olsen, E. D., and R. L. Poole. "Quick Removal of Alkali Metals from Quaternary Ammonium Bases during Ion Exchange," Anal. Chem. **37,** 1375 (1965).
456. Osterried, O. "Flame-Photometric Determination of Small Quantities of Cesium in Silicates after Enrichment on Zirconium Phosphate Ion Exchangers," Z. Anal. Chem. **199,** 260 (1964).

457. Pai, K. R., N. Krishnaswamy, and D. S. Datar. "Properties of Hydrous Oxide Inorganic Exchangers," Conference on Ion Exchange, London, 1969.
458. Pallmann. "Soil Research," Forsch. **6,** 30 (1938–39).
459. Parrish, J. R. "Measurement of Water Regain and Macropore Volume of Ion-Exchange Resins," J. Appl. Chem. **15,** 280 (1965).
460. Parrish, J. R. J. Chromatog. **18,** 535 (1965).
461. Parrish, J. R. "Packing of Spheres," Nature **190,** 800 (1961).
462. Pashkov, A. B., and V. S. Titov. "Principle Characteristics of Some Soviet Ion-Exchange Resins," Khim. Prom. 270–6 (1958).
463. Pashkov, A. B., *et al.* "Synthesis and Investigation of anion Exchangers Based on Vinyl and Alkylvinylpyridines," Conference on Ion Exchange, London, 1969.
464. Patel and Bafna. Nature **211,** 963 (1966).
465. Patt, P., and W. Winkler. "On the Preparation and Determination of Solanine by Ion Exchangers," Arch. Pharm. Ber. Deut. pharm. Ges. **293/65,** 846 (1960).
466. Paulson, J. C., *et al.* J. Am. Chem. Soc. **75,** 2039 (1953).
467. Pelzbauer, Z., and V. Forst. "The Electron Microscopic Evaluation of the Porosity of Ion Exchangers," Collection Czech. Chem. Commun. **31,** 2338 (1966).
468. Pepper, K. W., D. Reichenberg, and D. K. Hale. "Swelling and Shrinkage of Sulphonated Polystyrenes of Different Cross-Linking," J. Chem. Soc. **1952,** 3129.
469. Persoz, J. "Frontal Analysis in Ion Exchange: I. Theoretical Calculations; Application to the Determination of the Equivalent Theoretical Plate Height," Bull. Soc. Chim. France **1967,** 523.
470. Peterson, E. A., and H. A. Sober. Biochem. Prepn. **8,** 39, 43, 45, 47 (1961).
471. Peterson, E. A., and H. A. Sober. "Chromatography of Proteins, I. Cellulose Ion-Exchange Adsorbents," J. Am. Chem. **78,** 751 (1956).
472. Peterson, E. A., and H. A. Sober. J. Am. Chem. Soc. **78,** 756 (1956).
473. Pharmacia Fine Chemicals, Uppsala, Sweden.
474. Phipps, A. M. "Anion Exchange in Dimethyl Sulfoxide," Anal. Chem. **40,** 1769 (1968).
475. Piez *et al.* Biochem. **2,** 58 (1963).
476. Pilot, J., *et al.* "Reactivation of Phosphoric Acid and Anodising Liquors by Ion Exchange," Conference on Ion Exchange, London, 1969.
477. Plapp, F. W., and J. E. Casida. Anal. Chem. **30,** 1622 (1958).
478. Pollio, F. X. "Determination of Moisture in Ion Exchange Resins by Karl Fischer Reagent," Anal. Chem. **35,** 2164 (1963).
479. Polyanskii, N. G., *et al.* Kinetics Catalysis USSR Eng. Transl. **3,** 136 (1962).
480. Porath, J. Arkiv. Kemi. **11,** 259 (1957).
481. Porath, J., and P. Flodin. "Gel Filtration: a Method for Desalting and Group Separation," Nature **183,** 1657 (1959).

482. Porath, J., and E. B. Lindner. "Separation Methods Based on Molecular Sieving and Ion Exclusion," Nature **191,** 69 (1961).
483. Prout, W. E., E. R. Russel, and H. J. Groh. "Ion Exchange Absorption of Cesium by Potassium Hexacyanocobalt(II) ferrate(II)," J. Inorg. Nucl. Chem. **27,** 473 (1965).
484. Rachinskii, V. V. *The General Theory of Sorption Dynamics and Chromatography* (New York: Consultants Bureau, 1965). Russian.
485. Randerath, K. "A Simple Manufacturing Process for Cellulose Anion Exchangers and Anion Exchange Papers," Angew. Chem. **74,** 780 (1962).
486. Rane and Bhatki. Anal. Chem. **38,** 1598 (1966).
487. Rebek, M., and M. K. Semlitsch. "Preparation of Ultrapure Crystal Violet Base with the Aid of Ion Exchanger," Monatsh. Chem. **92,** 214 (1961).
488. Rebenfeld, L., and E. Pascu. J. Am. Chem. Soc. **75,** 4370 (1953).
489. Rebertus. Anal. Chem. **38,** 1089 (1966).
490. Reed, A., and A. E. Errington. "Ion Exchange in the Treatment of Effluent from Metal Finishing Processes," Conference on Ion Exchange, London, 1969.
491. Reed, T. B., and D. W. Breck. "Crystalline Zeolites: II. Crystal Structure of Synthetic Zeolite Type A," J. Am. Chem. Soc. **78,** 5972 (1956).
492. Reichenberg, D., and D. J. McCauley. "Properties of Ion Exchange Resins in Relation to their Structure: VII. Cation-Exchange Equilibria on Sulphonated Polystyrene Resins of Varying Degrees of Cross-Linking," J. Chem. Soc. **1955,** 2731.
493. Remond, J. "Ion Exchange Resins: its Use in Catalysis," Rev. Prod. Chim. **63,** 417, 421 (1960).
494. Rice, S. A., and F. E. Harris. "Polyelectrolyte Gels and Ion Exchange Reactions," Z. phys. Chem. (Frankfurt) **8,** 207 (1956).
495. Richter, A. "Attainable Residual Concentrations of Dissolved Materials in Water Treated with Ion Exchangers," Vom Wasser **26,** 241 (1959).
496. Riedel, H.-J. "Contribution to the Characterization of Cation Exchangers: Properties of Suitable Inorganic Materials," Ber. Kernforsch. (Jülich) **1962,** (32) CA 57: 10747g.
497. Riley, J. P., and D. Taylor. "The Use of Chelating Ion Exchange in the Determination of Molybdenum and Vanadium in Seawater," Anal. Chim. Acta **41,** 175 (1968).
498. Rombauts, W. A., and M. A. Raftery. "A Device for Automatic Gradient or Stepwise Chromatography," Anal. Chem. **37,** 1611 (1965).
499. Rosenberg, N. W., J. H. B. George, and W. D. Potter. "Electrochemical Properties of a Cation Transfer Membrane," J. Electro Chem. Soc. **104,** 111 (1957).
500. Rosenblum, P., A. S. Tombalakian, and W. F. Gordon. "Homogenous Ion-Exchange Membranes of Improved Flexibility," J. Polymer Sci. **A-14,** 1703 (1966).
501. Rowe, R. "How to Test Cation Resins for Stability," Power Eng. **66,** 68 (1962).

502. Ruch, R. R., *et al.* "Preconcentration of Trace Elements by Precipitation Ion Exchange using Organic–Hydrochloric Acid Systems," Anal. Chem. **37,** 1565 (1965).
503. Runge, F. Angew, Chem. **62,** 451 (1950).
504. Sailer, E., and K. Menli. "The Powdex-Process," Brennstoff-Waerme-Kraft **18,** 361 (1966).
505. St. John, C. V., *et al.* Anal. Chem. **23,** 1289 (1951).
506. Saldadze, K. M. "Structure and Properties of Ion Exchangers" Conference on Ion Exchange, London, 1969.
507. Saline Water Research and Development Progress Report 1963, p. 75.
508. Samuelson, O. *Ion Exchangers in Analytical Chemistry* (New York: John Wiley and Sons, Inc., 1953).
509. Samuelson, O. *Ion Exchange Separations in Analytical Chemistry* (Stockholm: Almqvist and Wiksell; New York: John Wiley and Sons, Inc., 1963).
510. Samuelson, O. "Ion Exchange Chromatography," Anal. Chim. Acta **38,** 163 (1967).
511. Samuelson, O. Tek. Tidskr. **76,** 561 (1946).
512. Samuelson, O., and B. Wallenius. J. Chromatog. **12,** 236 (1963).
513. Sato, *et al.* Bull. Chem. Soc. Japan **39,** 716 (1966).
514. Sattelmacher, P. G. "Are Nitrates Dangerous in Drinking Water?" Gas Wasserfach **104,** 1321 (1963).
515. Satterfield, C. N., and T. K. Sherwood. *The Role of Diffusion in Catalysis* (Reading, Mass.: Addison-Wesley Publ. Co., 1963).
516. Saunders, L. G. "An Ion Exchange Method for Preparing Water for Injection," Dechema Monograph. **55,** 275 (1964).
517. Saunders, L. G. *Modern Deionization: Ion Exchange Progress of Elga Products Ltd.* Special Publication, 1968.
518. Saunders, L. G., *et al.* "Preparation of Biologically Pure Water by Ion Exchange," Conference on Ion Exchange, London, 1969.
519. Schell, W. R., and J. V. Jordan. "Investigations of Anion Exchange of Pure Clays," Plant Soil (The Hague) **10**, 303 (1959).
520. Schiffers, A., and H. Schmidt. "Instrumentation and Control System for Decarbonation and Full Desalination," Chem.-Ztg. **89,** 379 (1965).
521. Schindewolf, U. "Liquid Ion Exchangers," Z. Elektrochem. **62,** 335 (1958).
522. Schlegel, H. "Pertinent Questions on the Application of Ion Exchangers in Waste Water Technology," Galvanotechnik **56,** 73 (1965).
523. Schlögl, R. *Mass Transport through Membranes* (Darmstadt: D. Steinkopf-Verlag, 1964).
524. Schlögl, R. Trans. Faraday Soc. **21,** 46 (1956).
525. Schlögl, R. Z. Phys. Chem. N.S. **1,** 305 (1954).
526. Schlögl, R. Z. Phys. Chem. N.S. **3,** 73 (1955).
527. Schlögl, R., and H. Schurig. "An Experimental Method for the Determination of Pore Sizes in Ion Exchangers," Z. Elektrochem. **65,** 863 (1961).
528. Schmid, G. "Theory of Electro-Osmosis," Chem. Ingr.-Tech. **37,** 616 (1965).

529. Schmidle, C. J., and R. C. Mansfield. Ind. Eng. Chem. **44,** 1388 (1952).
530. Schmidtmann, W. "Mixing Apparatus to Produce Salt and pH Gradients," Chem.-Ztg. **89,** 231 (1965).
531. Schröder, E. Plaste Kautschuk **6,** 325 (1959).
532. Schroeder, W. A., and Robberson. Anal. Chem. **37,** 1583 (1965).
533. Schtanikow, E. V. "Purification of Virus-Infested Water with the Aid of Ion Exchanger Resins," Gigiena i Sanit. **30,** 29 (1965).
534. Schuler, R. H., *et al.* J. Chem. Educ. **28,** 192 (1951).
535. Schultz, B. J., and E. H. Crook. "Ultrafine Ion Exchange Resins," Ind. Eng. Chem. Prod. Res. Dev. **7,** 120 (1968).
536. Seamster, A. H., and R. M. Wheaton. "Ion Exchange becomes Powerful Processing Tool," Chem. Eng. **67,** (17) 115 (1960).
537. Segal, H. L. Ann. N.Y. Acad. Sci. **57,** 308 (1953).
538. Segal, H. L., and L. L. Miller. Gastroenterology (Basel) **29,** 633 (1955).
539. Segal, H. L., *et al.,* Gastroenterology (Basel) **28,** 402 (1955).
540. Segal, H. L., *et al.* "Determination of Gastric Acidity without Intubation by Use of Cation Exchange Indicator Compounds," Proc. Soc. Exptl. Biol. Med. **74,** 218 (1950).
541. Seidl, J., J. Malinsky, K. Dusek, and W. Heitz. "Macroporous Styrene–Divinylbenzene Copolymers and their Use in Chromatography and for the Preparation of Ion Exchangers," Fortschr. Hochpolymer. Forsch. **5,** 113 (1967).
542. Seki, J. J. Biochem. (Tokyo) **45,** 855 (1958).
543. "Sephadex-Gel Filtration in Theory and Practice," Deutsche Pharmacia (1967/69).
544. Serva Development Laboratory. Heidelberg, Römerstr. 118.
545. Seugling, E. W., and E. P. Guth. "Study of the Cationic Exchange Properties of Acid Activated Bentonite: I. Effect of Structure Variation, pH, and Ionic Strength on the Exchange Rates of a Series of β-Phenyl-Ethylamines," J. Pharm. Sci. **50,** 929 (1961).
546. Seyb, E. "Need and Type of Water Treatment Prior to Ion Exchangers with Special Consideration of Radioactive Waste Effluents," Tech. Mitt. Haus Tech. Essen **52,** 424 (1959).
547. Sherma, J., and W. Rieman. Anal. Chim. Acta **18,** 214 (1958).
548. Sherry, H. S. "Ion Exchange Properties of the Synthetic Zeolite, Linde T," Conference on Ion Exchange, London, 1969.
549. Shimolin, P. E., and V. I. Bochkareva. "The Use of Ion-Exchangers for the Production of Deionized Nonpyrogenic Water from Tap Water," Probl. Gematol. i Pereliv. Krovi **11,** 54 (1966).
550. Shirato, S., *et al.* J. Ferment. Technol. **45,** 60 (1967).
551. Shulman, H. L., G. R. Youngquist, and J. R. Covert. "Development of a Continuous Countercurrent Fluid-Solids Contactor Ion Exchange," Ind. Eng. Chem. Process Design Develop. **5,** 257 (1966).
552. Siegel, S., *et al.* "Stabilization of Cold Tablets," J. Pharm. Sci. **51,** 1069 (1962).

553. Sjöström, E., and W. Rittner. "A Method for the Quantitative Determination of Alkaloid Salts by Cation Exchange and Subsequent Complexometric Titration," Z. Anal. Chem. **153,** 321 (1956).
554. Skelly. Anal. Chem. **33,** 271 (1961).
555. Skloss, *et al.* Anal. Chem. **37,** 1240 (1965).
556. Skogseid, A. Dissertation, Oslo, 1948.
557. Slater, M. J. "A Review of Continuous Countercurrent Contractors for Liquids and Particulate Solids," Brit. Chem. Eng. **14,** (1) 41 (1969).
558. Slater, M. J. "Comparison of the Hydrodynamics of two continuous Counter-Current Ion-Exchange Contactors," Conference on Ion Exchange, London, 1969.
559. Smit, J. van R. "Ammonium Salts of the Heteropoly Acids as Cation Exchangers," Nature **181,** 1530 (1958).
560. Smit, J. van R., J. J. Jacobs, and W. Robb. "Cation Exchange Properties of the Ammonium Heteropolyacid Salts," J. Inorg. Nucl. Chem. **12,** 95 (1959).
561. Sober. H. A., and E. A. Peterson. J. Am. Chem. Soc. **76,** 1171 (1954).
562. Sollner, K. "The Physical Chemistry of Ion Exchange Membranes," Svensk Kem. Tidskr. **70,** 267 (1958).
563. Specker, H., M. Kuchtner, and H. Hartkamp. "The Quantitative Separation of Inorganic Ions by Ion Exchange on Alginic Acid." Z. Anal. Chem. **141,** 33 (1954).
564. Speecke, A., and J. Hoste. Talanta **2,** 332 (1959).
565. Spes, H. "Catalytic Reactions in the Ion Exchange Columns with a Shift of the Chemical Equilibrium," Chem.-Ztg. **90,** 443 (1966).
566. Spillane and Scott. Lab. Pract. **17,** 352 (1968).
567. Stamberg, K. "Design: Technological Problems of Ion Exchanger Disposal," Paper before the Leipzig Symposium, June, 1968.
568. Steinbach, J., and H. Freiser. "Preparation of Standard Sodium Hydroxide Solutions by Use of a Strong Anion Exchange Resin," Anal. Chem. **24,** 1027 (1952).
569. Strelow, F. W. E. "An Ion Exchange Selectivity Scale of Cations Based on Equilibrium Distribution Coefficients," Anal. Chem. **32,** 1185 (1960).
570. Strelow, F. W. E. Anal. Chem. **40,** 928 (1968).
571. Strewlow, F. W. E., and C. J. C. Bothma. "Anion Exchange and Selectivity Sequence for Elements in the Sulfuric Acid Medium on a Strong Base Resin," Anal. Chem. **39,** 595 (1967).
572. Strelow, F. W. E., *et al.* "Ion Exchange Selectivity Scales for Cations in Nitric Acid and Sulfuric Acid Media with a Sulfonated Polystyrene Resin," Anal. Chem. **37,** 106 (1965).
573. Sugai, S., and J. Furuichi. "On Diffusions of Radioactive Ions in Ion Exchangers," J. Chem. Phys. **23,** 1181 (1955).
574. Sussman, S. Ind. Eng. Chem. **38,** 1228 (1946).
575. Sussman, S., F. C. Nachod, and W. Wood. "Metal Recovery by Anion Exchange," Ind. Eng. Chem. **37,** 618 (1945).

576. Swift, J. G. "Investigations on the Prolongation of Antihistamine Action by Complexing with Ion Exchangers," Arch. Intern. Pharmacodyn. **124,** 341 (1960).
577. Tabikh, A. A., J. Barshad, and R. Overstreet. "Cation Exchange Hystereses in Clay Minerals," Soil Sci. **90,** 219 (1960).
578. Tallmadge, J. A. "Ion Exchange Treatment of Mixed Electroplating Wastes," Ind. Eng. Chem. Process Design Develop. **6,** 419 (1967).
579. Talvitie, N. A., and R. J. deMint. "Radiochemical Determination of Strontium-90 in Water by Ion Exchange," Anal. Chem. **37,** 1605 (1965).
580. Taylor, P. B. J. Phys. Chem. **31,** 1478 (1927).
581. Symposium: Thirty Years of Synthetic Ion Exchanger Resins, Leipzig, 4–7 June, 1968.
582. Thomas, G. G., and C. W. Davis. Nature **159,** 372 (1947).
583. Thompson, C. M., and N. F. Kember. "Ion-Exchange Celluloses for Biochemical Separations," Proc. Biochem. **2,** 7 (1967).
584. Thompson, H. S. J. Royal Agricult. Soc. Engl. **11,** 68 (1950).
585. Titov, V. S., and A. B. Pashkov. "Innovations in the Synthesis, Production and Uses of Polymeric Ion-Exchange Materials," Khim. Prom. **1962,** 912–18.
586. Tombalakian, A. S., C. Y. Yeh, and W. F. Graydon. "Mass Transfer Coefficients Across Ion Exchange Membranes," Can. J. Chem. Eng. **42,** 61 (1964).
587. Tondeur, D., and G. Klein. "Multicomponent Ion Exchange in Fixed Beds: Constant-separation-factor Equilibrium," Ind. Eng. Chem. Fundamentals **6,** 351 (1967).
588. Torraca, E., *et al.* "Preparation and Properties of Inorganic Ion Exchangers of the Zirconium Phosphate Type," Conference on Ion Exchange, London, 1969.
589. Trandafilov, T., and L. Tomassini. "On the Adsorption Properties of Different Bulgarian Bentonites," Pharmazie **22,** 109 (1967).
590. Türkölmex, S. "New Methods of Waste Gas Purification: Deodorizing by Exchanging-Adsorption with Synthetic Ion Exchange Resins," Wasser Luft Betrieb **9,** 737 (1965).
591. Turner, J. B., *et al.* "Adsorption of Some Metals on Anion-Exchange Resins from Potassium Thiocyanate Solutions," Anal. Chim. Acta **26,** 94 (1962).
592. Ueda, T., and S. Harada. "Adsorption of Cationic Polysulfone on Bentonite," J. Appl. Polymer. Sci. **12,** 2395 (1968).
593. Umland, F. "The Interaction of Electrolyte Solutions and γ-Al_2O_3: IV. Development of a Formal Ion Exchange Theory for the Adsorption of Electrolytes from Aqueous Solution," Z. Elektrochem. **60,** 711 (1956).
594. Umland, F. "V. The Dependence of Electrolytic Adsorption on the Ratio of Adsorbent to Solution," Z. Elektrochem. **63,** 510 (1959).
595. Umland, F. "VI. Determination of Water Sorption from Aqueous Solution," Z. Anorg. Allgem. Chem. **317,** 129 (1962).

596. Unholzer, S. "Technical Tests for Automatic Control of Ion Exchangers in Full Desalination Installations by Conductivity-Difference Measurements," Chem. Tech. **21,** 160 (1969).
597. Urata, Y. "Substitution Reactions of Organic Halides with Anion Exchangers," J. Chem. Soc. Japan, Pure Chem. Sect. **83,** 932, 936, 1105 (1962).
598. Urbanek, B., *et. al.* "Application of Ion Exchangers in Blood Preservation: Principle and Conditions of Blood Preservation with Ion Exchangers." Cesk. Farm. **9,** 137 (1960).
599. U.S. Patent 2,366,007 (G. F. d'Alelio, General Electric Co.), 1945.
600. U.S. Patent 24,777,380 (S. J. Kreps and F. C. Nachod), 1949.
600a. U.S. Patent 2,593,417 (G. F. d'Alelio), 1952.
601. U.S. Patent 2,678,306 (A. H. Ferris), 1954.
602. U.S. Patent 2,678,307 (A. H. Ferris and W. R. Tyman), 1954.
603. U.S. Patent 2,764,563 (E. L. McMaster and W. K. Glesner), 1956.
604. U.S. Patent 2,891,007, Method of regenerating ion exchangers (P. H. Caskey and A. C. Reents), 1959.
605. U.S. Patent 3,005,815 (Merck and Co., Inc.), 1961.
606. U.S. Patent 3,041,292 (M. J. Hatch), 1962.
607. U.S. Patent 3,074,797 (Foremost Dairies Inc.), 1963.
608. U.S. Patent 3,092,617.
609. U.S. Patent 3,122,565 (Eisai Co., Ltd.), 1964.
610. U.S. Patent 3,139,401 (Hach Chemical Co.), 1962.
611. U.S. Patent 3,232,867 (Diamond Alkali Co.), 1966.
612. Vaisberg, E. S. "The Rate of Exchange of Large Organic Ions on Carboxylic Acid Cation-Exchange Resins: III. Variation of the Diffusion Coefficients of Streptomycin with the Composition of the Ion Exchanger," Russ. J. Phys. Chem. English Transl. **41,** 468 (1967).
613. Vasilev, A. A. "On the Structural Units of Sulfonated Ion Exchangers Based on Phenol–Formaldehyde Compounds," Z. Angew. Chem. Moscow, **41,** (5) 1099 (1968).
614. Vassiliou, B., and R. Kunin. "Fine Particle-Sized Ion Exchange Resins," Anal. Chem. **35,** 1328 (1963).
615. Vermeulen, T. "Separation by Adsorption Methods," Advan. Chem. Eng. **2,** 147 (1958).
616. VGB-Merkblatt. *Certification and Performance Control of Full Desalination Plants,* 2nd Ed. No. 10 (Essen: VGB-Dampftechnik GmbH., 1964).
617. Vincent, M. C., and M. J. Blake. "Analysis of Barbiturates in Different Dosage Forms by Ion Exchangers and Titration in a Nonaqueous Solvent," J. Am. Pharmacol. Assoc. Sci. Ed. **48,** 359 (1959).
618. Vincent, R. "Treatment of Condensates," Conference on Ion Exchange, London, 1969.
619. Volkin, E., and W. E. Cohn. J. Biol. Chem. **215,** 767 (1953).
620. Walker, G. T. "Ion Exchangers in the Cosmetics Industry," Seifen-Ole-Fette-Wachse **86,** 379 (1960).
621. Way, J. T. J. Royal Agricult. Soc. Engl. **11,** 313 (1850).

622. Webster, *et al.* Anal. Chim. Acta **38,** 193 (1967).
623. Weinstock, J., *et al.* J. Am. Chem. Soc. **75,** 2546 (1953).
624. Weiss, A. "On the Cation Exchange Capacity of Clay Minerals: I. Comparison of Experimental Methods," Z. Anorg. Allgem. Chem. **297,** 232 (1958).
625. Weiss, A. "On the Cation Exchange Capacity of Clay Minerals: II. Cation Exchange in Minerals of the Mica Vermiculite and Montmorillonite Groups," Z. Anorg. Allgem. Chem. **297,** 257 (1958).
626. Weiss, A. "On the Cation Exchange Capacity of Clay Minerals: III. Cation Exchange in Kaolinite," Z. Anorg. Allgem. Chem. **299,** 92 (1959).
627. Weiss, D. E., *et al.* Australian J. Chem. **19,** 561, 589, 765, 791 (1966).
628. *Whatman Advanced Ion-Exchange Cellulose Laboratory Manual* (Maidstone, Kent, England: W. & R. Balston Ltd., 1968).
629. Wheelwright, E. J. "A generic Ion-Exchange Process for the Recovery and Purification of Valuable Elements from the Nuclear Industry," Conference on Ion Exchange, London, 1969.
630. White, J. M., P. Kelly, and N. C. Li. "Dionyl Naphthalene Sulfonic Acid and Tri-*n*-Octylamine as Liquid Ion Exchangers for the Study of Fe(III) and In(III)-Chloride," J. Inorg. Nucl. Chem. **16,** 337 (1961).
631. Wickbold, R. "The Concentration of Very Small quantities of Silicic Acid by Ion Exchange," Z. Anal. Chem. **171,** 81 (1959).
632. Wieland, G. "New Developments in Countercurrent Regeneration of Ion Exchange Filters," Tech. Ueberwach. **10,** (2) 40 (1969).
633. Wiley, R. H., and G. Devenuto. "Irradiation Stability of Sulfonated Styrene Resins Crosslinked with Various Divinylbenzene Isomers and Mixtures Thereof." J. Appl. Polymer Sci. **9,** (6) 2001 (1965).
634. Wirth, L. F. "Sources and Detection of Impurities Affecting the Production of High Purity Water," Combustion **37,** 38 (1966).
635. Wirth, L. F., C. H. Feldt, and Odland. "Ion Exchanger Stability under the Influence of Chlorine," Ind. Eng. Chem. **53,** 638 (1961).
636. Wish, L. Anal. Chem. **31,** 326 (1959).
637. Woermann, D. "Investigations of Mosaic Membranes," Ber. Bunsenges. Physik. Chem. **71,** 87 (1967).
638. Wolf, F. "Theory and Application of Ion Exchange Catalysis: I. Kinetics of Ester Hydrolysis and Sugar Inversion in the Presence of Strongly Acidic Cation-Exchange Resins; II. Kinetics of Ester Hydrolysis in the Presence of Strongly Acidic Cation-Exchange Resins with Macroreticular Structure," Conference on Ion Exchange, London, 1969.
639. Wolf, F., and H. Mlytz. "On the Influence of the Bipolar Structure of Ion Exchange Resins (Snake-Cage Electrolytes) on the Separation of Electrolyte-Nonelectrolyte Mixtures," J. Chromatog. **34,** 59 (1968).
640. Wolf, F., and H. Schaaf. "On the Suitability of Strong Acid Cation Exchange Resins in the Presence of Gaseous Hydrogen Chloride as a Dehydration Catalyst," Z. Chem. **7,** 391 (1967).
641. Wolff, H. P. "Therapy with Cation Exchangers in the Light of a Five-Year Experience," Klin. Wochschr. **32,** 761 (1959).

642. Wolff, J. J. "Regeneration of Cation Exchangers in Counter-Current," Wasser Luft Betrieb **13,** (3) 83 (1969).
643. Wolniwiecz, E. "Treatment of Raw Water with the Use of Ion Exchangers, Especially Those with a Macroporous Structure," Textilindustrie **66,** 746 (1964).
644. Worsely, M., A. S. Tombalakian, and W. F. Graydon. "Cation Interchanges across Ion-Exchange Membranes," J. Phys. Chem. **69,** 883 (1965).
645. Wünsche, J.-J., and R. Henze. "The Softening of Very Hard and High-Alkaline Salt Water with Wofatit Resins by the Countercurrent Process," Chem. Tech. **20,** 219 (1968).
646. Wüsten, E, "Therapy of Hyperacidity with Cation Exchangers," Medizinische 1953, 763.
647. Wymore, C. E. "Cation Exchange Resins of the Sulfonic Acid Type as Siccatives," Ind. Eng. Chem. Prod. Res. Dev. **1,** 173 (1962).
648. Yang Jeng-Tsong. Compt. Rend. **231,** 952 (1950).
649. Yoshimura, J., and H. Waki. Japan Analyst **6,** 362 (1957).
650. Ziegler, M. Z. Anal. Chem. **180,** 351 (1961).
651. Zsigmond, A., and E. Gryllus. "Ion Exchanger Columns Operating in Countercurrent," Zucker **17,** 385 (1964).
652. Zundel, G., H. Noller, and G. M. Schwab. "Membranes of Polystyrene Sulfonic Acid and its Salts: I. Production and IR-Spectra and their Assignments," Z. Naturforsch. **16b,** 716 (1961).

Index

This book was typeset at SSPA Typesetting, Inc., Carmel, Indiana, in 10-point Caledonia, and it was printed and bound by LithoCrafters, Inc., Ann Arbor, Michigan